Achieving sustainable cultivation of bananas

Volume 1: Cultivation techniques

It is widely recognised that agriculture is a significant contributor to global warming and climate change. Agriculture needs to reduce its environmental impact and adapt to current climate change whilst still feeding a growing population, i.e. become more 'climate-smart'. Burleigh Dodds Science Publishing is playing its part in achieving this by bringing together key research on making the production of the world's most important crops and livestock products more sustainable.

Based on extensive research, our publications specifically target the challenge of climate-smart agriculture. In this way we are using 'smart publishing' to help achieve climate-smart agriculture.

Burleigh Dodds Science Publishing is an independent and innovative publisher delivering high quality customer-focused agricultural science content in both print and online formats for the academic and research communities. Our aim is to build a foundation of knowledge on which researchers can build to meet the challenge of climate-smart agriculture.

For more information about Burleigh Dodds Science Publishing simply call us on +44 (0) 1223 839365, email info@bdspublishing.com or alternatively please visit our website at www.bdspublishing.com.

Related titles:

Achieving sustainable cultivation of apples
Print (ISBN 978-1-78676-032-6) Online (ISBN 978-1-78676-034-0; 978-1-78676-035-7)

Achieving sustainable cultivation of mangoes
Print (ISBN 978-1-78676-132-3) Online (ISBN 978-1-78676-134-7; 978-1-78676-135-4)

Managing soil health for sustainable agriculture Volume 1: Fundamentals
Print (ISBN 978-1-78676-188-0) Online (ISBN 978-1-78676-190-3; 978-1-78676-191-0)

Managing soil health for sustainable agriculture Volume 2: Monitoring and management
Print (ISBN 978-1-78676-192-7) Online (ISBN 978-1-78676-194-1; 978-1-78676-195-8)

Water management for sustainable agriculture
Print (ISBN 978-1-78676-176-7) Online (ISBN 978-1-78676-178-1; 978-1-78676-179-8)

Chapters are available individually from our online bookshop: https://shop.bdspublishing.com

climate•SMART•publishing
IN AGRICULTURAL SCIENCE

BURLEIGH DODDS SERIES IN AGRICULTURAL SCIENCE

NUMBER 40

Achieving sustainable cultivation of bananas

Volume 1: Cultivation techniques

Edited by Professor Gert H. J. Kema, Wageningen University and Research, The Netherlands; and Professor André Drenth, University of Queensland, Australia

burleigh dodds
SCIENCE PUBLISHING

Published by Burleigh Dodds Science Publishing Limited
82 High Street, Sawston, Cambridge CB22 3HJ, UK
www.bdspublishing.com

Burleigh Dodds Science Publishing, 1518 Walnut Street, Suite 900, Philadelphia, PA 19102-3406, USA

First published 2018 by Burleigh Dodds Science Publishing Limited

Library of Congress Control Number: 2018956889

British Library Cataloguing in Publication Data
A catalogue record for this book is available from the British Library

ISBN 978-1-78676-156-9 (Print)
ISBN 978-1-78676-159-0 (PDF)
ISBN 978-1-78676-158-3 (ePub)
ISSN 2059-6936 (print)
ISSN 2059-6944 (online)

DOI 10.19103/AS.2017.0020

Typeset by Deanta Global Publishing Services, Chennai, India
Printed by Lightning Source

Contents

Series list x

Acknowledgements xv

Introduction xvi

Part 1 Banana production and its challenges

1 The origin, domestication and dispersal of bananas 3
Hugo Volkaert, National Center for Genetic Engineering and Biotechnology, Thailand
1 Introduction 3
2 Origins: morphological and molecular evidence 4
3 The role of *Musa balbisiana* 9
4 Domestication and early cultivation of bananas 10
5 Archaeological evidence for early banana cultivation 11
6 Dispersal of bananas into Oceania 12
7 Dispersal of bananas into Africa 13
8 Dispersal of bananas in South and Southeast Asia 14
9 Conclusion 15
10 Where to look for further information 16
11 References 16

2 Understanding the banana industry: monoculture and beyond 21
Dan Koeppel, Independent Journalist and Researcher, USA
1 Introduction 21
2 Banana varieties and the Cavendish monoculture 22
3 The development of a banana monoculture 23
4 The dynamics of the banana monoculture 24
5 Post-war developments: business as usual? 25
6 The Cavendish era 27
7 Cavendish in crisis and a new paradigm 28
8 The hunt for new varieties 29
9 Future trends and conclusion 30
10 Where to look for further information 31

3 Banana cultivation in Africa 33
W. K. Tushemereirwe and J. Kubiriba, National Agricultural Research Laboratories, Uganda
1 Introduction 33
2 Cultivar composition and distribution 34
3 Production systems 35
4 Banana production in Uganda 39
5 Nutrient deficiencies 40
6 Moisture stress 41
7 Pests and diseases affecting bananas 41

8 Improving banana varieties for management of pests and diseases 50
9 Marketing as a constraint to banana production in Uganda 54
10 Future trends and conclusion 55
11 Where to look for further information 56
12 References 57

4 Challenges and opportunities for smallholders in banana value chains 65
William Tinzaara, Bioversity International, Uganda; Dietmar Stoian, Bioversity International, France; Walter Ocimati, Enoch Kikulwe and Gloria Otieno, Bioversity International, Uganda; and Guy Blomme, Bioversity International, Ethiopia
1 Introduction 65
2 Challenges along the value chain 67
3 Enabling an environment for addressing the challenges 79
4 Future trends 81
5 Conclusion 84
6 Where to look for further information 86
7 References 86

Part 2 Improving practices across the banana value chain

5 A functional approach to bunch formation in banana 93
D. W. Turner, School of Agriculture and Environment, The University of Western Australia, Australia; and D. J. Gibbs, Consultant, Australia
1 Introduction 93
2 Gross morphology and bunch form 94
3 Phases of development in banana 95
4 The floral phase: overview 97
5 Inflorescence 99
6 Formation of hands and fruit: independent but closely coordinated processes 102
7 Growth of the aerial true stem 108
8 Case study: photosynthate availability and bunch formation in banana 110
9 Conclusion 113
10 Where to look for further information 114
11 References 114

6 Banana plant propagation methods 117
Sharon D. Hamill, Department of Agriculture and Fisheries, Australia
1 Introduction 117
2 Background to banana evolution 118
3 Banana morphology and vegetative propagation 118
4 Bits and suckers 119
5 Propagation using conventional planting material 122
6 Banana tissue culture 124
7 Liquid culture systems 132
8 Banana off-types and how to reduce them 132
9 Future trends and conclusion 136

10 Where to look for further information 137
11 References 138

7 Conserving banana germplasm through field genebanks 143
Mike Smith, Queensland Department of Agriculture and Fisheries, Australia
1 Introduction 143
2 Objectives of the field genebank 144
3 Choosing the site 145
4 Sourcing material 147
5 Planting 149
6 Crop management 150
7 Quality control 151
8 Dissemination of quality planting stock 152
9 Case study 152
10 Future trends and conclusion 155
11 Where to look for further information 156
12 Acknowledgements 157
13 References 157

8 Good agricultural practices: an end point or a starting point for more sustainable banana production? 159
Charles Staver, Bioversity International, France
1 Introduction 159
2 Why GAP in banana production? 161
3 GAP-related certification schemes in banana production 163
4 Is sustainable banana production being operationalized through GAP certification? 166
5 Improving GAP for more sustainable banana production 172
6 Scientific frontiers in bananas and implications for GAP 182
7 Conclusion: mutual lessons between GAP certification and approaches to sustainable banana production 187
8 Where to look for further information 189
9 References 189

9 Challenges in cultivation of bananas in the subtropics 195
Víctor Galán Saúco, Instituto Canario de Investigaciones Agrarias, Spain
1 Introduction 195
2 The influence of climate in banana production 196
3 Main problems for banana cultivation in the subtropics 200
4 Solving cultivation problems 200
5 Other cultivation practices relevant to the subtropics 207
6 Greenhouse banana cultivation 207
7 Banana cultivars for the subtropics 216
8 Plant protection 218
9 Access to market 218
10 Conclusion 219
11 Where to look for further information 219
12 References 220

10 Nutrition and soil management in banana cultivation 223
Jetse J. Stoorvogel, Wageningen University, The Netherlands; and Rafael A. Segura, CORBANA S.A., Costa Rica
1 Introduction 223
2 Assessing nutrient requirements 224
3 Developing fertilizer use recommendations 229
4 Precision application of nutrients 230
5 Conclusion 231
6 Where to look for further information 231
7 References 232

11 Harvesting and packaging of bananas 235
Juan José Aycart, Dole, Ecuador
1 Introduction 235
2 Criteria determining when to harvest bunches 237
3 Harvesting techniques 237
4 Bunch movement from the fields 239
5 Packing house operations 241
6 Fruit selection processes 247
7 Packing bananas 249
8 Future trends 252
9 Where to look for further information 253
10 References 253

12 Ripening systems for bananas 257
Frits Popma, Popma Fruit Expertise, The Netherlands
1 Introduction 257
2 Factors affecting banana ripening 258
3 Banana ripening cycles 259
4 The banana supply chain 261
5 Ripening technologies and rooms 266
6 Quality control 273
7 Summary 277
8 Future trends 277
9 About the author 278

Part 3 Assessing and improving sustainability

13 Life cycle assessment and carbon footprint of banana cultivation 281
Louis Bockel, Laure-Sophie Schiettecatte and Orane Debrune, Food and Agriculture Organization (FAO), Italy
1 Introduction 281
2 Assessing the environmental impact of banana production 286
3 An integrated socio-economic carbon footprint banana value chain tool 292
4 Future trends and conclusion 298
5 Where to look for further information 299
6 Abbreviations 299
7 References 299

14 Sustainable banana cultivation: from standards to multiple solutions 303
Sietze Vellema and Kees Jansen, Wageningen University, The Netherlands
1 Introduction 303
2 Sustainability and the range of standards in banana production 304
3 The limits of sustainability standards 307
4 Sustainability and diversity in banana production systems 308
5 Case study: the variety of banana producers in the Philippines 309
6 Conclusion 311
7 Where to look for further information 311
8 References 311

15 Organic banana cultivation and sustainability 317
Frans Wielemaker, Consultant (formerly Director of Research at Dole Fresh Fruit International), Costa Rica
1 Introduction 317
2 Organic banana production 318
3 Key requirements: location, soil and use of rotation 323
4 Soil fertility and fertilization 325
5 Disease management 331
6 Managing nematode and insect pests 337
7 Weed management 340
8 Discussion 342
9 Where to look for further information 344
10 Disclaimer 345
11 About the author 345
12 References 345

Index 349

Series list

Title	Series number
Achieving sustainable cultivation of maize - Vol 1 From improved varieties to local applications Edited by: Dr Dave Watson, CGIAR Maize Research Program Manager, CIMMYT, Mexico	001
Achieving sustainable cultivation of maize - Vol 2 Cultivation techniques, pest and disease control Edited by: Dr Dave Watson, CGIAR Maize Research Program Manager, CIMMYT, Mexico	002
Achieving sustainable cultivation of rice - Vol 1 Breeding for higher yield and quality Edited by: Prof. Takuji Sasaki, Tokyo University of Agriculture, Japan	003
Achieving sustainable cultivation of rice - Vol 2 Cultivation, pest and disease management Edited by: Prof. Takuji Sasaki, Tokyo University of Agriculture, Japan	004
Achieving sustainable cultivation of wheat - Vol 1 Breeding, quality traits, pests and diseases Edited by: Prof. Peter Langridge, The University of Adelaide, Australia	005
Achieving sustainable cultivation of wheat - Vol 2 Cultivation techniques Edited by: Prof. Peter Langridge, The University of Adelaide, Australia	006
Achieving sustainable cultivation of tomatoes Edited by: Dr Autar Mattoo, USDA-ARS, USA & Prof. Avtar Handa, Purdue University, USA	007
Achieving sustainable production of milk - Vol 1 Milk composition, genetics and breeding Edited by: Dr Nico van Belzen, International Dairy Federation (IDF), Belgium	008
Achieving sustainable production of milk - Vol 2 Safety, quality and sustainability Edited by: Dr Nico van Belzen, International Dairy Federation (IDF), Belgium	009
Achieving sustainable production of milk - Vol 3 Dairy herd management and welfare Edited by: Prof. John Webster, University of Bristol, UK	010
Ensuring safety and quality in the production of beef - Vol 1 Safety Edited by: Prof. Gary Acuff, Texas A&M University, USA & Prof. James Dickson, Iowa State University, USA	011
Ensuring safety and quality in the production of beef - Vol 2 Quality Edited by: Prof. Michael Dikeman, Kansas State University, USA	012
Achieving sustainable production of poultry meat - Vol 1 Safety, quality and sustainability Edited by: Prof. Steven C. Ricke, University of Arkansas, USA	013
Achieving sustainable production of poultry meat - Vol 2 Breeding and nutrition Edited by: Prof. Todd Applegate, University of Georgia, USA	014
Achieving sustainable production of poultry meat - Vol 3 Health and welfare Edited by: Prof. Todd Applegate, University of Georgia, USA	015

Achieving sustainable production of eggs - Vol 1 016
Safety and quality
Edited by: Prof. Julie Roberts, University of New England, Australia

Achieving sustainable production of eggs - Vol 2 017
Animal welfare and sustainability
Edited by: Prof. Julie Roberts, University of New England, Australia

Achieving sustainable cultivation of apples 018
Edited by: Dr Kate Evans, Washington State University, USA

Integrated disease management of wheat and barley 019
Edited by: Prof. Richard Oliver, Curtin University, Australia

Achieving sustainable cultivation of cassava - Vol 1 020
Cultivation techniques
Edited by: Dr Clair Hershey, formerly International Center for Tropical Agriculture (CIAT), Colombia

Achieving sustainable cultivation of cassava - Vol 2 021
Genetics, breeding, pests and diseases
Edited by: Dr Clair Hershey, formerly International Center for Tropical Agriculture (CIAT), Colombia

Achieving sustainable production of sheep 022
Edited by: Prof. Johan Greyling, University of the Free State, South Africa

Achieving sustainable production of pig meat - Vol 1 023
Safety, quality and sustainability
Edited by: Prof. Alan Mathew, Purdue University, USA

Achieving sustainable production of pig meat - Vol 2 024
Animal breeding and nutrition
Edited by: Prof. Julian Wiseman, University of Nottingham, UK

Achieving sustainable production of pig meat - Vol 3 025
Animal health and welfare
Edited by: Prof. Julian Wiseman, University of Nottingham, UK

Achieving sustainable cultivation of potatoes - Vol 1 026
Breeding improved varieties
Edited by: Prof. Gefu Wang-Pruski, Dalhousie University, Canada

Achieving sustainable cultivation of oil palm - Vol 1 027
Introduction, breeding and cultivation techniques
Edited by: Prof. Alain Rival, Center for International Cooperation in Agricultural Research for Development (CIRAD), France

Achieving sustainable cultivation of oil palm - Vol 2 028
Diseases, pests, quality and sustainability
Edited by: Prof. Alain Rival, Center for International Cooperation in Agricultural Research for Development (CIRAD), France

Achieving sustainable cultivation of soybeans - Vol 1 029
Breeding and cultivation techniques
Edited by: Prof. Henry T. Nguyen, University of Missouri, USA

Achieving sustainable cultivation of soybeans - Vol 2 030
Diseases, pests, food and non-food uses
Edited by: Prof. Henry T. Nguyen, University of Missouri, USA

Achieving sustainable cultivation of sorghum - Vol 1 031
Genetics, breeding and production techniques
Edited by: Prof. William Rooney, Texas A&M University, USA

Achieving sustainable cultivation of sorghum - Vol 2 032
Sorghum utilization around the world
Edited by: Prof. William Rooney, Texas A&M University, USA

Achieving sustainable cultivation of potatoes - Vol 2 033
Production, storage and crop protection
Edited by: Dr Stuart Wale, Potato Dynamics Ltd., UK

Achieving sustainable cultivation of mangoes 034
Edited by: Professor Víctor Galán Saúco, Instituto Canario de Investigaciones Agrarias (ICIA), Spain & Dr Ping Lu, Charles Darwin University, Australia

Achieving sustainable cultivation of grain legumes - Vol 1 035
Advances in breeding and cultivation techniques
Edited by: Dr Shoba Sivasankar et al., formerly International Crops Research Institute for the Semi-Arid Tropics (ICRISAT), India

Achieving sustainable cultivation of grain legumes - Vol 2 036
Improving cultivation of particular grain legumes
Edited by: Dr Shoba Sivasankar et al., formerly International Crops Research Institute for the Semi-Arid Tropics (ICRISAT), India

Achieving sustainable cultivation of sugarcane - Vol 1 037
Cultivation techniques, quality and sustainability
Edited by: Prof. Philippe Rott, University of Florida, USA

Achieving sustainable cultivation of sugarcane - Vol 2 038
Breeding, pests and diseases
Edited by: Prof. Philippe Rott, University of Florida, USA

Achieving sustainable cultivation of coffee 039
Edited by: Dr Philippe Lashermes, Institut de Recherche pour le Développement (IRD), France

Achieving sustainable cultivation of bananas - Vol 1 040
Cultivation techniques
Edited by Prof. Gert H. J. Kema, Wageningen University and Research, The Netherlands; and Prof. André Drenth, University of Queensland, Australia

Global Tea Science 041
Current status and future needs
Edited by: Dr V. S. Sharma, formerly UPASI Tea Research Institute, India & Dr M. T. Kumudini Gunasekare, Coordinating Secretariat for Science Technology and Innovation (COSTI), Sri Lanka

Integrated weed management 042
Edited by: Emeritus Prof. Rob Zimdahl, Colorado State University, USA

Achieving sustainable cultivation of cocoa 043
Edited by: Prof. Pathmanathan Umaharan, Cocoa Research Centre – The University of the West Indies, Trinidad and Tobago

Robotics and automation for improving agriculture 044
Edited by: Prof. John Billingsley, University of Southern Queensland, Australia

Water management for sustainable agriculture 045
Edited by: Prof. Theib Oweis, ICARDA, Jordan

Improving organic animal farming 046
Edited by: Dr Mette Vaarst, Aarhus University, Denmark & Dr Stephen Roderick, Duchy College, Cornwall, UK

Improving organic crop cultivation 047
Edited by: Prof. Ulrich Köpke, University of Bonn, Germany

Managing soil health for sustainable agriculture - Vol 1 048
Fundamentals
Edited by: Dr Don Reicosky, Soil Scientist Emeritus USDA-ARS and University of Minnesota, USA

Managing soil health for sustainable agriculture - Vol 2 049
Monitoring and management
Edited by: Dr Don Reicosky, Soil Scientist Emeritus USDA-ARS and University of Minnesota, USA

Rice insect pests and their management 050
E. A. Heinrichs, Francis E. Nwilene, Michael J. Stout, Buyung A. R. Hadi & Thais Freitas

Improving grassland and pasture management in temperate agriculture 051
Edited by: Prof. Athole Marshall and Dr Rosemary Collins, IBERS, Aberystwyth University, UK

Precision agriculture for sustainability 052
Edited by: Dr John Stafford, Silsoe Solutions, UK

Achieving sustainable cultivation of temperate zone tree fruit and berries – Vol 1 053
Physiology, genetics and cultivation
Edited by: Prof. Gregory Lang, Michigan State University, USA

Achieving sustainable cultivation of temperate zone tree fruit and berries – Vol 2 054
Case studies
Edited by: Prof. Gregory Lang, Michigan State University, USA

Agroforestry for sustainable agriculture 055
Edited by: Prof. María Mosquera-Losada, University of Santiago de Compostela, Spain & Dr Ravi Prabhu, World Agroforestry Centre (ICRAF), Kenya

Achieving sustainable cultivation of tree nuts 056
Edited by: Prof. Ümit Serdar, Ondokuz Mayis University, Turkey & Emeritus Prof. Dennis Fulbright, Michigan State University, USA

Assessing the environmental impact of agriculture 057
Edited by: Prof. Bo P. Weidema, Aalborg University/2.-0 LCA Consultants, Denmark

Critical issues in plant health: 50 years of research in African agriculture 058
Edited by: Dr Peter Neuenschwander and Dr Manuele Tamò, IITA, Benin

Achieving sustainable cultivation of vegetables 059
Edited by: Emeritus Prof. George Hochmuth, University of Florida, USA

Advances in breeding techniques for cereal crops 060
Edited by: Prof. Frank Ordon, Julius Kuhn Institute (JKI), Germany & Prof. Wolfgang Friedt, Justus-Liebig University of Giessen, Germany

Advances in Conservation Agriculture – Vol 1 061
Systems and science
Edited by: Prof. Amir Kassam, University of Reading, UK

Advances in Conservation Agriculture – Vol 2 062
Practice and benefits
Edited by: Prof. Amir Kassam, University of Reading, UK

Achieving sustainable greenhouse cultivation 063
Edited by: Prof. Leo Marcelis and Dr Ep Heuvelink, Wageningen University, The Netherlands

Achieving carbon-negative bioenergy systems from plant materials 064
Edited by: Dr Chris Saffron, Michigan State University, USA

Achieving sustainable cultivation of tropical fruits 065
Edited by: Prof. Elhadi Yahia, Universidad Autónoma de Querétaro, Mexico

Advances in postharvest management of horticultural produce 066
Edited by: Prof. Chris Watkins, Cornell University, USA

Pesticides and agriculture 067
Profit, politics and policy
Dave Watson

Integrated management of diseases and insect pests of tree fruit 068
Edited by: Prof. Xiangming Xu and Dr Michelle Fountain, NIAB-EMR, UK

Integrated management of insect pests: Current and future developments 069
Edited by: Emeritus Prof. Marcos Kogan, Oregon State University, USA & Prof. Leon Higley, University of Nebraska-Lincoln, USA

Preventing food losses and waste to achieve food security and sustainability 070
Edited by: Prof. Elhadi M. Yahia, Universidad Autónoma de Querétaro, Mexico

Achieving sustainable management of boreal and temperate forests 071
Edited by: Dr John Stanturf, Estonian University of Life Sciences (formerly US Forest Service), USA

Advances in breeding of dairy cattle 072
Edited by: Prof. Julius van der Werf, University of New England, Australia & Dr Jennie Pryce, DEDJTR-Victoria/La Trobe University, Australia

Improving gut health in poultry 073
Edited by: Prof. Steven C. Ricke, University of Arkansas, USA

Achieving sustainable cultivation of barley 074
Edited by: Dr Glen Fox, University of Queensland, Australia & Prof. Chengdao Li, Murdoch University, Australia

Advances in crop modelling for a sustainable agriculture 075
Edited by: Emeritus Prof. Ken Boote, University of Florida, USA

Achieving sustainable crop nutrition 076
Edited by: Prof. Zed Rengel, University of Western Australia

Achieving sustainable urban agriculture 077
Edited by: Prof Han Wiskerke, Wageningen University, The Netherlands

Acknowledgements

We wish to acknowledge the following for their help in reviewing particular chapters:

- Chapter 3: Dr Frank A. Shotkoski, Cornell University, USA
- Chapter 5: Dr I Ravi, ICAR-National Research Centre for Banana, India
- Chapter 6: Dr Michael Pillay, Vaal University of Technology, South Africa
- Chapter 7: Prof Benham Lockhart, University of Minnesota, USA; and Dr Hildegard Garming, Thűnen Institute, Germany
- Chapter 9: Dr Edgar Vinson, Auburn University, USA
- Chapter 11: Dr G. Manjunath, University of Horticultural Sciences – Bagalkot, India
- Chapter 12: Dr Scot Nelson, University of Hawaii, USA
- Chapter 13: Dr Laura Roibas, University Santiago De Compostela, Spain
- Chapter 14: Dr Charles Staver, Bioversity International, France
- Chapter 15: Dr Regina Angelina Sanderson Bellamy, Stockholm University, Sweden

Introduction

Bananas are produced in over 135 tropical and subtropical countries and, according to the Food and Agriculture Organization (FAO), represented an estimated combined economic value of US $52 Billion in 2017. Each year more than 100 billion bananas are consumed globally and over 400 million people rely on bananas for their staple food or as a source of income. A large portion (85%) of the 113 million tons of bananas are consumed locally while the remaining 15% is exported, mainly to countries in temperate regions. Banana is the world's most exported and valuable fruit and was, according to figures from the FAO, the sixth most valuable food crop after rice, maize, wheat, potato and soybean in 2016.

Despite their importance as a food crop, bananas are often considered an orphan crop when it comes to investment into making production more efficient and sustainable. Funds invested in research on bananas, in areas such as genetics, breeding, variety evaluation trials, physiology, pathology and entomology, have always been far less than the resources available to those researching major crops of comparable economic importance grown in temperate climes. Investment in interdisciplinary research involving environmental and social aspects of banana production is even further behind. Since no contemporary book exists on banana cultivation, we had no hesitation in agreeing to edit a three-volume book series on bananas. Volume 1 is focussed on the cultivation of bananas, Volume 2 on genetics and breeding, and Volume 3 on banana diseases. Each volume stands on its own but together, we hope, the three books will provide a comprehensive review of all the key aspects of research into making cultivation more sustainable in the long term.

Bananas were among the first crops cultivated by man. Their wide utility as building and packing material, fibre for making rope and string, and cooked food in the form of starch from the corm and pseudostem, male flowers and the seeded bananas, preceded eating the seedless fruit we are familiar with today. Early farmers cultivated seeded bananas for these purposes in Asia, from which they identified seedless parthenocarpic plants which they subsequently propagated vegetatively. Movement of these plants by man in Southeast Asia also facilitated hybridisation between different Musa species, giving rise to novel interspecific hybrids. It is interesting to note that almost all varieties of banana we use today were selected by farmers in antiquity. Thousands of such varieties were selected in the past and brought into cultivation through vegetative propagation, giving rise to our current global suite of germplasm of seedless bananas. Since many seedless bananas are intra- and inter- specific hybrids which vary in ploidy level, they are expected to have high levels of heterozygosity which is maintained by man through ongoing vegetative propagation. Selections of Cavendish for example can, under ideal growing conditions and best agricultural practice, produce up to 80 tons of fruit per ha. Although modern plant breeding has played no role in generating the current varieties, the genetic yield potential of these early farmer selections compares very favourably to modern breeding efforts in other fruit crops.

On the one hand, continuous vegetative propagation and high levels of sterility in banana varieties locks in high genetic potential but, on the other, they also hamper significant genetic improvement of bananas. After discovering a new sterile hybrid in the wild, any advance in productivity solely relies on improving cultivation methods. Therefore the first book of the banana trilogy deals with development of sustainable cultivation techniques as outlined in more detail below.

re-establish. This chapter briefly describes the basic anatomy of the banana plant required to understand the part of the plant used for propagation. The chapter then describes the various practical aspects of banana propagation using either conventional planting material or tissue culture plants and the benefits and disadvantages of each method. The chapter looks ahead to future research trends in the area of banana tissue culture.

Chapter 7 reviews how the availability of pest- and disease-free planting material is fundamental for successful banana cultivation. This chapter examines the key issues in the selection, establishment and management of a field germplasm collection, with a focus on how best to provide a reliable source of quality banana planting stock. The chapter discusses characterisation strategies for variety selection and includes a case study of banana cultivation in Queensland, Australia. The chapter addresses the sourcing of material for the germplasm collection and strategies to ensure freedom from pests and disease, and suggests future directions for research.

Chapter 8 reviews the main GAP-related certification schemes used in the sector: GlobalGap, Rainforest, EU Organic, Fair Trade and SA8000. While these schemes have made important improvements, growers have also seen them as a significant and growing bureaucratic burden. Some versions of standards risk being the certification of business as usual. The chapter shows how GAP certification procedures can be re-oriented to become more useful in benchmarking and improving productivity. It also argues that certification criteria should focus not just on specific inputs, but on broader ecological processes (such as soil ecosystem services). The chapter suggests using certification schemes to promote practices such as the use of cover crops and plant residues to protect soil, stimulate root activity and soil biology. As the chapter argues, a more holistic approach will help to achieve true sustainable intensification.

Chapter 9 highlights that fact that although a tropical climate is commonly thought to be more suited to banana production than a subtropical climate, banana yields obtained in the subtropics are excellent and, with good cultural practices, can be considered among the highest in the world. The relatively cool climate of the subtropics causes undesirable physiological phenomena and a long cropping cycle, but it has advantages as well. This chapter describes appropriate cultural practices designed to solve or minimize the climatic constraints on cultivation of bananas in the subtropics, including cultivation under greenhouse conditions.

Moving the focus onto a specific aspect of banana cultivation, Chapter 10 details the significant variation in banana production systems across the world, particularly at the level of soil management. However, as stated at the beginning of the Introduction, research into plant nutrition and soil management in relation to bananas has been limited. Bananas are unusual in their physiology and management, and the characteristics of the crop make it difficult to carry out standard fertilizer response experiments. The current scarcity of fertilization studies in bananas is perhaps surprising, given the significant economic consequences of fertilization costs. This chapter provides an overview of general aspects of soil management and then looks in more detail at crop nutrition. Focusing mostly on intensively managed systems where fertilizer use is highest, the chapter discusses a number of strategies for soil nutrition as well as advantages and disadvantages of each approach.

Chapter 11 highlights that bananas are one of the most widely traded crops globally, and with appropriate management, can be produced on a continuous basis. However, the journey from field to retail destination is a delicate, finely-tuned process which must be closely managed at every stage to ensure product safety and quality. This chapter describes each stage of this journey, evaluating different strategies and challenges. The chapter reviews criteria for determining when to harvest a bunch, followed by a description of different harvesting and transport techniques to avoid crop damage. Post-harvest losses can be reduced through effective techniques of dehanding, washing, sorting and packaging. The chapter concludes with a discussion of future trends in banana cultivation, balancing the need to minimise production costs with increasing consumer concern for the environmental and human impact of banana production.

Complementing the previous chapter's focus on harvesting, Chapter 12 highlights the importance of correct ripening for optimum taste, appearance, shelf life and consumer satisfaction. This chapter reviews the skills and technologies needed for successful ripening, together with innovations and recent developments. Advice on procedures and best practice throughout the supply chain is presented from the author's unique experience in this sector.

Part 3 Assessing and improving sustainability

The themes of chapters in the third part of the volume address how to assess and improve the sustainability of banana production. As one of the world's most traded fruits, bananas have attracted considerable attention for their social and environmental impact throughout the whole value chain. Chapter 13 describes concepts and tools for life cycle assessment (LCA) and carbon footprint (CFP) analysis as applied to the banana value chain. The chapter examines how LCA and CFP methods can be applied to analyse the environmental impact and sustainability of banana cultivation, and how it can inform banana labelling. It includes case studies of Ecuadorian and Peruvian banana value chains and looks ahead to future tools and practices that may be developed in this area.

The last two chapters of the book focus on ways of making cultivation of bananas more sustainable, with Chapter 14 highlighting that there is no single strategy for sustainable banana cultivation. Echoing Chapter 8, the chapter addresses the gap between implementation of global sustainability standards in the export industry based on a single banana variety, i.e., Cavendish, and the diversity of varieties, uses and production environments related to local food security and rural livelihoods. The gap between ecologically and socially-diverse production systems and management strategies relying on standardisation of agronomic practices complicates coordinated action working towards multiple solutions needed to improve sustainability. This chapter includes a case study of a major banana-producing region in the Philippines, representing a variety of banana production systems, which shows the importance of working towards a territorial approach that has the capacity to accommodate and connect multiple solutions to make banana cultivation more sustainable in the long term.

The last chapter of the volume, Chapter 15, addresses the fact that the cultivation of conventional bananas of the commercial Cavendish variety for export markets in the wet tropics requires a substantial amount of synthetic agrochemical input. Over the last 25 years organic banana production has soared in Latin America. This chapter describes organic banana production, identifying the key requirements for success, including location, soil quality and the use of rotations. The chapter also examines the importance of soil fertility and fertilization, disease management, weed management, and managing nematode and insect pests within organic systems.

Part 1

Banana production and its challenges

Chapter 1

The origin, domestication and dispersal of bananas

Hugo Volkaert, National Center for Genetic Engineering and Biotechnology, Thailand

1 Introduction
2 Origins: morphological and molecular evidence
3 The role of *Musa balbisiana*
4 Domestication and early cultivation of bananas
5 Archaeological evidence for early banana cultivation
6 Dispersal of bananas into Oceania
7 Dispersal of bananas into Africa
8 Dispersal of bananas in South and Southeast Asia
9 Conclusion
10 Where to look for further information
11 References

1 Introduction

The search for the origins of edible bananas probably started soon after European explorers brought home stories about the wonderful fruit they had come across in Africa. The wide variety in morphology of the banana plant and fruit initially created a profusion of names among taxonomists (Häkkinen and Väre 2008). However, since seeded banana species only occur in South and Southeast Asia and Oceania, the origin of the edible '*Musa sapientium*' and '*Musa paradisiaca*' had to be sought there. Currently about 40 seeded species in the genus *Musa* have been recognised. It gradually became clear that most edible banana varieties had closest affinities with the seeded species *Musa acuminata* Colla and *Musa balbisiana* Colla (Simmonds 1962). The Fe'i bananas cultivated by some Pacific islanders derive from *Musa* section 'Callimusa' (formerly 'Australimusa') species. The knowledge of these bananas is far too incomplete to even attempt to reconstruct a history of their domestication, and thus they will not be considered further here.

http://dx.doi.org/10.19103/AS.2018.0020.01

2 Origins: morphological and molecular evidence

Musa acuminata Colla is a highly polymorphic species complex with several recognised subspecies which occur in the wild from Northeastern India to the Philippine islands, Papua New Guinea and Northeastern Australia (Fig. 1). Morphological and molecular studies of the accessions of wild bananas have been able to distinguish five or more distinct clusters, corresponding to the ssp. *banksii* + *errans*, *zebrina*, *malaccensis*, *burmannica* + *burmannicoides* + *siamea* and *truncata* (e.g. Risterucci et al. 2009). The subspecies *microcarpa* is distinguished based on morphology but molecular genetic analysis would indicate it might be an admixed population (Perrier et al. 2009). However an additional 11 varieties from Indonesia alone were described by Nasution (1991) some of which like var. *alasensis* and var. *sumatrana* have clearly distinct characters and should be elevated to subspecies level.

Musa acuminata may be a rather promiscuous species, hybridising with other *Musa* species that occur in sympatry. Several edible banana cultivars are known hybrids of *M. acuminata* with *M. balbisiana* or *M. schizocarpa* or even with species from the section 'Callimusa' (D'Hont et al. 2000). Hybrids between *M. acuminata* and *M. rubra* Kurz (*M. laterita* Cheesman) can occasionally be found in the wild in Northern Thailand (Fig. 2a). A large population of wild bananas in Northern Thailand consists of apparently diploid and triploid hybrids between *M. acuminata* of both ssp. *malaccensis* and ssp. *burmannica* and an unknown species related to *M. sikkimensis* (Fig. 2b).

Musa balbisiana Colla occurs from Eastern and Northern India and Southern China to at least Central Indonesia but has also been reported from Papua New Guinea (Argent 1976). It is far less diverse in morphology and molecular genetics with only some recognisable varieties (Häkkinen and Väre 2008).

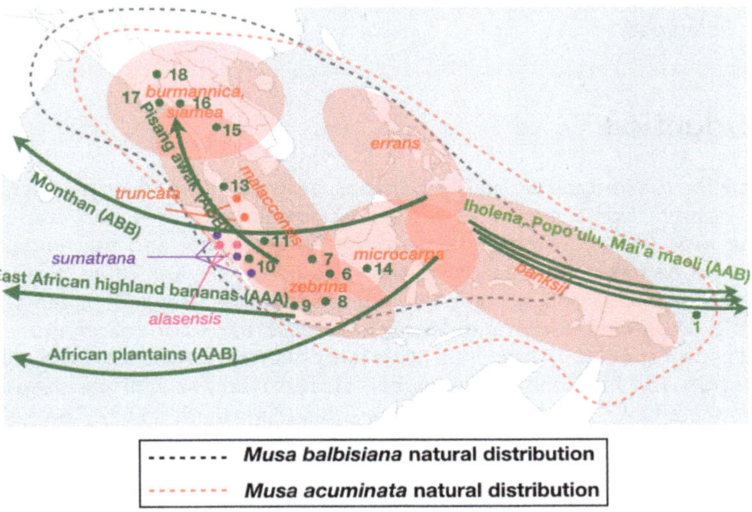

Figure 1 Geographical distribution of *Musa acuminata*, its subspecies and *Musa balbisiana*. Indicated are also the approximate origins for various banana cultivar groups. The numbers on the map refer to the numbers in Table 1.

Early studies placed the origin of cultivated bananas in Southeast Asia, more specifically in Malaysia as the presumed *M. acuminata* ancestral species is most diverse there (Simmonds 1962). Molecular analysis using isozyme markers (Lebot et al. 1993, 1994), cytoplasmic and nuclear DNA markers (Carreel et al. 1994, 2002), microsatellites (Perrier et al. 2009) and DArT (Risterucci et al. 2009) have indicated that almost all edible cultivars are highly heterozygous, most likely as the result of hybridisation between divergent parents. These studies also indicated the highest similarity between cultivars and wild *M. acuminata* ssp. *banksii* and *errans*. Since these two subspecies occupy a distinct geographic region, the Philippines, Eastern Indonesia and New Guinea, wild plants or primitive seeded cultivars are presumed to have been transported by humans in order to come into contact with other subspecies and hybridise, forming presumed heterozygous banana proto-cultivars.

Detailed analysis of RFLP markers (Raboin et al. 2005) and SSRs (Perrier et al. 2009) pointed to the origin of 'Gros Michel' and 'Cavendish' bananas from diploid parental varieties 'Sa' and 'Akondro Mainty', affiliated with the Mlali group of banana cultivars. From these molecular studies together with linguistic analysis and archaeological finds, a scenario was put forward wherein initial seeded proto-cultivars were taken along by humans on their voyage from their home range in Eastern Indonesia or Papua to Western Indonesia, Malesia and the Indian subcontinent. During the travel these fertile proto-cultivars hybridised with local wild bananas and at some point almost completely sterile

Figure 2 Examples of hybrids between *Musa acuminata* and other species that can be found in northern Thailand. (a) Hybrid with *Musa rubra/M. laterita*. The clump structure is quite loose, the rachis position is mostly supra-horizontal but never erect, the bud colour is a reddish orange, the bud shape is lanceolate as in *M. acuminata*. Fruits do rarely develop fully and the seed almost never mature. (b) Hybrid with an as yet unidentified species. Molecular analysis indicated that both diploid and larger triploid individuals occur. Very typical is the quick (diploids?) or more gradual (triploids?) decline of the male bud and termination of the apical meristem into a flower after which the male bud stage terminates. The fruits can be variable in shape and size, but mostly have pronounced ridges. The seed develop to maturity (scale bar = 5 mm).

hybrid diploid AA cultivars were generated. Because of the hybridisation between divergent genotypes, problems could have arisen during meiosis leading to the formation of diploid gametes which then after fertilisation produced triploid AAA genome cultivars. When primitive fertile cultivars were brought to areas where *Musa balbisiana* was present, interspecific hybrids were formed, which lead to the formation of AB, AAB and ABB cultivars (Perrier et al. 2011) through various crosses and back-crosses.

However, even though cultivated bananas clustered with the wild bananas from Papua New Guinea in multi-locus distance-based molecular genetic analyses, that does not amount to evidence that the cultivated bananas are derived from Papuan ssp. *banksii* genetic origins.

In a recent molecular genetic study, Sardos et al. (2016) used DArT markers to study genetic diversity and relationships among wild and cultivated bananas. While a nice separation of cultivars based on the B genome contribution was obtained, the clustering of the cultivars with *M. acuminata* subspecies or geographic origins was not resolved. The 'Cavendish' bananas did clearly not cluster together with the Papuan *M. acuminata banksii*, nor any of the other wild subspecies included in the analysis. A total of 117 distinct origins were estimated for the diploid AA and 78 for the triploid AAA cultivars. There was also a clear separation of the Papuan diploid cultivars from those in South and Southeast Asia.

In a study of SSR markers, Hippolyte et al. (2012) could assign parents for a few groups of cultivated bananas, but for most cultivars none of the studied wild bananas nor diploid cultivars matched the alleles present in the triploids. In another study (Volkaert 2018), a molecular analysis of alleles at five nuclear gene coding loci clearly specifies the diverse origins for edible bananas. This study confirmed that with rare exceptions all cultivars have admixed genotypes from a wide range of ancestral backgrounds. Some cultivars have indeed alleles very similar or identical to *M. a. banksii* commonly found in the Philippines and Papua New Guinea, notably the starchy ABB bananas such as 'Monthan' and 'Bluggoe' group, AAB African plantains and the AAB 'Popo'ulu', 'Iholena', 'Mai'a Maoli' groups. The 'Cavendish' and 'Gros Michel' bananas and most AA and AAA cultivars had no molecular genetic traces of a *M. a. banksii* origin, rather having alleles similar to those found in the ssp. *zebrina* and ssp. *malaccensis*, thus presumably a more Western Indonesian origin. From the study it is clear that the cultivated bananas have generally admixed (hybrid) genetic origins and come from a wide diversity of genetic backgrounds without a trace of a common ancestral homeland from where they have been postulated to be carried by people along their migrations.

As detailed in a few molecular studies (Perrier et al. 2009; Volkaert 2018), also some wild accessions of banana seem to have higher genetic diversity and could have derived from inter-subspecific hybrids. The accession of *M. a.* ssp. *microcarpa* from Borneo was more heterozygous and may be a hybrid with a *M. a. zebrina* parent (Perrier et al. 2009), but admixed individuals can also be found in other populations (Volkaert 2018).

Volkaert (2018) observed that a large number of alleles found in the cultivars were not present in the wild bananas studied, highlighting the poor representation of the ancestral wild bananas in the international germplasm collection. Thus the apparent closest relationship between cultivated bananas and *M. acuminata* ssp. *banksii* from Papua New Guinea is most likely an artefact of the poor representation of wild relatives in the germplasm collection and the very poor ability of distance-based clustering algorithms to resolve hybridisation and admixture events. An illustrative case are the cultivars belonging to the 'Pisang Jari Buaya' group which in a study of diversity at 19

Table 1 An attempt to indicate the putative region of origin and pre-modern region of distribution for some groups of banana cultivars

	Cultivar/group	Genome composition	Region of origin	Region of distribution	Remarks
1	Various local cultivars from Near Oceania, Bougainville	AA, AAA	Bismarck Archipelago	Near Oceania and Papua	Sachter-Smith (2011), Sachter-Smith et al. (2017), Sardos et al. (2016)
2	Iholena, Popo'ulu, Mai'a Maoli Laknau	AAB AAB	Eastern Indonesia, Bismarck Archipelago, Papua	Near and Remote Oceania Philippines	de Langhe et al. (2015), Kagy et al. (2016), Volkaert (2018)
3	African plantains	AAB	Eastern Indonesia, Philippines	West and Central African rainforest	de Langhe et al. (2015), Volkaert (2018)
4	Monthan Bluggoe	ABB	Eastern Indonesia	India	Volkaert (2018)
5	East African highland bananas	AAA	Java (?)	East African highlands	Volkaert (2018)
6	Cavendish Gros Michel Pome	AAA AAA AAB	Western Indonesia, Malayan peninsula	Indonesia. Annam, Southern China, Philippines Now globally	Kervegant (1933), Volkaert (2018)
7	Pisang Raja Bulu Mysore	AAB AAB	Western Indonesia, Malayan peninsula	Indo-Malaya, India	Volkaert (2018)
8	Pisang mas/Sucrier	AA	Western Indonesia, Malayan peninsula	Indo-Malaya Now globally	Volkaert (2018)
9	Kluai Leb Mu Nang	AA	Western Indonesia, Malayan peninsula	Indo-Malaya, Thailand	Volkaert (2018)
10	Pisang Jari Buaya	AA	Sumatra	Indo-Malaya, Philppines, Papua	Volkaert (2018)
11	Red	AAA	Western Indonesia, Malayan peninsula	Indo-Malaya, India Now globally	Volkaert (2018)

(Continued)

Table 1 (*Continued*)

	Cultivar/group	Genome composition	Region of origin	Region of distribution	Remarks
12	Pisang Awak/Kluai NamWa	ABB	Western Indonesia, Malayan peninsula	Indo-Malaya, Thailand	Volkaert (2018)
13	Kluai Khai Thong Reuang	AAA	Malayan peninsula, Thailand	Thailand, recent introduction to Central Africa (Yangambi km[5])	Volkaert (2018)
14	Pisang Batu	ABB	Western Indonesia, Malayan peninsula	Philippines (Saba), Indo-Malaya, Thailand (KluaiHin)	Volkaert (in prep)
15	Kluai Teen Tao	AAA	Thailand	Thailand	Volkaert (in prep)
16	Kluai Nang Klaay Kluai Teeb (Mukdahan)	AAB ABB	Thailand, Myanmar	Thailand	Volkaert (in prep)
17	Kluai Theparod Kluai Leb Chaang Kud	ABB/BBB ABB/BBB	Thailand, Myanmar	Thailand	Volkaert (2018, in prep)
18	Kluai New Mu Nang	AXB	Thailand, Myanmar	Thailand	Volkaert (in prep)

SSR markers by Christelová et al. (2017) were clustered together with the *M. acuminata*. ssp. *burmannica*, while for each of the five gene coding regions studied by Volkaert (2018) clearly demonstrated these cultivars belong to a (possibly F1) hybrid between an admixed ssp. *zebrina* and ssp. *malaccensis* plant and a unique highly divergent genotype belonging to an as yet unknown subspecies or lineage in *M. acuminata* not included in any germplasm collection.

The conclusion that has to be drawn is that the cultivated bananas originated from a wide diversity of genetic backgrounds presumably at several locations all over island and mainland Southeast Asia and at different times. Though the involvement of human transportation of germplasm and subsequent hybridisation cannot be excluded, it clearly was not a prerequisite for the generation of seedless banana cultivars.

It is clear however that the wild *Musa acuminata* accessions in germplasm collections across the world lack representative germplasm from a large section of the wild diversity in Indonesia, Borneo and Philippines. As long as that gap has not been filled, any deductions about the genetic and geographical origin of the different banana cultivars remain highly speculative. Nevertheless I have made an attempt to indicate the region of origin for several groups of banana cultivars in Table 1 and Fig. 1.

3 The role of *Musa balbisiana*

From the study of Volkaert (2018) it also follows that the hybridisation between *M. acuminata* and *M. balbisiana* happened at several locations with Eastern Indonesian *M. a. banksii – errans* ('Monthan' (ABB), African plantains (AAB), Pacific plantains (AAB)), Western Indonesian *M. a. zebrina – malaccensis* ('Pisang Awak' (ABB), 'Pome' (AAB), 'Silk' (AAB), 'Mysore' (AAB)) and Mainland SE Asian *M. a. burmannica – siamea* ('Kluai Nang Klaay' (AAB), 'Kluai Teeb' (ABB), 'Kluai Leb Chang Kud' (ABB)).

It has been assumed that the natural distribution of *Musa balbisiana* is clearly north of the *M. acuminata* range (Perrier et al. 2011) and the *M. balbisiana* found in Southeast Asia has been brought there by human migration in the opposite direction of the assumed banana proto-cultivar spread (de Langhe et al. 2015). However, the presumed anthropogenic presence of *M. balbisiana* in Thailand, Malaysia, Indonesia and the Philippines is based on the observation by Simmonds (1956) and others since, that the species grows rather sparsely distributed, mostly close to villages and is sometimes cultivated. *M. balbisiana* is the preferred plant for non-food purposes such as the leaves being used as food wrapper and decorations and for religious and cultural ceremonies.

However, since there are many regions in Southeast Asia where *M. balbisiana* does not occur at all, neither in the villages nor in the wild, people living in those places use other banana species as a wrapper or for religious and cultural ceremonies (see Nasution 1991), thus refuting the hypothesis that *M. balbisiana* is linked to human presence through its importance in traditional practices. The linkage of *M. balbisiana* to non-food usage might be explained differently. In many locations *M. balbisiana* naturally grows in environments that are also favoured by people to establish villages which accounts for its proximity to human settlements where the plants are tolerated, rather than actively nurtured. Since cutting leaves (and even more so cutting the pseudostem) from cultivated banana plants delays and/or diminishes the fruit production, people preferred to use any other banana plant for non-food uses. Thus, as the wild species growing closest to the village was *M. balbisiana*, it became the preferred species for these uses and was linked to the traditions through story telling. If no *M. balbisiana* was present close to the village, then other species (*Musa acuminata*, *M. itinerans*, but also many other plants such as *Nelumbo nucifera*) were used for wrapping, cooking, fibre or other purposes.

4 Domestication and early cultivation of bananas

Whether humans had an active role in the transportation of germplasm or not, genetic admixture has probably been a driving force in the genesis of edible bananas. After admixture the interplay between alleles at interacting loci might lead to problems in embryo development resulting in (hybrid) sterility. When additional alleles for parthenocarpic fruit development are present, seedless diploid plants were generated. Interactions of alleles among some other loci might have led to problems during meiosis, producing unreduced gametes and setting the stage for the origin of triploid cultivars. The existence of such interacting loci is the basis of the Bateson–Dobzhansky–Muller model and variations thereof to explain hybrid incompatibility in interspecific hybrids (e.g. Scopece et al. 2010). In rice, at least 50 loci controlling fertility have been identified in *indica* × *japonica* hybrids, some causing female gamete abortion, others causing pollen sterility and some that cause both (Ouyang and Zhang 2013). In *Arabidopsis* duplication of genes followed by sub-functionalisation has been reported to lead to incompatibility in progeny that receive both the non-functional copies of the genes (Bikard et al. 2009; Vlad et al. 2010; Durand et al. 2012). Repeated origins of triploids combined with asexual reproduction have been observed in *Arabis holboellii* (Sharbel and Mitchell-Olds 2001) and other plant species. Such 'hybrid incompatibilities' do not necessarily have to happen in the F1 generation, but can be expressed at any generation of back-crossing and intermating in heterogeneous populations. The several subspecies within *Musa acuminata* may have evolved in partial isolation at times but became reconnected at other times during the glacial–interglacial climate cycles and ensuing changes in seawater level. This probably set the stage for multiple sets of mutually incompatible alleles to occur in the large and diverse populations across mainland and insular Southeast Asia.

The seedless banana cultivars were brought into a system of repeated vegetative propagation by early agricultural societies and taken along by people on their long-distance travels across the oceans to faraway places. During the multiple cycles of vegetative reproduction, epigenetic changes or somaclonal mutations may have occurred, changing the expression of some genes, generating morphologically distinct cultivars. This resulted in an impressively large number of varieties among certain cultivar groups such as the African plantains (AAB), 'Popo'ulu', 'Iholena', 'Mai'a Maoli' (AAB) East African highland bananas (AAA), 'Cavendish' (AAA) and 'Monthan' (ABB). Stature, fruit colouration, persistence of the male bud and other traits are variable in several of the cultivar groups. In most of these cases the geographical region with the highest diversity is a long distance from the region where the cultivars were formed. Thus it may even be that the physiological stress induced by long (ocean) travel followed by multiplication from a few surviving suckers after establishment in a new environment resulted in the release of pre-existing or induced somaclonal or epigenetic diversity in the surviving meristematic cells. This epigenetic or somaclonal diversity seems also higher in triploid cultivars than among the diploid cultivars.

5 Archaeological evidence for early banana cultivation

As the banana plant hardly contains any durable tissues, vegetative material does not preserve at all, especially in tropical conditions. Thus for long time the archaeological study of banana cultivation and use has been hampered by an almost complete absence of plant remains. However, the study of phytoliths has yielded some surprising results in recent decades.

Archaeobotanical evidence from Kuk Swamp in Papua New Guinea has established that plant exploitation and possibly cultivation started in the high valleys of Papua as early as 10 000 years before present (YBP). High frequencies of Musaceae phytoliths were recorded in feature fills associated with mound cultivation in an anthropic landscape along the wetland margin leading to the hypothesis that bananas were not only present, but probably cultivated starting from at least 6500 YBP (Denham et al. 2003).

Since molecular studies at the time had indicated the closest similarity between cultivated bananas and wild bananas from Papua New Guinea, it had been surmised that the traces of banana cultivation at Kuk Swamp implied an early origin and start of domestication of bananas in or around New Guinea and a subsequent spread from there. However the latest molecular studies by Sardos et al. (2016) and Volkaert (2018) indicate that this conclusion may have been premature. There is good evidence for an early and independent origin of plant cultivation in the Papua region and bananas might still have been cultivated early at Kuk Swamp, but it is unknown whether they were cultivated for the edible fruits, for the starchy corm or for other non-food uses of banana. Even if they were cultivated for the edible fruits, there is no evidence that the Papuan bananas were transported westward throughout insular Southeast Asia by people. Papua New Guinea and Near Oceania have a large number of indigenous edible AA and AAA cultivars which are not found elsewhere (Arnaud and Horry 1997; Sachter-Smith 2011; Sachter-Smith et al. 2017; Sardos et al. 2016).

Archaeobotanical evidence for early banana cultivation in Southeast Asia is ambiguous. As the wild banana populations are ubiquitously present it is almost impossible to attribute phytoliths found in archaeological contexts to cultivation. Only phytoliths of larger size, derived from triploid plants, would be indicative for cultivation in the region. Thus only a detailed molecular study of the contemporary wild populations of *Musa acuminata* and *Musa balbisiana* can shed light on the different locations of the origin for each cultivar group.

Phytolith finds in South Asia are still few and far between. The typical 'volcaniform' (Ball et al. 2006) phytoliths of *Musa* have only been reported in an archaeological context from Kot Diji, a Harappan culture site in Pakistan (Fuller and Madella 2001). The size of the recovered phytoliths suggest that they were derived from triploid bananas. This would imply that triploid bananas were cultivated in South Asia around 4000 YBP (Fuller and Madella 2009). Although the importance and scale of the banana cultivation in Harappan is unclear, the observation that banana cultivars might have spread beyond the assumed natural distribution of the wild bananas at such early date is significant.

Linguistic evidence has been integrated with molecular genetic and archaeobotanical finds into a hypothesis on the origin and spread of bananas (Perrier et al. 2011). Words used to name bananas among the populations of Southeast Asia were studied to extract information about the relationships among languages. Languages spoken by people show similarities among related populations and can be analysed in ways similar to the evolution of genetic diversity. However, as human cultures and languages can be transmitted both vertically (from ancestry populations to descendant populations) but frequently also horizontally (from donor culture to receptor culture), relationships among languages do not always reflect relationships among the human populations that speak the languages. A relevant case in point is the complex linguistic and genetic ancestry of modern-day Remote Oceanians (Posth et al. 2018; Lipson et al. 2018). Complicating the matters more is the fact that the word used to describe bananas usually refers to both edible and wild bananas while additional designators are used to specify the status as wild or a particular cultivar. Thus the relationships traced through the study of the words for banana do not necessarily trace the sharing of banana cultivars and cultivation, especially in regions where both wild and cultivated bananas co-occur. It is only in those instances where the linguistic relationships for the word 'banana' differs from the general pattern that there might be some information related to bananas specifically, although languages are replete with serendipity that could foil any attempt to extract information on banana domestication from languages.

6 Dispersal of bananas into Oceania

The Pacific plantains (AAB) share a genetic ancestry with the wild *M. acuminata banksii* from Eastern Indonesia and Papua (de Langhe et al. 2015; Volkaert 2018). Morphological analysis (Kepler and Rust 2011) and molecular studies (Kagy et al. 2016; Volkaert 2018) have indicated that there are at least three distinct types of Pacific plantains outside Papua New Guinea: 'Popo'ulu', 'Iholena' and 'Mai'a Maoli'. These traditional varieties spread throughout the Pacific islands following the South- and Eastward movement of Austronesian speaking people from East Asia (Skoglund et al. 2016) as deduced from anthropological, archaeological and linguistic sources (Kirch Patrick 2010; Kepler and Rust 2011). Even though at present all three groups of cultivars are cultivated throughout most of Remote Oceania, it is unclear whether they were part of the same wave of Austronesian migration as other people followed behind (Posth et al. 2018; Lipson et al. 2018). Within Papua New Guinea itself these cultivars are essentially unknown, though some other similar AAB cultivars do occur.

As archaeobotanical evidence for an early independent agricultural origin on Papua New Guinea is becoming stronger, and as the presence of phytoliths implicates the cultivation of Eumusa bananas in several Near and Remote Oceanian sites (e.g. Lentfer and Green 2004), a plausible scenario for the Pacific plantains would start with the discovery of edible hybrid bananas in wild populations in Eastern Indonesia or Papua, their local cultivation and gradual spread. When migrating Austronesian or Papuan people arrived in the region and added the bananas to their culture, the spread of these cultivars from Near Oceania to Remote Oceania became possible. This migration started more than 3000 years ago (Bellwood 1995; Sand 2010; Petchey et al. 2014) and ended when settlers reached Hawaii about 1200 years ago. During that time the cultivars underwent a diversification process resulting in the wide variety of forms within each of the cultivar groups that is observed today.

7 Dispersal of bananas into Africa

Traditional banana cultivation systems in Africa can be separated in three types. In West and Central African rainforest environments, the African plantains (AAB) have been cultivated since prehistorical times. In the higher elevation regions of Eastern Africa, AAA varieties have been cultivated as a staple food. Along the coastal regions of East Africa a more diverse set of AA, AAA, AAB and ABB bananas can be found.

The discovery of *Musa* phytolith evidence at Munsa, Uganda, dating to 5000 YBP (Lejju et al. 2006) and at Nkang, Cameroon, dating to 2500 YBP (Mbida et al. 2001) would suggest a very early arrival of banana cultivation in Africa. Even though there has been controversy over the early date for Cameroon (see Vansina 2003 and response by Mbida et al. 2005) and the phytolith identification of the Ugandan site has been called into question (Neumann and Hildebrand 2009), it is clear that bananas were already cultivated all over the African tropical region long before European explorers arrived along the coasts.

The East African highland bananas have been shown to be genetically highly homogeneous arising most likely as the progeny from a single cross but having undergone an epigenetic/somaclonal diversification over many centuries of vegetative propagation (Kitavi et al. 2016; Volkaert 2018). These bananas form the staple food for the people living around the African Great Lakes and are used in cooking, baking and brewing. The more than 200 cultivars can be grouped into five morphological clone sets (Karamura et al. 1999). The African plantains are cultivated in the humid forest lowlands of tropical Central and Western Africa and are an important component of the local food and culture (Blench 2009). They also form a large group of morphologically distinct cultivars, again with a highly homogeneous, likely single, genetic origin (Noyer et al. 2005; Volkaert 2018), the cultivar diversity thus generated by epigenetic or somaclonal changes. There is no or only minimal overlap of the areas of cultivation for these two groups of bananas.

It is still highly speculative how these banana cultivars arrived in Africa. From archaebotanical evidence, bananas might have arrived in Eastern Africa at a very early date (Lejju et al. 2006). Though the phytolith evidence will never be able to clarify the particular type, one could assume that it belonged to an ancestral East African highland banana. Genetically, the A genomes in these cultivars (AAA) are related to *M. acuminata* ssp. *zebrina*, and thus most likely from a Western Indonesian (Javanese var. *rutilifes*? (Nasution 1991)) origin. A prehistorical direct link between Java and the East African highlands can most likely be excluded.

However prehistorical contacts between Java and Southern India are highly likely. Early contacts between Southern India and Arabia, and between Arabia and the East African coast have also been documented (Fuller et al. 2011). If the East African highland bananas derive from a Western Indonesian origin and were carried by humans to the West in several stages, then the presence of EAH bananas in intermediate regions has been erased as climatic and socio-agricultural environments changed (cf. disappearance of any banana cultivation after the Harappan civilisation which yielded the Kot Diji phytoliths). Even in its presumed origin region of Western Indonesia this type of bananas is no longer cultivated. Somehow these bananas could adapt well to the East African highlands and became part of the local societies. Later on the sporadic contacts between the East African highland people and African East coast, Arabia and India may have ceased, failing to bring in other cultivars until recently.

The presence of the West African rainforest plantains is even more difficult to explain. The A genome of West African plantains is similar to the A genomes of wild bananas in Central and East Indonesia (Borneo? Sulawesi? Moluccas?). Based on linguistic analysis and cultural associations, Blench (2009) postulated that the African plantains started their diversification and expansion from West and Central Africa. Given the large number of distinct cultivars all originating from possibly one single genotype, a very long time of vegetative propagation has been assumed. The finding of banana phytoliths in Cameroon at a very early date might fit with the hypothesis that plantain cultivation together with taro (*Colocasia esculenta*) and greater yam (*Dioscorea alata*) contributed to a societal change and propelled the expansion of Bantu-speaking people out of Western Africa.

If plantains would have been brought overland to Western Africa, they might well have passed through Eastern Africa where banana cultivation might well already have been established as the archaeobotanical evidence from Uganda predates the Cameroonian. Plantains are however not part of the traditional banana cultivation in Eastern Africa. If plantains would have been found on Madagascar, that might have been an indication that Austronesian ocean farers brought along plantains from their supposed homeland in Southern Borneo (Brucato et al. 2016; Kusuma et al. 2016) to Africa. However, plantains are not part of the traditional cultivation systems on Madagascar either (see also Crowther et al. 2016). In addition, the arrival of Austronesian people in Indian Ocean islands and Madagascar (2300 YBP, Burney et al. 2004; 2000–3000 YBP, Pierron et al. 2017) might well post-date the archaeobotanical evidence of *Musa* phytoliths in Cameroon (2500 YBP). Plantains are known though rarely cultivated in Southeast Asia, but some plantains are important cultivars in South India. However, the morphological diversity of these plantains is far less than in Africa.

More archaeobotanical research is definitely needed to establish the period when and the location(s) where the different bananas arrived in Africa and the routes by which they spread throughout the continent.

8 Dispersal of bananas in South and Southeast Asia

The story of banana cultivation in South and Southeast Asia is very complex as cultivars with highly divergent genome compositions can be grown together even in the same backyard garden. Still some interesting observations about the origin and spread of particular groups of cultivars can be made. A large diversity of ABB genome cultivars belonging to the 'Monthan'/'Bontha' group can be found in Southern India (Saraswathi et al. 2011). Genetically the A genome in 'Monthan' and 'Bluggoe' is related to the ssp. *banksii* from Papua and Eastern Indonesia. The large morphological diversity in India is thus not an indication of the local genetic origin of the cultivars, but it might actually indicate that they were originally introduced from a faraway location and went through a process of repeated vegetative propagation in new environments, not unlike the bananas in the Pacific and the African plantains with which they share genetic similarities.

The diversity of 'Pisang Awak'/'Kluai Nam Wa' in Thailand similarly might indicate a process of vegetative propagation and diversification in new environments. Genetically the 'Pisang Awak' A genome is related to the wild *M. a. malaccensis* and *M. a. zebrina* from Western Indonesia. While 'Pisang Awak' nowadays is a commonly planted cultivar in Indonesia and Malaysia, it seems to display rather limited morphological diversity. In

Thailand, there are several stature variants, variants with more of less wax on leaves and pseudostem, a brown fruit peel variant and variants in fruit pulp colour. None of these would be indicative of a genetic origin within Thailand though. As the names for the 'Pisang Awak' in Indonesia indicate, cultivars may even have been re-introduced from mainland Southeast Asia: 'Pisang Siem' might indicate that the cultivar came from Siam, the old name for Thailand and 'Unti Belanda' from South Sulawesi might indicate that the cultivar was introduced during the Dutch colonial era.

Similarly, the diversity of the commercially important 'Cavendish' and 'Gros Michel' cultivar groups might be greatest outside its region of origin. Genetically the affinities of these cultivars are mostly with *M. a. zebrina* and *M. a. malaccensis*, thus Western Indonesia and Malaya. From their original location, these cultivars spread throughout Southeast Asia and diversified into different morphological forms. Different types were introduced to the Caribbean from Southern China and Annam in the early nineteenth century (Kervegant 1933) where they diversified even further.

Within Southeast Asia and Near Oceania one also encounters a large diversity of diploid banana cultivars most of which until recently did not spread far. The morphological diversity within each of those cultivars is apparently not large in contrast to the diversity found in the triploid cultivars. This might be because of molecular mechanisms such as outbreeding depression (Campbell and Waser 1987; Fischer and Matthies 1997; Schierup and Christiansen 1996) as the result of crossing between very divergent parent plants might have limited their potential to adapt to different environments, while dosage compensation (Birchler and Veitia 2012; Yao et al. 2013) alleviated this burden in triploid bananas. Alternatively, it could be explained by the larger size and higher productivity of triploid bananas that made them clear favourites of the local people which lead to their long-distance spread and large-scale vegetative propagation that released the inherent epigenetic variation.

9 Conclusion

The origins of the banana are as varied as the colour, shape and taste of its fruits. Molecular genetic analysis indicates a wide variety of genetic and geographic origins and most likely also different times. The banana fruits were readily adopted by early agriculturalists that may already have been familiar with the cultivation of banana plants for the edible corm. The edible bananas were carried along on the long-distance migrations undertaken by some of these people and readily adopted by people already present when they arrived in Africa and in India.

To establish the original homelands for the different cultivars a thorough exploration of the wild diversity of *Musa acuminata* and *M. balbisiana* will be needed all through their native ranges. Once a more approximate location of the genetic origins can be established, one could attempt to trace back the banana cultivars and link them to particular human populations that have lived there in the past.

Naturally occurring hybridising populations of *M. acuminata* such as in Northern Thailand should be studied to understand the genetic basis for the origin of diploid and triploid bananas. There is also a need for a systematic search for microfossils (phytoliths and starch grains from corm and fruits) in archaeological contexts in Southeast Asia, South Asia, Africa and across Oceania.

10 Where to look for further information

A good introduction to the bananas and their diversity and evolution is still N. W. Simmonds' 1962 book '*The Evolution of the Bananas*'.

A starting point for information on banana diversity, collection and conservation is the Bioversity International programme on bananas (https://www.bioversityinternational.org/research-portfolio/banana-genetic-resources-and-management-systems/) and the related webpages of the International Transit Centre (https://www.bioversityinternational.org/banana-genebank/) and ProMusa (http://www.promusa.org/). These sites include links to further information resources and key contacts.

The archaeobotanical and anthropological information on banana domestication is highly distributed as it has links to archaeology in the Palaeotropics, anthropology of ancient seafarers and modern human population genetics projects. As explained in the chapter, most of these projects may rely on the premature or even flawed conclusion that bananas originated mainly in the Papua region and were greatly influenced by human interventions.

For background on the Kuk Swamp archaeobotanical site I would like to refer to the Terra Australis vol 46 special issue: 'Ten Thousand Years of Cultivation at Kuk Swamp in the Highlands of Papua New Guinea' (https://press.anu.edu.au/publications/series/terra-australis/ten-thousand-years-cultivation-kuk-swamp-highlands-papua-new/download). Several critical evaluations of early banana cultivation in Africa were published in a special issue of *Ethnobotany Research and Applications* in 2009 (http://journals.sfu.ca/era/index.php/era/). Archaeobotanical research is also ongoing in Sri Lanka (Belilena cave and Fa Hien rock shelter) and in Southeast Asia. For background on ancient links between Southeast Asia, India, Arabia and Africa across the Indian Ocean, see Fuller and Boivin (2009).

11 References

Argent, G. C. G. 1976. The wild bananas of Papua New Guinea. *Notes from the Royal Botanic Garden, Edinburgh* 35: 77–114.

Arnaud, E. and Horry, J. P. (Eds). 1997. *Musalogue: A Catalogue of Musa germplasm. Papua New Guinea Collecting Missions, 1988–1989*. International Network for the Improvement of Banana and Plantain, Montpellier, France, 126pp.

Ball, T. B., Vrydaghs, L., van den Houwe, I., Manwarig, J. and De Langhe, E. 2006. Differentiating banana phytoliths, wild and edible: *Musa acuminata* and *Musa balbisiana*. *Journal of Archaeological Science* 33: 1228–36.

Bellwood, P. 1995. Austronesian prehistory in Southeast Asia: Homeland, expansion and transformation. In: Bellwood, P., Fox, J. J. and Tryon, D. (Eds), *The Austronesians: Historical and Comparative Perspectives*. The Australian National University, Canberra, pp. 96–111.

Bikard, D., Patel, D., le Metté, C., Giorgi, V., Camilleri, C., Bennett, M. J. and Loudet, O. 2009. Divergent evolution of duplicate genes leads to genetic incompatibilities within *A. thaliana*. *Science* 323: 623–6.

Birchler, J. A. and Veitia, R. A. 2012. Gene balance hypothesis: Connecting issues of dosage sensitivity across biological disciplines. *Proceedings of the National Academy of Sciences USA* 109: 14746–53. doi:10.1073/pnas.1207726109.

Blench, R. 2009. Bananas and plantains in Africa: Re-interpreting the linguistic evidence. *Ethnobotany Research & Applications* 7: 363–80.

Brucato, N., Kusuma, P., Cox, M. P., Pierron, D., Purnomo, G. A., Adelaar, A., Kivisild, T., Letellier, T., Herawati, S. and Ricaut, F.-X. 2016. Malagasy genetic ancestry comes from an historical Malay trading post in Southeast Borneo. *Molecular Biology and Evolution* 33: 2396–400.

Burney, D. A., Burney, L. P., Godfrey, L. R., Jungers, W. L., Goodman, S. M., Wright, H. T. and Jull, A. J. 2004. A chronology for late prehistoric Madagascar. *Journal of Human Evolution* 47: 25–63. doi:10.1016/j.jhevol.2004.05.005.

Campbell, D. R. and Waser, N. M. 1987. The evolution of plant mating systems: Multilocus simulations of pollen dispersal. *American Naturalist* 129: 593–609.

Carreel, F., Fauré, S., González de León, D., Lagoda, P. J. L., Perrier, X., Bakry, F., Tézenas du Montcel, H., Lanaud, C. and Horry, J. P. 1994. Évaluation de la diversité génétique chez les bananiers diploïdes (*Musa* sp.). *Genetics Selection Evolution* 26 suppl 1: s125–36.

Carreel, F., Gonzalez de Leon, D., Lagoda, P., Lanaud, C., Jenny, C., Horry, J. P. and Tézenas du Montcel, H. 2002. Ascertaining maternal and paternal lineage within *Musa* by chloroplast and mitochondrial DNA RFLP analyses. *Genome* 45: 679–92.

Christelová, P., de Langhe, E., Hřibová, E., Čížková, J., Sardos, J., Hušáková, M., van den Houwe, I., Sutanto, A., Kepler, A. K., Swennen, R., Roux, N. and Doležel, J. 2017. Molecular and cytological characterization of the global *Musa* germplasm collection provides insights into the treasure of banana diversity. *Biodiversity and Conservation* 26: 801–24.

Crowther, A., Lucas, L., Helm, R., Horton, M., Shipton, C., Wright, H. T., Walshaw, S., Pawlowicz, M., Radimilahy, C., Douka, K., Picornell-Gelaber, L., Fuller, D. Q. and Boivin, N. 2016. Ancient crops provide first archaeological signature of the westward Austronesian expansion. *Proceedings of the National Academy of Sciences USA* 113: 6635–40.

de Langhe, E., Perrier, X., Donohue, M. and Denham, T. 2015. The original banana split: Multi-disciplinary implications of the generation of African and Pacific plantains in island Southeast Asia. *Ethnobotany Research & Applications* 14: 299–312.

Denham, T., Haberle, S. G., Lentfer, C., Fullagar, R., Field, J., Porch, N., Therin, M., Winsborough, B. and Golson, J. 2003. Multi-disciplinary evidence for the origins of agriculture from 6950–6440 Cal BP at Kuk Swamp in the highlands of New Guinea. *Science* 301: 189–93.

D'Hont, A., Paget-Goy, A., Escoute, J. and Carreel, F. 2000. The interspecific genome structure of cultivated banana, *Musa* spp, revealed by genomic DNA *in situ* hybridization. *Theoretical and Applied Genetics* 100: 177–83.

Durand, S., Bouché, N., Perez Strand, E., Loudet, O. and Camilleri, C. 2012. Rapid establishment of genetic incompatibility through natural epigenetic variation. *Current Biology* 22: 326–31.

Fischer, M. and Matthies, D. 1997. Mating structure and inbreeding and outbreeding depression in the rare plant *Gentianella germanica* (Gentianaceae). *American Journal of Botany* 84: 1685–92.

Fuller, D. Q. and Boivin, N. 2009. Crops, cattle and commensals across the Indian Ocean. *Études Océan Indien*, 42–3. URL: http://oceanindien.revues.org/698; doi:10.4000/oceanindien.698.

Fuller, D. Q. and Madella, M. 2001. Issues in Harappan archaeobotany: Retrospect and prospect. In: Settar, S. and Korisettar, R. (Eds), *Indian Archaeology in Retrospect. Volume II. Protohistory.* Publications of the Indian Council for Historical Research, Manohar, New Dehli, pp. 317–90.

Fuller, D. Q. and Madella, M. 2009. Banana cultivation in South Asia and East Asia: A review of the evidence from archaeology and linguistics. *Ethnobotany Research & Applications* 7: 333–51.

Fuller, D. Q., Boivin, N., Hoogervorst, T. and Allaby, R. 2011. Across the Indian Ocean: The prehistoric movement of plants and animals. *Antiquity* 85: 544–58.

Häkkinen, M. and Väre, H. 2008. Typification and check-list of *Musa* L. names (*Musaceae*) with nomenclatural notes. *Adansonia* 30: 63–112.

Hippolyte, I., Jenny, C., Gardes, L., Bakry, F., Rivallan, R., Pomies, V., Cubry, P., Tomekpe, K., Risterucci, A. M., Roux, N., Rouard, M., Arnaud, E., Kolesnikova-Allen, M. and Perrier, X. 2012. Foundation characteristics of edible *Musa* triploids revealed from allelic distribution of SSR markers. *Annals of Botany* 109: 937–51.

Kagy, V., Wong, M., Vandenbroucke, H., Jenny, C., Dubois, C., Ollivier, A., Cardi, C., Mournet, P., Tuia, V., Roux, N., Doležel, J. and Perrier, X. 2016. Traditional banana diversity in Oceania: An endangered heritage. *PLoS ONE* 1: e0151208.

Karamura, D. 1999. *Numerical Taxonomic Studies of the East African Highland Bananas (Musa AAA-East Africa) in Uganda*. International Network for the Improvement of Banana and Plantain, Montpellier, France, 192pp.
Kepler, A. K. and Rust, F. G. 2011. *The World of Bananas in Hawaii: Then and Now*. Pali-o-waipi'o Press, Hawaii, USA.
Kervegant, D. 1933. Introduction à l'étude des variétés de bananiers à fruits comestibles de la Martinique (Suite). *Revue de Botanique Appliquée et d'Agriculture Coloniale* 13(bulletin n°141): 337–46. doi:10.3406/jatba.1933.5235.
Kirch Patrick, V. 2010. Peopling of the Pacific: A holistic anthropological perspective. *Annual Review of Anthropology* 39: 131–48. doi:10.1146/annurev.anthro.012809.104936.
Kitavi, M., Downing, T., Lorenzen, J., Karamura, D., Onyango, M., Nyine, M., Ferguson, M. and Spillane, C. 2016. The triploid East African highland banana (EAHB) genepool is genetically uniform arising from a single ancestral clone that underwent population expansion by vegetative propagation. *Theoretical and Applied Genetics* 129: 547–61. doi:10.1007/s00122-015-2647-1.
Kusuma, P., Brucato, N., Cox, M. P., Pierron, D., Razafindrazaka, H., Adelaar, A., Sudoyo, H., Letellier, T. and Ricaut, F.-X. 2016. Contrasting linguistic and genetic origins of the Asian source populations of Malagasy. *Science Reports* 6: 26066. doi:10.1038/srep26066.
Lebot, V., Aradhya, K. M., Manshardt, R. and Meilleur, B. 1993. Genetic relationships among cultivated banans and plantains from Adia and the Pacific. *Euphytica* 67: 163–75.
Lebot, V., Meilleur, B. and Manshardt, R. 1994. Genetic diversity in Eastern Polynesian Eumusa bananas. *Pacific Science* 48: 16–31.
Lejju, J. B., Robertshaw, P. and Taylor, D. 2006. Africa's earliest bananas? *Journal of Archaeological Science* 33: 102–13.
Lentfer, C. J. and Green, R. C. 2004. Phytoliths and the evidence for banana cultivation at the Lapita Reber-Rakival site on Watom Island, Papua New Guinea. *Records of the Australian Museum, Supplements* 29: 75–88.
Lipson, M., Skoglund, P., Spriggs, M., Valentin, F., Bedford, S., Shing, R., Buckley, H., Phillip, I., Ward, G. K., Mallick, S., Rohland, N., Broomandkhoshbacht, N., Cheronet, O., Ferry, M., Harper, T. K., Michel, M., Oppenheimer, J., Sirak, K., Stewardson, K., Auckland, K., Hill, A. V. S., Maitland, K., Oppenheimer, S. J., Parks, T., Robson, K., Williams, T. N., Kennett, D. J., Mentzer, A. J., Pinhasi, R. and Reich, D. 2018. Population turnover in Remote Oceania shortly after Initial settlement. *Current Biology* 28: 1157–65. doi:10.1016/j.cub.2018.02.051.
Mbida, C., Doutrelepont, H., Vrydaghs, L., Swennen, R. L., Beeckman, H., de Langhe, E. and de Maret, P. 2001. First archaeological evidence of banana cultivation in central Africa during the third millennium before present. *Vegetation History and Archaeobotany* 10: 1–6.
Mbida, C., Doutrelepont, H., Vrydaghs, L., Swennen, R. O., Swennen, R. U., Beeckman, H., de Langhe, E. and de Maret, P. 2005. The initial history of bananas in Africa. A reply to Jan Vansina, Azania, 2003. *Azania* 40: 128–35. doi:10.1080/00672700509480419.
Nasution, R. E. 1991. A taxonomic study of the species *Musa acuminata* Colla with its intraspecific taxa in Indonesia. *Memoirs of the Tokyo University of Agriculture* 32: 1–122.
Neumann, K. and Hildebrand, E. 2009. Early bananas in Africa: The state of the art. *Ethnobotany Research & Applications* 7: 353–62.
Noyer, J. L., Causse, S., Tomekpe, K., Bouet, A. and Baurens, F. C. 2005. A new image of plantain diversity assessed by SSR, AFLP and MSAP markers. *Genetica* 124: 61–9. doi:10.1007/s10709-004-7319-z.
Ouyang, Y. and Zhang, Q. 2013. Understanding reproductive isolation based on the rice model. *Annual Reviews in Pant Biology* 64: 111–35.
Perrier, X., Bakry, F., Carreel, F., Jenny, C., Horry, J.-P., Lebot, V. and Hippolyte, I. 2009. Combining biological approaches to shed light on the evolution of edible bananas. *Ethnobotany Research & Applications* 7:199–216.
Perrier, X., De Langhe, E., Donohue, M., Lentfer, C., Vrydaghs, L., Bakry, F., Carreel, F., Hippolyte, I., Horry, J.-P., Jenny, C., Lebot, V., Risterucci, A.-M., Tomekpe, K., Doutrelepont, H., Ball, T., Manwaring, J., de Maret, P. and Denham, T. 2011. Multidisciplinary perspectives on banana (*Musa*

spp.) domestication. *Proceedings of the National Academy of Sciences USA* 108: 11311–18. https://doi.org/10.1073/pnas.1102001108

Petchey, F. J., Spriggs, M., Bedford, S., Valentin, F. and Buckley, H. 2014. Radiocarbon dating of burials from the Teouma Lapita cemetery, Efate, Vanuatu. *Journal of Archaeological Sciences* 50: 227–42.

Pierron, D., Heiske, M., Razafindrazaka, H., Rakoto, I., Rabetokotany, N., Ravololomanga, B., Rakotozafy, L. M.-A., Rakotomalala, M. M., Razafiarivony, M., Rasoarifetra, B., Andriamampianina Raharijesy, M., Razafindralambo, L., Ramilisonina,, Fanony, F., Lejamble, S., Thomas, O., Abdallah, A. M., Rocher, C., Arachiche, A., Tonaso, L., Pereda-lotha, V., Schiavinato, S., Brucato, N., Ricaut, F.-X., Kusuma, P., Sudoyo, H., Ni, S., Boland, A., Deleuze, J.-F., Beaujard, P., Grange, P., Adelaar, S., Stoneking, M., Rakotoarisoa, J.-A., Radimilahy, C. and Letellier, T. 2017. Genomic landscape of human diversity across Madagascar. *Proceedings of the National Academy of Sciences USA* 114: 6498–506. doi:10.1073/pnas.1704906114.

Posth, C., Nägele, K., Colleran, H., Valentin, F., Bedford, S., Kami, K. W., Shing, R., Buckley, H., Kinaston, R., Walworth, M., Clark, G. R., Reepmeyer, C., Flexner, J., Maric, T., Moser, J., Gresky, J., Kiko, L., Robson, K. J., Auckland, K., Oppenheimer, S. J., Hill, A. V. S., Mentzer, A. J., Zech, J., Petchey, F., Roberts, P., Jeong, C., Gray, R. D., Krause, J. and Powell, A. 2018. Language continuity despite population replacement in Remote Oceania. *Nature Ecology & Evolution* 2: 731–40. doi:10.1038/s41559-018-0498-22018.

Raboin, L.-M., Carreel, F., Noyer, J.-I., Baurens, F.-C., Horry, J.-P., Bakry, F., Tezenas Du Montcel, H., Ganry, J., Lanaud, C. and Lagoda, P. J. L. 2005. Diploid ancestors of triploid export banana cultivars: Molecular identification of 2n restitution gamete donors and n gamete donors. *Molecular Breeding* 16: 333–41.

Risterucci, A. M., Hippolyte, I., Perrier, X., Xia, L., Caig, V., Evers, M., Huttner, E., Kilian, A. and Glaszmann, J.-C. 2009. Development and assessment of diversity arrays technology for high-throughput DNA analyses in *Musa*. *Theoretical and Applied Genetics* 119: 1093–103.

Sachter-Smith, G., 2011. *Bananas of the Solomon Islands: A Catalog of Banana Plants Seen in the Islands of Guadalcanal, Makira and Malaita, Papua New Guinea*. Bioversity International, 54pp. www.musalit.org/seeMore.php?id=15133

Sachter-Smith, G., Paofa, J., Rauka, G., Sardos, J. and Janssens, S. 2017. *Bananas of the Autonomous Region of Bougainville: A Catalog of Banana Diversity Seen on the Islands of Bougainville and Buka, Papua New Guinea*. Bioversity International, 97pp. www.musalit.org/seeMore.php?id=17006

Sand, C. 2010. Lapita Calédonien. Archéologie d'un premier peuplement insulaire Océanien. Société des Océanistes, Travaux & Documents Océanistes 2, 296pp.

Saraswathi, M. S., Uma, S., Vadivel, E., Durai, P., Siva, S. A., Rajagopal, G. and Sathiamoorthy, S. 2011. Diversity analysis in Indian cooking bananas (*Musa*, ABB) through morphotaxonomic and molecular characterisation. *Acta Horticulturae – Proceedings of the International ISHS-ProMusa Symposium on Global Perspectives on Asian Challenges* (Guangzhou, China, 14–18 September 2009) no 897, pp. 123–31. ISBN: 978-90-6605-138-6.

Sardos, J., Perrier, X., Doležel, J., Christelova, P., Van den houwe, I., Kilian, A. and Roux, N. 2016. DArT whole genome profiling provides insights on the evolution and taxonomy of edible banana (*Musa* spp.). *Annals of Botany* 118: 1269–78.

Schierup, M. H. and Christiansen, F. B. 1996. Inbreeding depression and outbreeding depression in plants. *Heredity* 77: 461–8. doi:10.1038/hdy.1996.172.

Scopece, G., Lexer, C., Widmer, A. and Cozzolino, S. 2010. Polymorphism of postmating reproductive isolation within plant species. *Taxon* 59: 1367–74.

Sharbel, T. F. and Mitchell-Olds, T. 2001. Recurrent polyploid origins and chloroplast phylogeography in the *Arabis holboellii* complex (Brassicaceae). *Heredity* 87: 59–68.

Simmonds, N. W. 1956. Botanical results of the banana collecting expedition 1954–55. *Kew Bulletin* 11: 463–89. doi:10.2307/4109131.

Simmonds,, N. W. 1962. *The Evolution of the Bananas. Tropical Science Series*. Longmans, London (GBR), 170p.

Skoglund, P., Posth, C., Sirak, K., Spriggs, M., Valentin, F., Bedford, S., Clark, G. R., Reepmeyer, C., Petchey, F., Fernandes, D., Fu, Q., Harney, E., Lipson, M., Mallick, S., Novak, M., Rohland, N., Stewardson, K., Abdullah, S., Cox, M. P., Friedlaender, F. R., Friedlaender, J. S., Kivisild, T., Koki, G., Kusuma, P., Merriwether, D. A., Ricaut, F.-X., Wee, J. T. S., Patterson, N., Krause, J., Pinhasi, R. and Reich, D. 2016. Genomic insights into the peopling of the Southwest Pacific. *Nature* 538: 510–13. doi:10.1038/nature19844.

Vansina, J. 2003. Bananas in Cameroon c. 500 BCE? Not proven. *Azania* 38: 174–6.

Vlad, D., Rappaport, F., Simon, M. and Loudet, O. 2010. Gene transposition causing natural variation for growth in *Arabidopsis thaliana*. *PLoS Genetics* 6: e1000945.

Volkaert, H. 2018. DNA sequence analysis of diversity reveals multiple genetic and geographic origins of cultivated bananas. *New Phytologist* (*submitted*).

Yao, H., Gray, A. D., Auger, D. L. and Birchler, J. A. 2013 Genome dosage effects on heterosis in triploid maize. *Proceedings of the National Academy of Sciences USA* 119: 2665–9. doi:10.1073/pnas.1221966110.

Chapter 2

Understanding the banana industry: monoculture and beyond

Dan Koeppel, Independent Journalist and Researcher, USA

1 Introduction
2 Banana varieties and the Cavendish monoculture
3 The development of a banana monoculture
4 The dynamics of the banana monoculture
5 Post-war developments: business as usual?
6 The Cavendish era
7 Cavendish in crisis and a new paradigm
8 The hunt for new varieties
9 Future trends and conclusion
10 Where to look for further information

1 Introduction

It was March 2008, and Chiquita – the world's leading marketer of the world's most popular fruit – was expanding operations to a new continent. A farm in Mozambique, a nation of 25 million people located on Africa's southeast coast, would soon account for as much as 30% of the company's export business to Europe, according to the banana company's then chief executive officer Fernando Aguirre. That would keep banana prices and supplies stable, and would also, as the company promised, provide huge local benefits, adding as many as 3000 jobs to the regional economy.

However, that expectation failed to materialize. A decade later, that effort, abandoned early on by Chiquita, but continued, with similar ambitions, by regional entrepreneurs, has become not an example of sustainable, profitable expansion of the commercial banana industry, but the opposite: a possible glimpse into that industry's destruction. With bananas in Mozambique came banana disease, Fusarium wilt tropical race 4 (Foc-TR4), an incurable malady never before seen in Africa, where the fruit is a subsistence staple for millions of people.

How did the disease appear thousands of miles from where it had previously been seen? There is a chance it was brought to the region by banana growers themselves, perhaps on tools and equipment; perhaps on clothing worn by imported workers or visitors; or

http://dx.doi.org/10.19103/AS.2017.0020.04

perhaps – somehow – in the starter plants used to launch the Mozambique operation. The fact is that it came, and from far away, which is a testament to the pathogen's ability to traverse oceans.

The troubles related to the fruit offer both a blueprint for a long-predicted disaster and a reminder that history offers lessons. Foc-TR4 had surfaced in Malaysia in the late 1980s and spread to a few Asian nations, including Taiwan and Indonesia. But it soon reached the Chinese mainland. It spread throughout the islands that straddle the confluence of the Indian and Pacific Oceans, through the banana-dependent nation of the Philippines, and then, jumping a thousand miles of ocean, appeared in Australia. In all those places where it arrived, Foc-TR4 devastated the production of Cavendish in plantations. Dispersal of this pathogen happens through contaminated soil, irrigation water and planting material at the local level. The intercontinental move has been especially disturbing, because it showed how easily the pathogen could be transmitted. Foc-TR4 has as yet not hit Latin America, where nearly all bananas consumed by shoppers in the United States are grown. But scientists have predicted it will, eventually. 'Given today's modes of travel, there's almost no doubt it will hit Cavendish crops', said Randy Ploetz, the University of Florida's plant pathologist who identified the original Malaysian strain of the fungus.

2 Banana varieties and the Cavendish monoculture

A Cavendish wipeout would have consequences beyond deaccessorized corn flake breakfasts. The banana is not only the world's most popular fruit, but also the world's most *important* fruit. While most Americans and Europeans eat an average of three or four bananas per person per week, the fruit is relied on as a main course in places like Asia and Africa. In Uganda, Rwanda, Burundi and Tanzania, along with parts of Kenya and the Democratic Republic of Congo, individual consumption ranges up to over 1 kg of bananas a day. Bananas are an important source of energy and are rich in certain minerals as well as vitamins A, C and B6. Without bananas, people would starve.

It is important to realize that there is more than one kind of banana. In fact, the Cavendish – the breed we import, the breed stricken in Mozambique, and just about the only banana type consumed in parts where the fruit is not grown locally – is just one of more than a thousand cultivated varieties of the fruit. Despite this apparent diversity, the Cavendish banana, as known in the Western world, is remarkably uniform. This variety remains statistically the sole global export banana because most other types are too delicate to withstand the ocean voyage required to bring the fruit to non-tropical ports in sellable condition. The global expansion of the Cavendish monoculture, propelled by the drive to sell bananas to residents of distant cities and countries, has begun to squeeze out local varieties, many of which might contain genetic material that could be used in breeding disease-resistant fruit.

The loss of such varieties is critical because bananas – they are seedless and sterile, reproducing via cuttings – cannot be archived in traditional 'seed vaults' currently being operated as protectors of biodiversity. Instead, banana conservation has traditionally been accomplished via growing representative plants in field collections. If these vanishing local varieties are disease resistant, it seems especially ill advised to replace them with a variety that is susceptible to a range of maladies, although that is found to be the current practice.

The breadth of the Cavendish monoculture in banana-exporting countries is astonishing. Why, with so many delicious alternative varieties, did we only get this single version of the fruit? Moreover, why did we get a variety so bland that people in India – the world's largest banana producer, with over 600 recognized cultivars, some of which are so delicious and nutritionally valuable that they are viewed as sacred – call Cavendish 'The Hotel Banana', fit only for timid and unknowing tourists? In the past three decades, Western supermarkets have seen riotous introductions of multiple apple and citrus varieties, along with a growing consumer embrace of exotic fruits – kiwi, açai berries, pomegranate – once considered too unfamiliar for our palates. However, the banana trade has not embraced diversification of its offering to consumers and instead is persisting mostly with the one variety, Cavendish, that it knows how to produce, transport and ripen.

3 The development of a banana monoculture

Looking back into recent history, we learn that Cavendish is the modern era's *second* commercial banana type. When the business was launched by a few dedicated and ruthless entrepreneurs in the late nineteenth century, they grew an entirely different variety, called the *Gros Michel*, or 'Big Mike'. But not long after companies like United Fruit and Standard Fruit – today known as Chiquita and Dole, respectively – broke ground on banana farms in Central and South America, a disease began to hit those plantations. The malady, caused by a fungus, was named 'Panama disease' after the country where it had the earliest, most devastating effects. It moved swiftly from nation to nation, rendering banana fields fallow, never capable of growing Gros Michel again. The spread of Panama wilt disease made previous plantations unviable to such an extent that the banana companies were continuously looking for new land to establish and exploit new plantations.

That was devastating enough.

But the real cost of the bananas – the monoculture, the business model and the way banana infrastructure was created to serve those two elements – was not paid just in fallowed farms and lost profits. Reliance on a single 'world banana', the devastating spread of Panama wilt and the need to expand banana cultivation led to a series of historically significant events which had far-reaching consequences. For some, the insistence on a single banana variety to supply foreign markets was paid in blood, as the following example illustrates.

Just before Christmas, 1929, the town of Ciénaga – in the banana-growing region of Magdalena, Colombia – had been tense for months. Workers there had been demanding better conditions for most of the year, and when they went on strike against the biggest banana company in the world, the action had spread across the nation. They were protesting for basic human rights: healthcare when they were injured, improvements in cramped housing, protection from toxic pesticides and the right to be paid in cash, rather than the scrip issued by the United Fruit Company.

In Latin America, United Fruit was nicknamed 'El Pulpo' – the Octopus – for the way it seemed to control everything. The company's many-tentacled approach was never more evident than during the strike. In the United States, the banana giant launched a media campaign. It was the height of the depression, and public sympathies were with workers. A *New York Times* article assured readers that the protestors belonged to a 'subversive movement', and were 'not representatives of any established body of laborers'. In Colombia, the company asked for – and was granted – a clampdown. Rights to free assembly and free speech were suspended. The move met with the approval of the US government. In a telegram sent to the US Secretary of State Frank Billings Kellogg, ambassador Jefferson Caffery wrote: 'I have been following the . . . strike through United Fruit Company representative here'. The note assured Washington that the Colombian government would give 'protection to American interests'.

But the strike continued and expanded. By autumn, 32 000 workers had left their plantations. On 5 December, martial law was declared.

The next day was a Sunday. Banana workers and their families had just emerged from church, and were crowded in the town square, where they were waiting to hear a speech by the regional governor. Local troops – their actions unnoticed by workers and their families, who were listening to their pastor give a sermon – set up machine gun positions above the square. An order was given. The area was to be cleared in five minutes. Packed together as it was, the crowd was unable to disperse, following which the troops opened fire.

Ambassador Caffery relayed the news to his superiors in Washington. 'I have the honor to report', Caffery wrote, 'that the Bogotá representative of the United Fruit Company told me yesterday that the total number of strikers killed by the Colombian military exceeded one thousand'. The bodies of the slain men, women and children were, reportedly, dumped in the sea; the sequence of events was famously fictionalized by the author Gabriel Garcia Marquez in his Nobel-prize-winning novel, 'One Hundred Years of Solitude'. (With over 25 such banana-led interventions in twentieth-century Latin American history and the actions of the banana industry so deeply woven into regional history, the corruption of banana barons has become a literary meme; name a mid-century Central or South American author or poet, and there is a good chance he or she will, at one point, have taken up the topic of bananas.)

The Colombian banana strike was over, and Americans – though they did not know why – were once again guaranteed that their favourite fruit would remain their most affordable fruit. The corporation that orchestrated the bloodshed, United Fruit, would continue to thrive. Today, it is known as Chiquita.

4 The dynamics of the banana monoculture

Cheap bananas were – and remain – the centrepiece of the strategy that makes the fruit more popular, even today, than apples and oranges combined. A moment of consideration, and the outrageousness of that idea settles in. Bananas are grown in tropical areas and must be shipped thousands of miles. They go bad within a couple of weeks. Apples are grown within a day's drive of nearly every US city. They last for months if they are stored properly.

Under those circumstances, how could bananas possibly be such a bargain?

Shipping and storage were essentially fixed costs. But land and labour could be gotten strategically. Terrain for plantations did not have to be bought at fair prices, nor did workers have to be paid or treated well, if governments cooperated. Money could ensure that cooperation. If money did not work, force was always an option. Over and over, often with US soldiers at their side, banana companies could shoot their way through the Central American fields, guaranteeing that their 'formula' remained intact and that the bananas they had taught us to love came to our tables cheaply and reliably. Overthrows led by the fruit companies became so frequent that the nations of Latin America gained a nickname that is still used when describing a nation ruled by a puppet government today: they were called 'Banana Republics'.

Another factor keeping banana prices down – and that demonstrated the brilliance of the banana entrepreneurs – was sameness. Even with over a thousand banana varieties found around the world, but United Fruit and its smaller competitors – from their very beginnings in the 1880s – focused on just one, the Gros Michel. This standardization ensured that growing and shipping methods were kept the same throughout the plantation world, and it guaranteed that consumers would find the fruit to be always reliable, with consistent ripening and taste. It was the same formula Henry Ford used for his Model T decades later, and that McDonald's would later adopt for hamburgers.

But bananas are a product of nature, not of industry. The banana companies' belief that the latter could overcome the former would lead to even more bloodshed. As we have seen, undermining the entire industry was the unstoppable fungus causing the 'Panama disease', after the place it was first identified. Nearly every Gros Michel it touched perished. Contaminated soil could never again grow the fruit. There was no cure. Plantation by plantation, bananas disappeared. In the 1920s, *The New York Times* visited one wrecked banana farm, where plants lay fallen and blackened amidst abandoned box cars and torn-up rail beds, and described it as a 'leper colony'. At home, the shortages were beginning to be felt, though in ways that almost seem coded.

In the spring of 1923, Tin Pan Alley composers Frank Silver and Irving Cohn wrote a song that became, up until then, the biggest musical hit ever recorded. It described a real-life event: a request for fruit denied by a fruit vendor who was inexplicably out of stock. The title phrase of the song was uttered to Silver by his girlfriend's younger brother, who had heard it from a grocer named Jimmy Costas – an immigrant who had not quite mastered the English language: 'Yes', Costas told his disappointed potential customer. 'We have no bananas'.

But supplies did not decline – they simply dipped, and then rose again. The reason was simple: there was plenty of land in Central and South America. It just had to be taken. Panama disease rampaged on, and so did the banana companies. When bananas died, blood was spilled, new plantations were cut from the forests and jungles, and the fruit at home remained cheap and plentiful. The only respite came in the form of World War II. United Fruit's ships were called into duty for troop and materiel transport. Several were sunk. Since Americans could not get as many bananas, production decreased.

5 Post-war developments: business as usual?

When the war ended, there were two banana companies left: United Fruit – which was about to introduce a mascot called 'Miss Chiquita' – and Standard Fruit (today's Dole). Both knew that their banana was dying. They could stick to the old ways – take more land, trample more people and deny the inevitable – or find a new banana variety to replace Gros Michel. But doing so would require completely reinventing the industry, which was so custom-tailored to the single existing variety it sold that switching to another fruit would almost require the boldness to be willing to start from scratch. Everything, from the way plantations were organized to the way bananas were harvested, shipped, distributed and ripened needed to change. Standard Fruit began experimenting to accomplish just that. Chiquita, with more to lose, refused – and the path they took led to the greatest human rights disaster the hemisphere had seen since the slave trade of the colonial era and the genocide of Native Americans during the westward expansion of the United States. After World War II and the defeat of fascism, the men and women of Latin America believed that they could now have a greater say in their destinies. Once again, Chiquita was terrified of the idea. Even now, with Panama disease having rendered billions of acres fallow, it would not give up one square inch.

But the politicians of the Banana Republics were becoming less pliable – and one, a Guatemalan named Jacobo Arbenz, was determined to test the banana companies. Arbenz was elected in 1950. His goal, he said, was to turn his country from a 'dependent nation with a semi-colonial economy, to an economically independent country'.

The first request Arbenz made, in a meeting with a United Fruit representative, was simple: he asked that the banana company obey existing laws – something it had never done. The result was an explosive argument; reporting back to corporate headquarters, the United Fruit board met, and the company's retired chairman – Samuel Zemurray – took charge.

Zemurray was in his 70s, and at this point in his career, he was better known as a benign philanthropist than an orchestrator of revolutions. He had made generous donations to schools and universities in both the United States and Latin America. But 'Sam the Banana Man' had begun his career with one of the industry's most audacious escapades. In 1910, using his personal yacht, he accompanied a pair of hired thugs (with the colourful names of Lee Christmas and Machine Gun Maloney) to Honduras, where local authorities had been hostile to his presence. The trio launched a coup, and six weeks later, a new president, hand-picked by the banana entrepreneur, was in charge. His first act was to grant his patron thousands of acres of land, on which he could grow bananas tax-free.

With Guatemala, Zemurray returned to form. For the next three years, he personally orchestrated a campaign against Arbenz. Dozens of articles – fuelled by luxurious press junkets, and planted by public relations specialists paid by Zemurray – appeared in newspapers and magazines, including *The New York Times*, *Time*, *Newsweek* and *The Christian Science Monitor*. All portrayed Arbenz as a dangerous radical, whose aim, as one report put it, was to 'engender hatred' of American business.

Arbenz responded by pushing his reforms through. In 1952, Arbenz issued an order called 'Decree 900'. It allowed the government to take possession of any plot of agricultural land over 670 acres – with one qualifier: the land had to be unused. The scheme specifically targeted banana fields made unproductive by Panama disease. Zemurray reached out to even higher contacts. The US Secretary of State, John Foster Dulles, had been a partner at United Fruit's law firm. Dulles' brother Allen was head of the CIA, and a former member of the United Fruit Board of Directors. Zemurray was on a first-name basis with both. Ed Whitman, the banana company's chief of public relations, was married to President Eisenhower's private secretary. The result could have been predicted: in 1953, Eisenhower authorized the CIA to overthrow Arbenz.

The effort was a masterpiece of stagecraft, complete with false reports of political turmoil and defections, all broadcast via the vast radio network United Fruit had strung across Latin America. As US aircraft buzzed Guatemala City, and photographs of mutilated corpses began appearing in American newspapers – a United Fruit executive, decades later, admitted the pictures were of earthquake victims – loyal commanders in the field began finding it difficult to convince local farmers to take up arms; they were certain that they would head into the jungles and be slaughtered by the US soldiers who, the radio said, were waiting for them. (Many of the broadcasts featured realistic-sounding battles, complete with canned gunfire and bomb blasts.)

Six weeks later, the campaign was over, and Arbenz surrendered. He was brought, in chains, to the airport, where he was stripped to his underwear and paraded before the press before boarding a flight to Mexico. Guatemala would once again become banana-friendly – but the price to the Guatemalan people would be beyond high. The country would not see political stability until the 1990s, as a succession of dictatorships plunged the nation into chaos that included disappearances, death squads and genocide that saw over 200 000 native Mayans killed.

As for Arbenz, he wandered, stateless, for fifteen years. No country would accept him. In 1971 – the circumstances are unclear – he was found drowned in the bathtub of his Mexico City hotel room. He was only redeemed in 1995, when his body was finally returned home; over 100 000 people watched as his coffin was borne to the national cemetery, where he was interred in a white, pyramid-shaped mausoleum that is still visited daily by hundreds of pilgrims, who see the fallen leader as the Lincoln or Kennedy of their nation.

6 The Cavendish era

Guatemala's blood did nothing to slow the demise of the banana. The variety – the Gros Michel – is now functionally extinct as a commercial banana variety, growing only in a few isolated patches. By 1960, Chiquita was on the verge of bankruptcy, and Zemurray, stricken with Parkinson's disease, was dying. Its rival, Standard Fruit, had adopted a new variety – called the Cavendish – but Chiquita, which preferred military adventure to product development, was far behind.

The problem with Cavendish was that Chiquita did not believe in it. There were good reasons for that. It was, in almost every way, inferior to the fruit it now had to replace. Cavendish 'broke' the banana equation. It was not tough. It needed expensive bagging and boxing. It ripened poorly, which meant special gases needed to be pumped into temperature-controlled cargo holds to keep the fruit green during transport. Worst of all, the Cavendish simply was not delicious. Compared to the Gros Michel, Chiquita believed that the housewives of America would not accept such a radically different banana.

But now, there was no choice. The banana industry had to change. It did so by modifying the way it handled and grew bananas, tearing apart a plantation system it had spent fifty years building. Assembly lines were introduced to plantations. For the first time, women were employed on a wide scale, used to clean and separate the fruit. As they did, they added something new to the mix: stickers, denoting a brand name, on every bunch.

With regard to the taste issue, the banana industry got lucky: it was, after all, the beginning of the suburban era. The Western world embraced convenience above anything else and it was the dawn of the fast food age. The Western palate was dumbing down at just the right time for a blander banana to take hold.

Cavendish's other advantage was that it could be grown in more places. All that fallow land could support bananas again. Labour still had to be cheap. Skullduggery became more mild, but it continued. The pliable governments of Guatemala, Costa Rica and Honduras made sure that workers did not make trouble, though the countries occasionally attempted to band together and take a little bit bigger piece of the banana pie. (One such attempt was broken up in the early 1970s by a Chiquita bribe to the sitting Honduran president. When the payments were revealed, Eli Black – the fruit giant's CEO – threw himself out the 43rd storey window of the company's New York headquarters.)

Lately, the banana industry has moved south: it is now growing far more fruit in Ecuador and Peru than it does in Central America. That is because these countries are bigger, and less susceptible to hurricanes, which can destroy an entire nation's crop, as happened in Honduras in 1998, but also because those countries have much weaker labour laws. Banana companies have adopted improved labour and environmental standards, but the system of land and plantation ownership has mostly dissolved into a web of subcontractors and independent suppliers, which makes it difficult for the banana companies and governments to enforce those rules.

7 Cavendish in crisis and a new paradigm

Over time, we have come to know, even love the humble Cavendish. But Panama disease is back, and this time, Cavendish is not immune. The new strain, Foc-TR4, has now spread across tens of thousands of square miles, and has destroyed millions of hectares of plantations, in China, the Philippines, Australia and Mozambique. It has yet to hit Latin America, but it is generally agreed that it will. When? It is a matter of years, or perhaps decades. Nobody can say for sure. But this pathogen is very good at reproducing and all it will take is a single clump of dirt to begin its spread to a new continent, and it will then be unstoppable.

Publicly, the banana companies have expressed little concern, saying that the disease will not come, will come in the distant future or can be easily controlled. Privately, the industry is concerned. This can be seen in Chiquita's experiment with plantations in Mozambique. Chiquita has said that these were built mostly to serve Middle Eastern consumers. This is certainly true, but if you look at banana history, you will see another possibility: that Africa – almost devoid of a commercial banana industry and devoid of Panama disease – might have been being groomed as a replacement to Latin America, just as banana growers moved from nation to nation as the original incarnation of the Panama wilt pursued Gros Michel. But, in 2010, Chiquita disassociated itself from the project. (The company was undergoing an overall repositioning of its business, shifting more towards bananas and away from a failed diversification into packaged foods. That included a move of its corporate headquarters and the hiring of a new CEO; the Africa exit was, according to company investor filings, a result of difficulties in getting quality fruit out of Mozambique.)

Both Chiquita and Dole have begun, over the last few years, to work on Cavendish replacement strategies. The problem is that there is no obvious banana waiting to stand in the Cavendish, as Cavendish was waiting to replace Gros Michel. Of the hundreds and hundreds of banana varieties, none natively possesses the combination of qualities needed for large-scale monoculture. High productivity, short production cycle, ease of shipping, slowness of ripening and the ability to grow on vast plantations are the factors that make the banana equation that has existed since the industry's nineteenth-century beginnings: sell billions of fruit cheaply.

There are alternative bananas, but they will require alternative thinking, and travel beyond the shores of our traditional banana lands. Our 'next' banana may look and taste far different than the one we love today. Possible origin points include India, where more than half the world's banana varieties grow. In that country, nearly every village has its own banana type, and nearly every type is delicious, with fruits that offer multiple shades of taste, sour then sweet, some as creamy as ice cream, others tart, like apples. The problem is that some of these new varieties are also susceptible to Foc-TR4, and, as well, may not meet 'the export industries equation'.

The answer, then, could be to throw the equation away. The adoption of those new bananas from far away could finally transform the way the industry operates in Latin America. The current containerization of the world makes transport of different varieties under different optimal conditions a lot more feasible compared to the reefer ships which were built to only carry one variety. A new banana world could include more than one kind of fruit, priced at various levels, similar to the way apples, chocolate or coffee are now sold. Coffee is a good example: today, you can buy 'Fair Trade' coffee that offers not just higher quality, but guarantees that workers growing it are provided with a just share of the profits. Fair Trade bananas exist, but because bananas are so cheap, the extra few pennies that trickle down – compared to the five-dollar premium per pound ethical coffee garners – buys much less. A spectrum of supermarket bananas, based on fruit transplanted from Asia, could change all this. With the disappearance of their banana, the banana republics could be delivered into the hands of those who have bled, toiled and wept for the fruit we have loved – and never truly understood – for over a century.

8 The hunt for new varieties

As we have seen, Chiquita has been rethinking its position – at least in private – on Panama disease. In 2011, after a visit to the Democratic Republic of Congo, I spoke to a banana breeder who told me he had been hired by the company to come up with a list of varieties that might replace the Cavendish. The company also contracted with the *Fundación Hondureña De Investigación Agrícola* – FHIA – to begin developing new banana varieties, using conventional breeding methods that may result in disease-resistant Cavendish-like varieties. The FHIA arrangement was ironic in that the facility, independently operated since the 1980s, had once been owned by the banana company, operating as Chiquita's primary research outpost in Latin America for decades.

In the non-commercial part of the banana world, efforts to build a Panama-disease-resistant commodity variety were yielding more public and tangible success. James Dale, operating with a grant from the Gates Foundation, has developed Foc-resistant cultivars based on ladyfinger, a variety of the fruit popular in Australia. Ladyfinger has an apple-like taste and is more fragrant. However, it is more difficult to grow, and its yield is not as high and thus there is a genuine question as to whether it would stand in as a pure Cavendish replacement, but even if it could – or if the technology was transferrable to Cavendish or Gros Michel, as Dale hopes it will be – there is another problem: many consumers are not willing to purchase any genetically modified banana. Dale has recently begun testing his lab-bred bananas in actual commercial breeding areas, a huge step. A second set of GMO experiments, conducted in Africa by Rony Swennen, a Belgian scientist, reports that he has successfully created 'transformed' plantains – those are the for-cooking bananas that constitute the west African subsistence diet – that could, if adopted, play a major role in securing food supplies for that continent.

As the banana world began, slowly, to embrace the idea of new varieties, its 'original' newcomer, Cavendish, was getting sicker – especially in Mozambique. Chiquita's former plantations were experiencing big challenges. One of the biggest was water. On multiple plots in the Mozambique farms, plants were sharing water supplies. Inundation from common irrigation sources was one of the primary ways Panama wilt destroyed the Gros Michel plantations in the mid-twentieth century. Although some quarantine and clean farming measures were instituted after the pathogen arrived, the disease has continued to spread rapidly; the operation's annual production has been declining since 2014, a result of being 'severely impacted by the onset of Panama disease', according to the company's annual report.

9 Future trends and conclusion

The appearance of banana disease in three new countries, and on a new continent, set off alarm bells in late 2013. The public was wondering if yet another banana apocalypse had begun. The pathogen will continue to move to new areas, and recently, it has been reported from Israel and several countries in Southeast Asia.

The presence of Foc-TR4 in every banana-growing continent except Latin America is further proof that the disease is coming to that region; look at a map, and you can clearly see that the fungus is on an approach, and one that appears to be less lackadaisical than a decade ago.

The best thing that could happen in the banana world is variety, for the monoculture finally to be broken. It would be good for the environment, good for consumers and good for workers, who – if the business were structured differently – might actually be able to reap benefits from a market that is not constantly in a race to the commodity-littered bottom. The banana industry of the future will likely include a commodity variety; it likely will be genetically modified; and it will include multiple 'gourmet' or heirloom types, as well, sold at higher prices with greater profit margins.

How likely is this to happen? There are hopeful signs, listed below in no particular order.

The first is that parts of the banana industry are beginning to understand that variety is needed – and when that understanding is coupled with a desire to finally change the economic balance between banana workers and banana barons, there is great potential. Bananas, because of their chequered history, have always been a key target for human rights advocates, and the Fair Trade movement – which seeks to provide agricultural workers in supplier companies with living wages and basic human rights – began to take hold in Europe in the 1980s. A similar movement was launched in the United States by a Boston-based company called Equal Exchange. Distributing Cavendish bananas through food co-ops in many parts of the United States, Equal Exchange has made an impact on banana farmers in Ecuador and Peru. These initiatives provide a workable model for the bigger banana companies.

But even the most ardent fair traders of bananas will acknowledge that their work remains as much a reminder of potential than a present-day reality. The problem is scale – and margins. Fair Trade coffee and chocolate have been fairly successful because those product categories allow for wide variations at premium prices. Bananas – because of the commodity-based Cavendish monoculture – have very low margins; consumers in the United States who buy Fair Trade bananas have expressed a willingness to pay a dollar a pound or more for such fruit, but with the average price of conventional bananas in supermarkets at 30%–40% less than that – and with consumers expecting bananas to be cheap, so conditioned are they to the industry's fundamentally unchanged in a century business model – there is a real question as to whether a Fair Trade model based on continuing the banana monoculture could ever be popular enough to make an actual difference in the lives of banana workers.

Similar efforts are underway elsewhere. In 2013, a little-noticed shipment of bananas arrived at the port of Los Angeles. It came from Mindanao, Philippines – the farthest anyone has ever shipped a commercial banana to a major port. Thanks to new container technology, the fruit, despite having spent an extra week at sea, arrived intact. This time, the fruit contained was Cavendish, but with the Philippines battling Panama disease – and having an alternative variety, called Lacatan, that is absolutely delicious and potentially exportable – the Pilipino Banana Growers and Exporters Association says that shipments of alternate varieties may happen before 2020.

The Philippines are also home to an experiment in replacing Cavendish with a conventionally bred version of the fruit – created by Taiwanese researchers following painstaking decades of selecting and reselecting the few Cavendish that survive a Panama disease infection in their fields – that is said to show levels of tolerance to the malady. Similar efforts with conventionally bred modified Cavendish are said to be underway in Brazilian and Japanese market. How that Cavendish would perform in large-scale exports is still undetermined.

There is also a greater understanding that replacing one monoculture with another is not a way forward. Anecdotally, consumers and chefs want to taste a plump and creamy Indian *Rasthali*; a complexly flavoured – tart, then sweet – African *Ybota Ybota;* or a candy-like, vitamin-A-rich Micronesian *f'ei*. One small grower in Santa Barbara, California, north of Los Angeles, is now farming such bananas at his small, seaside plantation; he routinely sells out his stock whenever he appears at local farmers' markets. Can such desire and initiative be scaled? Small businesses have the ambition and a willingness to think unconventionally and be innovative, something the commodity-driven mainstream banana industry has been unable to do.

If Chiquita does not do it, then there is an even more fascinating possibility, one that emerged from the headlines on global business pages in June 2017, when it was announced that Amazon – the global online marketplace – was buying the US-based Whole Foods supermarket chain. For years, Whole Foods has promoted a more sustainable (though still Cavendish-based) business model. Representatives of that company communicated a desire to start importing more varieties, possibly grown on company-run farms in Costa Rica, but that the logistics are still too daunting. One could speculate that being owned by the world's largest mail order distributor of goods could help solve that problem, and that finally all the complex pieces are coming together, creating a path towards giving the world's favourite fruit a genuine future; to make the banana not just genetically or agriculturally sustainable, but also sustainable for people, both where bananas are grown, and wherever they are consumed.

10 Where to look for further information

Though there has been quite a bit of academic research on all aspects of banana production, there has not been much in the way of integrated, multidisciplinary approaches to the topic. So these suggestions are for those who wish to understand not just the banana as an organism, but as a product of humanity whose history is filled with tragedy and triumph (but mostly tragedy).

The author's own book, *Banana: The Fate of the Fruit that Changed the World* has been published in six languages, and is used frequently in educational settings. (Published in 2007 in the United States by Penguin; the book remains in print and is widely available.)

Over the years, there have been other histories of the banana industry. Two of the earliest – they are not difficult to find in online used book outlets – is Charles Morrow Wilson's *Empire in Green and Gold*, written in 1947 by Charles Morrow Wilson. Another particularly prophetic history is *Conquest of the Tropics*, by Fredrick Upham Davis, published in 1914. It was one of the first warnings – ignored – issued to the banana industry about the dangers of monoculture.

One way the banana story has been most engagingly told is through fiction. The most famous account of the industry comes in Gabriel Garcia Marquez's *One Hundred Years of Solitude*. It is considered the Nobel-prize winning author's greatest and most important work, and it largely concerns the events – in fictional form – of an actual massacre of striking banana workers that happened in Colombia in late 1929.

Marquez's work is just the beginning of an extensive body of Latin American banana literature. Most notable (and massive) among them is Miguel Angel Asturias' *Banana Trilogy*, consisting of *Strong Wind* (1950); *The Green Pope*, 1954; and *The Eyes of the Interred* (1960).

Finally, there is Pablo Neruda's book of poetry, *Canto General* (1950) which acts as a history of the banana industry and Latin America. This is a passage from the fourth canto, called 'The Liberators':

> 'Our land - wide land, wilderness -
> was filled with murmurs, arms, mouths.
> A mute syllable kept burning,
> congregating the clandestine rose,
> until the meadows shook,
> trampled by metals and gallops
>
> Truth was as hard as a ploughshare.
>
> It broke the earth, established desire, sank in its germinal propaganda, and was born in the secret springtime'.

Chapter 3

Banana cultivation in Africa

W. K. Tushemereirwe and J. Kubiriba, National Agricultural Research Laboratories, Uganda

1 Introduction
2 Cultivar composition and distribution
3 Production systems
4 Banana production in Uganda
5 Nutrient deficiencies
6 Moisture stress
7 Pests and diseases affecting bananas
8 Improving banana varieties for management of pests and diseases
9 Marketing as a constraint to banana production in Uganda
10 Future trends and conclusion
11 Where to look for further information
12 References

1 Introduction

Bananas and plantains (*Musa* spp) are a staple food for rural and urban consumers and are important sources of income for smallholders across banana-growing African countries. They are important in the humid and sub-Saharan Africa, providing 25% of carbohydrates and 10% of calorie intake to about 70 million people. Per capita consumption ranges from 150 kg in southwest Cameroon to 460 kg in East and Central African regions. This is the highest consumption globally and is indicative of the population's dependency on *Musa* for food and cash (Kanmegne et al., 2006; Karamura et al., 1998).

Plantain production in Africa is estimated to be 37.2 million tonnes, which is more than 50% of global production (FAOSTAT, 1990, 2012). Plantains, therefore, are among the most important sources of food energy throughout the African lowland humid forest zone (Swennen, 1990) (Table 1). They are relatively high-value products, like other horticultural crops, compared to other staples such as maize, rice, cassava and sweet potato.

http://dx.doi.org/10.19103/AS.2017.0020.09

Table 1 Importance of plantain in West and Central Africa

Country	Rank among other crops	Reference
Liberia	First staple for women and third for men after cacao and rubber	English (2008)
Ghana's humid forest zone	Joint second with maize after cassava	Chamberlin (2007)
Nigeria	Third after cassava and yam	Akinyemi et al. (2010)
Kivu Province, DRC	Third after cassava and cooking banana	Ekesa et al. (2012)
Southern Cameroon	Favourite staple and demand projected to increase	Dury et al. (2002)

Bananas and plantain form an integral part of the farming systems in tropical Africa. Where bananas and plantains are well established, there are stable micro-ecologies contributing to more sustainable natural resource management, almost like the forest trees. The *Musa* spp., therefore, have been key in reducing soil erosion in hilly areas such as those of the Democratic Republic of Congo (DRC), Burundi and mountainous areas of Uganda.

In the Great Lakes Region of Africa (GLRA), banana is considered one of the most important food security crops. This is because the plant's all year round fruiting habits coupled with moderately high yield ensure continuous supply of dietary carbohydrates. Furthermore, the perennial nature of the crop makes it require less labour and least affected by social instability and drought (Tushemereirwe et al., 2001).

Despite its importance, banana production in Africa faces two big challenges; low productivity and short plantation life. Most of the bananas are produced by small subsistence farmers who apply minimal management techniques and rarely replenish the nutrients removed when banana bunches are harvested. They compensate for the nutrient loss by shifting to new plots whenever a great decline in productivity is observed. However, as the population grows, the land available becomes more limiting, forcing farmers to shift to plots where bananas have already been grown. Such plots are no longer productive, forcing farmers to prefer annual crops which lack the advantages that banana possesses.

The short life span of banana plantations is another big concern faced by producers of, especially, East African highland banana (EAHB) and plantains. These banana types are the most predominant in Africa where they are main sources of carbohydrate. During diagnostic surveys of banana production systems in Uganda, farmers claimed their plantations used to last more than 30 years and indeed in relatively recent production areas of southwestern Uganda, there are many plantations that have lasted for more than 50 years which are still highly productive. However, in most banana production areas across Africa, plantations last three to five years. Reduced longevity of plantations is largely attributed to banana pests (weevils, nematodes) and diseases (black Sigatoka, Fusarium wilt, banana xanthomonas wilt, bunchy top, banana streak virus), the control and management of which are of utmost priority to people involved in research and development in the major banana-producing African countries.

2 Cultivar composition and distribution

Plantain is one of the most important staple crops in West and Central Africa (Table 1). It is grown in the humid forest and derived, moist savannah agroecosystems, ranging from

Guinea Bissau in the west of the region to the DRC in the south-east (Jalloh et al., 2012). The plantains are predominant in West and Central Africa where they account for 73% of the Musa crop. The dessert bananas (Cavendish and Gros Michel) form 21% and a negligible portion (6%) of cooking bananas.

In the east of Africa, the *Musa* cropping system is dominated by East African highland banana (AAA-EAHB) or simply highland banana, which makes up 71% of the *Musa* spp. grown (http://banana.mappr.Info). Other types include dessert bananas (Cavendish, Gros Michel and others) (17%) and plantain (11%). Highland bananas occupy mainly the Great Lakes areas of Uganda, Northern Tanzania, Rwanda, Burundi and eastern DRC. The group is said to be endemic to this region with no clear analogue elsewhere in the world. A key factor responsible for the group's distribution is the altitude.

The dessert bananas (AAA) include Dwarf Cavendish, Lacatan, Red Banana and Gros Michel. The largest concentration of Cavendish dessert bananas is generally found in the low-lying coastal regions (below 800 m asl) – in South Africa, Mozambique, Tanzania, Kenya Somalia and Ethiopia (Karamura et al., 1998).

2.1 East African highland bananas

East African highland bananas (AAA-EAHB) are a distinct subgroup of bananas, domesticated in the Great Lakes region of East and Central Africa. East African highland bananas are mainly cooking bananas steamed and consumed as a starchy food. They are also used to make juice and rarely used as a dessert. Their ancestors are believed to have arrived in the region about 4000 years ago. The bulk of the subgroup is grown in Uganda with about 120 varieties (Karamura and Pickersgill, 1999). The region, therefore, is considered a secondary home of banana diversity.

It is a subgroup classified into five clonesets (Karamura and Pickersgill, 1999):

- Nfuka refers to changing and morphologically unstable bananas having the greatest numbers of cultivars.
- Musakala means lax, and refers to the spacing between the fruit, and includes high-yielding cultivars commercially grown for the urban markets.
- Nakabululu refers to the short and plump variety; bunch is compact with short fruit.
- Nakitembe refers to ekitembe and is a Ugandan name for *Ensete*.
- Mbidde refers to beer and has plenty of latex which is sticky.

Although the various cultivars of the AAA-EAHBs look different, they are relatively genetically uniform. This is exhibited in their reaction to various stresses. For example, most are similarly susceptible to black Sigatoka (Tushemereirwe, 1996); to nematodes (Kashaija et al., 1994) and more recently to banana xanthomonas wilt (Ssekiwoko et al., 2006). They are also resistant to Fusarium wilt (FOC) race 1 (Kangire, 1998). It is for this reason that the EAHB breeding programme agreed between the National Agricultural Research Organisation (NARO) and the International Institute of Tropical Agriculture (IITA) was set up to increase its genetic base in the region and to ensure resilience of the cropping system.

3 Production systems

Five common types of plantain systems were distinguished across West and Central Africa, adapted after Akinyemi et al. (2010): (1) food intercropping systems; (2) home garden (compound) systems; (3) plantain–cacao systems; (4) other agroforestry systems and (5) monocropping systems.

In Eastern Africa, banana is an important component of production systems on-farm. It is intercropped with perennial and annual crops, especially, legumes which contribute to balancing diet for the farmers. *Musa* production systems in East and Southern Africa are complex because of variable agroecological and socio-economic conditions. Despite the complexity, the systems can be divided into three subsystems, namely backyard garden, subsistence and commercial (Karamura et al., 1998).

The subsystems are characterized by the management levels. The crop management practices include planting materials, irrigation, pest control and cropping system. The subsystems are further characterized by yields, incomes attained, utilization by end users as well as farmers, perceptions, priorities, extension support and profitability.

Generally, most bananas in Africa are grown by smallholder farmers in three main cropping systems. These, in their order of decreasing importance, include backyards, intercrops and monocrops (Swennen and Vuylsteke, 2001).

3.1 Banana backyard systems

Banana backyard systems were described by Akinyemi et al. (2010) as a home garden (compound) and by Karamura et al. (1998) as a backyard garden. It is a characteristically low-input system. It is usually found in the peri-urban areas, where land is a key determinant of farm size (a few mats to 0.25 ha), or where other bananas/plantains crops are not important as in southern Ethiopia, Malawi, Mozambique and Zambia (Karamura et al., 1998).

Plantain is also grown near homesteads in home garden systems and are commonly found in highly populated areas for example, Nigeria (Nweke et al., 1988; Gobin et al., 2001), the Ashante region of Ghana (Drechsel et al., 2006) and in central Cameroon. The home gardens also comprise multipurpose shrubs with a range of annual crops and vegetables characterized by high levels of organic inputs, such as small livestock manure and kitchen waste.

In this system, the crop is not well managed as it only supplements for other food sources or is kept for non-food uses. In Uganda, for example, it is used to provide green leaves for wrapping and for cooking the popular starchy banana known as matooke. Highland bananas are found around the homesteads in southern Ethiopia mainly for windbreaks to protect *Ensete* crops. Since the banana backyard garden systems are minimally managed, they tend to be pest/disease foci and a source of infection for other neighbourhood bananas.

3.2 Banana intercrop system

Karamura et al. (1998) referred to banana intercrop systems in East and Central Africa as banana subsistence systems. Akinyemi et al. (2010) classified it as plantain–agroforestry systems in Central and West Africa. This agroforestry system consists of plantain as a shade crop for cacao saplings and is common in Liberia (English, 2008), Ghana (Ngeleza et al., 2011), SW Nigeria (Akinyemi et al., 2010) and central Cameroon (Sonwa et al., 2007).

This is the most dominant system for growing bananas and plantains in the tropical world, covering over 87% of global banana/plantain production (INIBAP, 1996). There are perennial, low-input, small (0.25–5 ha) and rural-based systems in Africa, Asia and Central

America. This system is key for food security with some surplus sold to rapidly expanding local and urban markets.

Another characteristic feature of this system is its complexity, in terms of cultivars grown, soils and terrain, pest/disease communities, management skills as well as crop uses, even in one ecoregion. Karamura et al. (1996) reported an average of 12 banana cultivars per farm, along with a mixture of other intercrops in Uganda. Due to limited land and financial resources, farmers try to meet all their dietary needs from the same part of land by mixing crops and sometimes crops and animals. These practices ensure food security and a balanced diet to rural communities, but are easily improvable using modern technologies such as mechanization. This complexity (ecological and socio-economic) needs to be considered during technology development and promotion for the productivity of the system to be meaningfully improved.

3.3 Banana monocrop systems

Although Karamura et al. (1998) described the banana monocrop system as being similar to the Cavendish-dominated commercial systems, there are a few highland banana monocrops that may be included in this system. These systems are less widespread in West Africa but are reported to be predominantly found in Nigeria (Akinyemi et al. (2010) and Southern Cameroon (Carrière, 2003). It is the least complex uniform management. They account for most of the dessert banana exported into global markets. It is an intensively managed system, from selection of planting materials, land preparation, crop establishment and plantation management to marketing and/or processing (Robinson, 1996).

An informed farming community that ensures intensive crop management and highest returns to investment is largely responsible for the differences between this system and the other two. While commercial banana systems are advantageous financially, they do not support cultivar diversity, being largely monocultivar. The significance of this becomes apparent in the face of a marauding disease such as Fusarium wilt. A case in question is the Race 4 (*Fusarium oxysporum* f. sp. cubense) which has led many farmers in South Africa to abandon banana farming. This is in contrast to the subsistence systems where cultivar diversity ensures that farmers have food security as not all cultivars succumb to the disease en masse.

Another consideration to be taken into account relates to the long-term effects on the environment. The use of chemical additives and heavy equipment to improve yields may have adverse effects on the environment if due care is not taken. Use of pesticides may alter the biotic composition of the specific ecoregion, while the use of heavy equipment may alter soil structures. In areas where irrigation is based on rivers, soil/water borne diseases are readily spread downstream.

3.4 Productivity of *Musa* cropping systems

The *Musa* sp. cropping systems have gained a lot of research and development attention, particularly with regard to pest management. However, not much success has been recorded to date. Pests and diseases have intensified in some areas and in extreme cases eliminated whole system of susceptible cultivars altogether. As a result, yields have continued to remain below 10t/ha except for Sudan and South Africa, whose systems are dominated by the commercially grown Cavendish bananas (Table 2).

Table 2 Production and productivity 2002–12

		Production (million tons)		Productivity (ton/ha)	
	Country/Year	2002	2012	2002	2012
Plantain system	Ghana	2.28	3.56	8.2	10.5
	Cameroon	1.24	3.45	5.3	12.5
	Nigeria	2.13	2.80	5.0	6.1
	DRCongo	1.20	0.51	4.8	1.8
	Central African Republic	0.08	0.09	2.9	2.8
	Guinea Bissau	0.04	0.05	3.0	3.1
	Equatorial Guinea	0.03	0.04	4.8	6.1
	Sierra Leone	0.03	0.04	5.4	5.6
East African highland banana system	Uganda	10.50	9.77	5.9	5.3
	Rwanda	2.78	3.22	8.0	9.2
	Tanzania	2.81	3.26	4.1	4.5
	DRCongo	1.51	0.83	4.4	2.3
	Malawi	0.57	0.74	11.8	14.8
	Mozambique	0.11	0.47	6.4	7.6
	Sudan	0.33	0.75	27.1	28.6
	South Africa	0.39	0.39	35.7	50.9

Source FAOSTAT (2002, 2012).

Banana yields in Uganda ranged from 1.9 to 31 t/ha/yr in central Uganda, 3.8 to 36 t/ha/yr in south Uganda and 4.6 to 37 t/ha/yr in southwestern Uganda (Wairegi and Van Asten, 2010). This implies that it is possible to increase banana productivity on-farm using practices that are already in use by other farmers. The proportion of farmers likely to adopt improved soil nutrition and soil water practices was estimated at 12.23% for Uganda (Edmeades et al., 2007).

Lescot and Ganry (2010) stated that, in Africa, 76.5% of plantain production was from smallholder intercropping systems, 13.5% from smallholder monocropping systems and only 10% of production was from intensified systems. They gave yield ranges of 0.1–3.0 t/ha on poor soils and 1.0–4.0 t/ha on fertile soils for smallholder intercropping systems (planting density of 800/ha). In southern and central Cameroon near the Atlantic coast, plantains yield 4.0–4.8 Mg/ha/yr in intercropping systems (Kanmegne et al., 2006; Mutsaers et al., 1981).

Average yields in East African highland banana cropping systems do not exceed 30% of the crop's potential of over 60–70 t/ha/yr (van Asten et al., 2005). Over the past 50 years, attempts have been made to increase production by increasing the land area under banana cultivation. Therefore, intensive approaches are indispensable for closing the yield gap that is crucial for maintaining food self-sufficiency, income and livelihoods of Africa's smallholder banana farmers.

However, in the Cavendish-dominated monocrop production systems, yields range between 40 and 60 t/ha and profit is the ultimate objective. Consequently, production and income, as well as consumer/market requirements, are predictable. This level of management ensures relatively high yields through quality control of planting material and timely application of inputs.

4 Banana production in Uganda

Banana is one of the five major starchy food crops in Uganda, with the other food crops being maize, cassava, sweet potatoes and sorghum (Table 3). It is considered the most important food security crop in the country.

Almost all the bananas produced are consumed locally in the region, though recently, some have been exported fresh to regional and international markets on a small scale. In terms of rural revenue and returns to family labour, banana is one of the best crops in Uganda (Embrechts et al., 1996; Bagamba, 1994).

Drastic decline in the traditional banana-growing areas has led to replacement of bananas with annual crops. At the same time, the crop's major production area has shifted from central Uganda to the country's southwest where productivity is already declining. Data from recent studies indicate productivity estimates of 12.0 t/ha for central Uganda and 20.0 t/ha for the relatively high production areas of southwestern Uganda (Wairegi and Van Asten, 2010). This is considered a great decline in comparison with 70 t/ha and above attainable in a research plot (van Estein et al., 2005).

This huge yield gap has been attributed to severe abiotic and biotic constraints. The most key abiotic stresses include nutrient deficiencies, notably potassium (K) and nitrogen (N) (Nyombi et al., 2010; Wairegi and Van Asten, 2010), and drought (Okech et al., 2004; Taulya, 2015). On the other hand, the biotic stresses include pests, mainly weevils and nematodes (Rukazambuga et al., 1998; Barekye et al., 2000; Kashaija et al., 2004), and diseases including black Sigatoka, Fusarium wilt, banana streak virus (BSV), banana bunchy top and banana Xanthomonas wilt (Tushemereirwe et al., 1993, 2006a; Tushemereirwe, 1996; Sharma, 1987; Magnaye, 1994).

Table 3 Production of major crops in Uganda, 2008/9

Crop	Area (ha)	Metric tons (MT)	Yield (MT/ha)	Total plots number	Mean plot size (ha)
Banana	929 274	4 297 345	5.0	3 173 467	0.3
Cassava	871 389	2 894 311	3.3	3 147 389	0.3
Maize	1 014 260	2 361 956	2.3	2 873 654	0.4
Sweet potatoes	440 256	1 818 769	4.1	2 277 083	0.2
Sorghum	399 252	375 795	0.9	1 212 563	0.3

Data Source Uganda Bureau of Statistics (UBOS) (2010).

Since 1991, the National Agricultural Research Organization, Uganda, Makerere University, Bioversity International and the International Institute of Tropical Agriculture (IITA) have been investigating these stresses and have generated information that has been instrumental in their elucidation and management. Most of the studies were based in Uganda but the management options derived have been adapted across banana-growing areas of East and Central Africa through regional networks such as Association for strengthening Agricultural Research in East and Central Africa and the Banana Research Network for Southern and Eastern Africa (Kubiriba et al., 2014; Kubiriba and Tushemereirwe, 2014).

5 Nutrient deficiencies

Banana bunches are being increasingly moved from rural areas to urban centres, leading to continuous loss of nutrients, especially potassium (K), which is considered the most limiting nutrient to banana yield. Studies conducted on highland bananas in Uganda indicated that K interaction with rainfall was the main driver of bunch yield and its components (Nyombi, et al., 2010; Van Asten et al., 2011; Taulya, 2015). Bananas are reported to uptake more K than all other essential plant nutrients combined and its main purpose is translocation of assimilates from leaves to fruits and growing points of plants (Pettigrew, 2008). Potassium is also reported to be important in banana dry matter production and allocation to bunches.

The other key nutrient is nitrogen (N). Wairegi and Van Asten (2010) reported highland banana yield increases of 10 t/ha/yr in response to NPK fertilizer application in southern Uganda, which they attributed to N deficiency. Their observation was in correlation with the results of studies conducted in India, which reported yield increases in response to N for Robusta bananas (Hedge, 1988; Singhi, 2004). This study dispels reports from Guadeloupe (Damour et al., 2012), Sri Lanka (Weerasinghe et al., 2004) and Nigeria (Baiyeri, 2002), which suggested that N level has no effect on banana yields. Waireigi and Van Asten (2011) concluded that nutrients interact with other factors to influence highland banana yield.

5.1 Nutrient replenishment

Use of organic materials: Animal manure, household refuse and mulch have been promoted in Uganda to remedy the decline in soil fertility since the 1940s. Survey studies showed that Ugandan farmers apply manure and mulch in varying proportions to replenish the nutrients (Wairegi, 2010). However, the majority utilize a smaller amount of these materials to increase yield significantly, for which they put forth high cost and unavailability of the materials as the reasons.

Use of inorganic fertilizer: Highland banana response to inorganic fertilization has been variable, site specific and in some cases negligible. Smithson et al. (2001) reported good banana response to inorganic fertilizer (100N + 25P + 100K) at Rubale in southwestern Uganda, which was attributed to low pest damage, the traditional mulching practice and rainfall above long-term average. On the other hand, the same inorganic fertilizer application failed to cause significant response at Kabulasoke in central Uganda. The low response was attributed to high pest damage observed in the plantations, insufficient and absence of mulch to ameliorate the moisture stress. It has thus been recommended that inorganic fertilizers should be mixed with mulch and its usage be combined with effective pest control measures to get their full benefit.

Less than 1% of farmers in Uganda have used inorganic fertilizers in their banana plantations though many of them are aware of their great effect on yield. Fertilizer usage is hampered by high prices as the biggest factor, followed by unavailability of an appropriate fertilizer in nearby shops and a perception that fertilizers spoil soil quality (Wairegi, 2010).

6 Moisture stress

A recent study (Taulya, 2015) ranked moisture stress as the most important banana production constraint in Uganda. Water was identified as the most important input for highland banana growth and yield, followed by K. Soil moisture is required for movement of nutrients, especially K and N, to the roots for uptake. Nutrient availability and uptake are impaired under moisture stress (Duman, 2012). Adequate soil water is also required to keep the stomata open and hence ensure the supply of carbon dioxide for photosynthesis to occur (Taulya, 2015). Low annual rainfall (678 mm) at a research station in southwestern Uganda (Mbarara) reduced banana yields by 50% (Okech et al., 2004). Bananas require about 25 mm of rainfall per week for satisfactory growth, which corresponds to 1300 mm per annum.

Moisture stress has previously been reported to reduce banana dry matter production (Firth et al., 2003) and yield (van Asten et al., 2011). However, Taulya (2015) observed that K increased total banana dry matter in early growth stages of banana plants irrespective of whether wet or dry conditions prevailed. The surprising observation was attributed to the role K plays in osmotic adjustment. This is a mechanism by which plants adapt to moisture stress conditions (Cutler and Loomis, 1977).

Highland bananas were reported to exhibit dry matter allocation plasticity in favour of belowground biomass structures in response to K and N deficiency but not to water stress (Taulya, 2015). They are more sensitive to K than to moisture stress.

6.1 Moisture stress management

Use of mulch: In subsistence production systems mulching is the only strategy farmers rely on to manage moisture stress. High yields of 66%, attributable to moisture retention as a result of mulch, were reported for plots close to homesteads (Bananuka and Rubaihayo, 1994). However, use of mulch is hampered by its ever increasing shortage and high cost.

Irrigation: Commercial farms are able to apply irrigation when and where necessary. However, the measure is unknown in subsistence production systems. With supplemental irrigation high yield of up to 59 t/ha was achieved at Kabanyolo in central Uganda (Bananuka, 2001).

7 Pests and diseases affecting bananas

7.1 Pests of bananas

Several pests are known to attack banana, but only nematodes and weevils are of economic importance in Uganda (Speijer et al., 1994b). They cause significant loss of the crop yield as they progressively build up through successive ratoons (Speijer et al., 1994b).

Their management is quite difficult especially for the resource-poor farmers who cannot afford chemical pesticides. The two pests appear to be responsible for the shortening of plantation life from over 50 to 3–5 years in central Uganda.

7.1.1 Banana nematodes

Plant parasitic nematodes are an important constraint to banana production worldwide, and are among the causes of banana production decline in Uganda (Gowen and Qencherve, 1990; Speijer et al., 1994a). Nematode infestations, leading to widespread toppling of bananas, were reported in Masaka and Rakai districts in mid-1980s. Nematodes attack and damage banana roots, leading to yield loss through stunting of plants, delayed maturation, slow ratooning, reduced bunch weight and toppling.

Diagnostic surveys conducted in the major banana-growing areas of Uganda in 1991–93 revealed eight nematode species that infected banana roots; namely *Radopholus simillis* (Cobb) Thorne, *Pratylenchus goodeyi* Sher & Allen, *Pratylenchus coffeae* (Zimmermann) Filipjev, Schuurmans & Stekhoven, *Pratylenchus zeae* (Graham), *Helicotylenchus multicinctus* (Cobb) Golden, *Helicotylenchus dihystera* (Colbran), *Helicotylenchus pseudorobustus* (Golden) and *Meloidogyne* spp (Kashaija et al., 1994).

Data from the surveys suggested that root damage (% root necrotic area and dead roots) and the resultant plant toppling were influenced by different factors, including banana cultivar and elevation (Speijer et al., 1994a; Kashaija, 1996). Generally, the East African highland bananas were more susceptible to nematode damage than introduced varieties such as Kayinja and Bogoya (AAA) groups (Kashaija, 1996). More toppling was observed at low elevation (1000–1350 m asl) than at higher elevations (1400–1900 m asl) (Speijer et al., 1994a)

Pratylenchus goodeyi was the most abundant nematode with occurrence of 96% followed by *H. multicinctus* at 83% of the survey sites. *R. simillis* and *Meloidogyne* spp were each found in 53% of the survey sites. The other four species were localized and rare, and were reported to be found on bananas in Uganda for the first time (Kashaija et al., 1994).

Pratylenchus goodeyi (the lesion nematode) is found in both low and high-elevation zones, but more abundant at (1400–1800 m asl) altitudes (Kashaija et al., 1994). *H. multicinctus* (the spiral nematode) is more common and dominant at low elevations in Uganda (Kashaija et al., 1994). The species is often found in association with *R. similis* (Gowen, 1995), a burrowing nematode first reported on bananas in Fiji (Cob). It was found almost wherever bananas are cultivated by 1993 (Bridge, 1993), except in Canary Islands, Egypt and Taiwan (Gowen, 1995). Its widespread occurrence is attributed to movement of infested planting materials in export cultivars.

Elevation has a major influence on distribution of the nematodes, and temperature seems to be the main underlying factor (Kashaija, 1996). Other factors that may influence occurrence and distribution of the nematodes include soils, cropping systems and banana cultivars. The surveys further revealed that damage by nematodes is often not recognized by farmers and is often attributed to the readily seen pest, the weevil (Gold et al., 1993).

7.1.2 Weevils

The banana weevil *Cosmopolites sordidus* (Germar) (Curculionidae) is rated the most devastating pest of bananas (Gowen, 1995; Gold, 1995). It was first reported in Uganda in the early 1900s and by the late 1950s was considered present throughout all banana-growing areas of the country.

Data from on-station trials on East African highland bananas revealed that the loss caused increased from 10% in the first ratoon to 48% in the fourth ratoon (Rukasambuga, 1996). This observation was earlier supported by reports that in older plantations with high weevil infestations, snapping was quite extensive, leading to 50–100% yield losses (and reduced life span of plantations) (Gold et al., 1993).

The East African highland bananas (AAA) and the plantains (AAB) are more susceptible to banana weevil attack than the introduced banana such as Kisubi (AB), Kayinja (ABB) cultivars and Gros Michel (Gold et al., 1994). Introduced bananas had peripheral damage that did not penetrate into the inner corm.

The diagnostic survey carried out in Uganda in 1991–93 revealed that the banana weevil was present in all survey sites below 1700 m asl, but was not present in Kabale (1976 m asl) and Kapchoprwa (1830 m asl) (Gold et al., 1994).

7.2 Pest management

Management of the two key banana pests is based on the disruption of their life cycles (Abera et al., 2000). This is achieved through integration of use of preventive measures (clean planting materials), destruction of breeding sites (post-harvest residues), denial of the conducive ecological conditions for breeding and multiplication (and good crop husbandry), reduction in populations to below threshold levels (trapping, chemical control) and host plant resistance.

7.2.1 Use of clean planting material

The spread of nematodes and weevils across and within regions is attributed to transfer of infested planting materials (Bridge, 1993; Abera et al., 2000). This could have been avoided by using nematode and weevil-free conventional or *in vitro* micro-propagated planting material. Field suckers are cleaned by corm paring followed either by exposure to sunlight or by chemical treatment (Gowen, 1995).

Clean planting materials may have little effect on attack by weevils from neighbouring fields, but they minimize the spread of the weevils into new plantations and therefore delay weevil infestation. Farmers obtain them through any of the following means as recommended by NARO (Tushemereirwe et al., 2003a):

- Clean suckers may be obtained from non-infested fields for instance from areas where the occurrence of weevils has not been reported.
- Planting materials from infested fields may be cleaned by paring the corm to remove eggs and superficial larvae, and by dipping the suckers in a recommended pesticide to kill the internally imbedded larvae.
- Tissue cultured plants are normally free of weevils.

7.2.2 Destruction of post-harvest residues

Within an established plantation, it is very difficult to control nematodes using cultural practices, which have been effective in weevil management. Data collected on-farm in Uganda for three years showed that it was possible to reduce banana weevil population from 52 000 to 13 000/ha, with associated reduced corm damage of 41% and increase in yield of 70% (Masanza, 2003). This was achieved through digging up corms and chopping the corms together with harvested pseudostems into small pieces which are spread out to dry quickly. This reduces the hiding and breeding sites of adult weevils and exposes weevil eggs and larvae to desiccation. This, if consistently effected, may effectively manage the weevil problem, even without trapping them, as the adults within the plantation eventually die and movement from the surrounding fields is not significant except by planting materials (Masanza, 2003).

7.2.3 Weevil trapping

Adult weevils are trapped using any of the two methods below as described by Gold and Bagabe (1994).

- The pseudostem trap: made from pseudostem pieces split longitudinally into halves and placed against a banana plant with split surface on the ground.
- The disc–on-stump trap: made by cutting a harvested stump 15–25 cm above ground level and then placing a pseudostem sheath or banana leaves on top of the stump.

Although trapping of weevils has been recommended, it has not been widely adopted by farmers.

7.2.4 Crop rotation

Crop rotation is based on a mechanism of denying the pest an appropriate host in which to complete the life cycle. It has, however, only been used for nematode management where nematodes are starved to death and, therefore, causing decline in nematode population in a field (Namaganda et al., 2016). The next banana crop then is started with a low initial inoculum. The method is efficient where bananas are replanted frequently. This has not been a common practice in Uganda. However, a few farmers in areas where banana production has severely declined as in central Uganda often uproot the less productive plantations and replant annual crops such as cassava.

7.2.5 Chemical control

Pesticides offer the best and most effective option in controlling nematodes and weevils in already established plantations. Increase in yields and productivity due to application of nematicides was recorded in heavy infested banana fields (Gowen and Quencherve, 1990; Gowen, 1995). Organophosphates or carbamate insecticides known to be less persistent than organochlorides are most commonly used. Carbofuran (Furadan), Primiphosethyl (primicid) and Dusban are applied at rates recommended by the manufacturers. Normally, insecticides are applied to soil around the base of the banana stool using a simple shaker in the case of granules or a sprayer for the liquid formulations. The pesticides also greatly reduce the weevil populations (Night et al., 2010). However, the chemicals are not affordable to many smallholder farmers. Chemical control is therefore used by relatively few farmers and as a last resort.

7.2.6 Biological control of banana weevil

The first attempt to control C. *sordidus* using introduced predators dates back to 1934. A Histerid beetle, *Plaesius javanus* Erichs was imported from Java and released on Kibibi island in Lake Victoria in 1935. Subsequent surveys in 1937 and 1945 showed that the predator had managed to get established in the area (Harris, 1947; Ingram, 1970). More recently, use of indigenous fungal pathogens *Beauveria bassaiana* (Balsamo) Vuillemin and *Metarhizium anisolpliae* (Mwetchnikoff) Sorokin was initiated in 1991 (Nankinga, 1994). *B. bassiana* was the most promising (Nankinga and OgengaLatigo, 1996). Banana fields treated with *B. bassiana* showed significantly reduced weevil infestation. However, the delivery system for effective spread of the biocontrol agent in a banana plantation was not efficient.

7.3 Banana diseases

Bananas and plantains in Uganda are severely affected by four economically important diseases, namely black Sigatoka, Fusarium wilt, Xanthomonas wilt and BSV.

7.3.1 Black Sigatoka

The commonly observed leaf spot of East African highland bananas is caused by a mixture of air-borne fungal diseases including yellow Sigatoka (*Mycosphaerella musicola* Leach), leaf speckle (*Periconiella sapientumicola* Siboe) and black Sigatoka (*Mycosphaerella fijiensis* (Morelet)) which is by far the most destructive disease. Black and yellow Sigatoka produce similar symptoms and are hard to distinguish in the field although they can be readily separated from leaf speckle. Sometimes the same plant is affected by all the three diseases together (Tushemereirwe, 1996).

Black Sigatoka was found to be first reported in central Uganda in 1989 (Tushemereirwe and Waller, 1993) and is the most widespread of the three leaf spots (Tushemereirwe, 1996). On the other hand, yellow Sigatoka which was found in Uganda much earlier in 1938 (Stover, 1962) and leaf speckle have always been considered as minor in economic importance (Leakey, 1970). The arrival of black Sigatoka made the leaf spot complex economically important. A trial conducted at Kawanda to establish pest status of the leaf spot complex driven by black Sigatoka on Mbwazirume highland banana cultivar revealed yield loss of 37% (Tushemerweirwe, 1996). The loss was mainly due to incomplete fruit filling as a result of premature drying of leaves.

By 1993, black Sigatoka was reported to be present across the banana-growing areas of Uganda below 1450 m asl. (Tushemereirwe et al., 1993), or where mean minimum temperatures exceeded 150°C (Tushemereirwe, 1996). Where present, incidence of black Sigatoka on highland banana cultivars was always 100%. On the other hand, yellow Sigatoka was present on highland banana only at elevations above 1450 m asl. with incidence ranging from 5 to 10% (Tushemerweirwe, 1996). Leaf speckle was present at all survey sites with 100% incidence on highland banana but affecting ageing leaves.

7.3.2 *Fusarium* wilt (Panama disease)

Fusarium wilt of banana caused by *Fusarium oxysporum* Schlect. f.sp. *cubense* (E.F. Smith) Snyd.and Hans (FOC) was first reported to be found on bananas in Uganda in 1952 (Leakey, 1970). Recent work has shown that FOC is only pathogenic to four exotic dessert cultivars grown in Uganda, namely Gros Michel, Sukari Ndizi, Kisubi and Kayinja and does not affect highland bananas.

Fusarium wilt, also called Panama disease (Brandes, 1919), is the most destructive disease of non-indigenous bananas in Uganda. This is the major reason why the introduced bananas comprise less than 8% of the cultivar on-farm (Karamura et al., 1998). Three races of the pathogen are known to infect edible bananas, races 1, 2 (Stover and Waite, 1960) and 4 (Su et al., 1986). Race 4 is mainly found in the subtropics with low temperatures, whereas races 1 and 2 are widely distributed wherever susceptible hosts are found. In Uganda, both race 1 and 2 differentials, Gros Michel and Bluggoe, respectively, are widely grown. Bluggoe is not affected by the disease, while Gros Michel suggests the presence of only race 1 in Uganda.

FOC race 1 is responsible for causing the disease on the four known susceptible cultivars in Uganda: Kayinja (Pisang awak-ABB), Bogoya (Gros Michel – AAA), Kisubi and Sukari Ndizi (AB). The disease has been found all over Uganda's banana-growing areas.

7.3.3 Banana streak virus disease (BSV)

While the other banana viral diseases are not important in Uganda, BSV disease appears to be a problem (Tushemereiwe et al., 1996). The casual organism of banana streak disease, the BSV, was first isolated in 1986 (Lockhart, 1986). It was later recorded in India, Morocco, Nigeria, Central America, Rwanda, Jordan, Tanzania, China, Mauritius, South Africa, Madagascar, Ecuador (Jones, 1993) and Uganda (Tushemereiwe et al., 1996).

Jones (1993) reported up to 90% yield losses on 'Poyo' plants with severe BSV symptoms. In severely infected areas in Uganda, plantations have had almost 100% loss in saleable yield (Tushemereirwe et al., 1996). Apart from the effect of BSV on the yield (quantity and quality) and plant growth of the banana, it also hinders germplasm exchange (Lockhart, 1986). The disease is mainly spread through planting material and mealybug vectors (Kubiriba et al., 2001). Spatial–temporal spread analysis of BSV in Uganda (Kubiriba et al., 2013) showed that it is a slow spreading disease, but has sufficient time in the perennial cropping system to increase to epidemic levels.

Like other *Badnavirus*, BSV DNA sequences are integrated into the *Musa* genome (Harper et al., 1999). Harper et al. (2005) reported that supergroup I contains BSV species that have integrated sequences that could be activated to give episomal (pathogenic) virus in cultivars with the B genome. Supergroup II contains integrated sequences that are so perturbed that they are considered not to be activatable ('dead'). These sequences mainly occur in the A genome. No evidence was found for activatable sequences integrated in East African highland bananas (AAA-EAHB) (Harper et al., 2005). Therefore, BSV spread in East African highland bananas seems to be vector-borne rather than arising from genomic BSV sequences. A study carried out by Harper et al. (2005) revealed that there were ten 'species' of BSV in Uganda.

The high variability of BSV both genomically and serologically poses challenges for developing polymerase chain reaction (PCR)– and antibody-based diagnostics. Immunocapture (IC)-PCR, currently used as a 'Gold standard' BSV diagnostic tool, was developed to circumvent the detection of integrated DNA (James et al., 2011). This, however, is still problematic since the antiserum (even with a cocktail) cannot capture all BSV isolates, and use of several primer sets is unlikely to detect the entirety of the BSV sequence diversity since some species are not characterized. Rolling-circle amplification (RCA) is a diagnostic test recently developed by James et al. (2011). RCA using bacteriophage Phi29 DNA polymerase, as a sequence-independent protocol, is therefore able to detect all circular badnavirus viral particles, discriminating against the integrated sequences.

7.3.4 Banana *Xanthomonas* wilt

A bacterial wilt disease caused by *Xanthomonas campestris* pv. *musacearum* (Xcm) was first reported in Ethiopia on Enset cultivars (*Ensete ventricosum*), a relative of bananas (Yirgou and Bradbury, 1968), and later on bananas (Yirgou and Bradbury, 1974). In 2001, the disease was reported in Uganda, where it had a devastating effect (Tushemereirwe et al., 2004). This was the first time the disease was reported outside Ethiopia.

Bacterial wilt diseases of bananas are considered to be of little importance in commercial plantations where avoidance measures are easy to implement. However, in Uganda where smallholder farmers intensively cultivate the bananas, the disease was initially devastating. Within a period of four years, the disease rapidly spread from a site in central Uganda where it was first reported to all banana-growing areas of the country (Tushemereirwe et al., 2006b) (Fig. 1).

Whenever a plantation is affected by the disease, all banana plants rapidly get affected, reaching an incidence of 100% within a year. Extensive studies that followed the arrival of the disease in Uganda revealed the means by which it was spreading and this led to formulation of management measures (Tushemereirwe et al., 2006a). It was established that one of the primary means of spread of the disease causing bacteria was by contaminated foraging insects, mainly stingless bees (*Plebeinadenoiti*), fruit flies (Drosophilidae), grass flies (Chloropidae), birds and bats (Tinzaara et al., 2006; Buregyeya et al., 2014).

Healthy plants are infected when the bacteria are carried by insects from oozing peduncles to fresh cushions on the peduncle from which male flowers have recently dehisced (Buddenhagen and Elsasser, 1962). The banana plant continuously drops off flowers from the peduncle, leaving a fresh wound that serves as a source of inoculum if the plant is infected or as a point of pathogen entry for uninfected plants. It was concluded that timely removal of the male inflorescence (male bud) would interrupt the disease spread by insects. It was recommended that the male buds should be removed with forked sticks to avoid transmission by tools.

Survey studies conducted between 2002 and 2005 revealed that tools used in plantation management (machetes, pruning knives, hoes) were major means of disease transmission, an observation supported by findings in Ethiopia (Tushemereirwe et al., 2003b; Yirgou and Bradbury, 1974). It was concluded that avoiding the use of contaminated tools prevents the transmission by tools.

The studies further revealed that all banana cultivars in Uganda were susceptible to the disease (Tushemereirwe et al., 2006b; Ssekiwoko et al., 2006). Apart from *Musa balbisiana,* which showed a high level of tolerance, all wild bananas and their relatives were also susceptible to the disease (Ssekiwoko et al., 2006). A study investigating survival of the Xanthomonas wilt pathogen outside the host revealed that the pathogen does not survive beyond 35 days in debris of both soil and banana (Mwebaze et al., 2006).

7.4 Management of banana diseases

7.4.1 Avoiding introduction of diseases into uninfected fields

Introduction of pathogens to new locations and fields may be caused by the use of pre-infected planting material or movement of vectors from remote infected sources (Cabaleiro and Segura, 1997). The soil-borne Fusarium wilt disease will only be effectively controlled

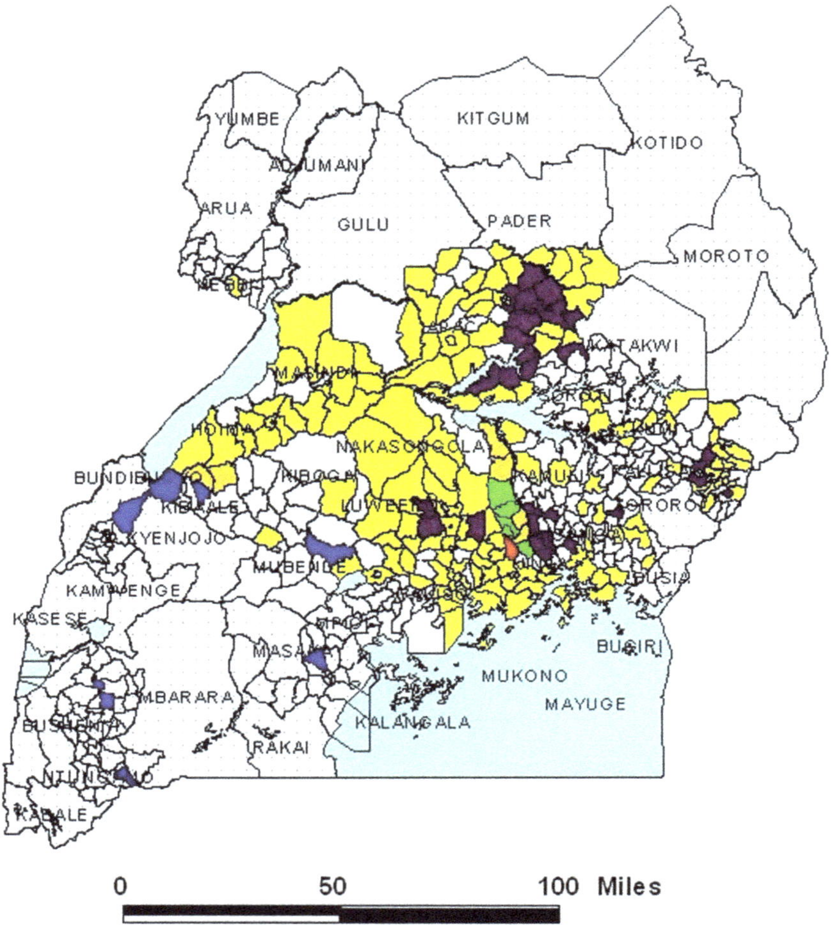

Figure 1 Distribution of banana bacterial wilt in Uganda as at April 2005. Darker colours show early outbreaks in 2000–2001 with lighter colours showing the spread of the disease by 2005. *Source* BXW NARO report to Kilimo Trust.

using planting materials of susceptible cultivars if the non-infected suckers are planted in non-infested soil. The planting materials are produced for farmers in the tissue culture laboratories or suckers obtained from plots which have never had the disease.

One other way of avoiding entry of the pathogen into new geographical area/field is through management of vector-spread epidemic. Early removal of the male bud is an effective means of preventing the spread of BXW by insect vectors (Tinzaara et al., 2006). Removal of male buds is recommended for both affected and unaffected fields. This technique effectively blocks the spread of BXW infection via insect vectors from within and outside the field (Tushemereirwe et al., 2006b).

7.4.2 Management of inoculum from the banana field

Roguing the whole affected stool was recommended at initial stages of the BXW epidemic in Uganda (Tushemereirwe et al., 2004). Then, single stem removal, that is, removing only infected plants from the stool, was recommended. This was based on the fact that BXW infection starts from the upper parts of the plant on the inflorescence and takes some time to get to the lower parts of the plant (Ssekiwoko et al., 2006). Single stem removal has also been evaluated and found to be effective in control of BXW in Kenya and DRC (Kubiriba et al., 2014). Studies showed that no bacteria were isolated from refuse of infected banana plants after five days even when they were left on the ground (Mwebaze et al., 2006). Preventing the use of cutting tools in infected fields during the three months was an effective alternative to cleaning tools as farmers worked in their plantations (Tushemereirwe et al., 2006a). Disinfection of cutting tools used for pruning leaves off the plants or after removing infected plants limits mechanical spread of BXW. Xcm strains were detected on cutting tools kept at room temperature up to 22 days after use (Buregyeya, 2010).

Removal of infected plants is also reported to effectively control the so-called 'clowd' diseases where new infections occur at places nearer to those where old infections had occurred than at places that are far away (Thresh and Owusu, 1986). Examples of such diseases are cocoa swollen shoot virus and BSV. It is important to remove the infected plants at the onset of the disease so that the epidemic is delayed (Kubiriba et al. 2013). Once the disease epidemic has advanced, the rate of removal of the infected planted should be higher than the rate of disease prevalence (Cabaleiro and Segura, 1997).

7.4.3 Good crop husbandry

Good husbandry of the banana plant considerably reduces the impact of leaf spot diseases and BSV (Tushemereirwe, 1996; Murekezi, 2003). The effect is mediated by increasing the rate of leaf production which in turn avails more leaf tissues to carry out photosynthesis (Tushemerweirwe, 1996).

7.4.4 Legislation to enforce disease management

There has been a strategy of helping the farmers constitute rules within their communities, including the enforcement procedure used in BXW control in Uganda (Kubiriba and Tushemereirwe, 2014). With support from the local government in enforcing the bye-laws formulated by affected communities, disease prevalence was reduced from 45 to 75% at its peak in 2012 to below 5% by 2015 (Table 4) (Kubiriba et al., 2016).

Table 4 BXW incidence in Uganda between 2013 and 2015

	% affected farms	
Region	November 2013	November 2015
Central	56.0	2.8
East	66.7	1.7
Mid-west	54.6	1.1
South-west	29.1	4.2
Over all	51.1	1.9

8 Improving banana varieties for management of pests and diseases

8.1 Conventional breeding

Host plant resistance is the most effective and most ecologically sustainable plant protection measure for controlling pests and diseases. Consequently, in the priority setting process of research areas in Uganda in the early 1990s, improving the highland bananas through breeding was identified as the most appropriate strategy for addressing the problems of pest and diseases. Farmer participatory appraisals of banana cropping systems in Uganda that followed revealed that most of the recommended cultural control practices were considered highly labour demanding and expensive by the farmers (Gold et al., 1993). Chemical control measures existed, but were economically inappropriate in the framework of the resource-poor smallholders who grow the bulk of crops in Uganda.

In 1994, the National Agriculture Research Organization (NARO), Uganda, in partnership with the International Institute of Tropical Agriculture (IITA), initiated a banana breeding programme with the aim of improving highland bananas for resistance to black Sigatoka disease and banana pests. Black Sigatoka, which had been reported to be found in Uganda in the same period (Tushemereirwe et al., 1993), triggered the need for banana breeding.

The breeding programme started with a medium-term objective of introducing elite lines from the established international breeding centres and evaluating them for consumer acceptability and resistance of pests and diseases in Uganda. Through this approach, two varieties, FHIA 17 and FHIA 23, were selected after a series of evaluations have been progressively adopted by farmers as replacements of Gros Michel (Bogoya), a dessert banana that is highly susceptible to Fusarium wilt and black Sigatoka (Tushemereirwe et al., 2000).

The programme had a long-term objective of developing consumer-acceptable East African highland banana hybrids with higher yield and resistance to pests and diseases. The breeding process involved cross-pollination of the susceptible female-fertile highland banana clones using improved black Sigatoka-resistant diploid male parents, the method already applied successfully in Honduras and at the IITA Onne station in Nigeria (Rowe, 1984; Vuylsteke et al., 1997).

The process started with evaluation of the highland banana cultivars (3× triploids) for female fertility by crossing them with Calcutta IV, a highly male-fertile wild banana (2× diploid). This resulted in tetraploid (4×) hybrids and some diploids (2×). The most seed-fertile clone sets were Nfuka and Nakabululu (Vuylsteke et al., 1996). Seed set rates also varied greatly among clones within clone sets, with means ranging from 1 to 20/bunch.

Selected tetraploids from the evaluation process that followed were then crossed with improved diploids accessed from IITA and the resultant progeny were taken through a series of evaluations (early evaluation, agronomic evaluation, on-farm evaluation and pest evaluation). Elite hybrids, KABANA 6H and KABANA 7H, that are resistant to black Sigatoka, highly tolerant to weevil and nematodes and fairly acceptable to consumers were finally selected and availed to farmers (Nowakunda et al., 2015). These hybrids are being rapidly adopted in low-elevation areas of Uganda where farmers have stopped cultivating banana due to pests and diseases.

Table 5 Performance of improved highland banana hybrids pests and diseases

Genotype	Youngest leaf spotted (YLS)	Bunch weight (kg)	% Cross section (weevil damage)	% Root necrosis (nematode damage)	Consumer acceptability score (1-6; worst to best)
12571S (M20)	9.3 ± 0.1	20.4 ± 0.5	6.8 ± 2.4	2.3a*	4.5 ± 1.3
1260S-1 (M19)	8.7 ± 0.2	19.2 ± 0.4	8.8 ± 1.7	3.3a	3.5 ± 1.3
M9	9.2 ± 0.2	21.3 ± 0.8	25.2 ± 1.7	6.1a	4.0 ± 1.1
M2	10.9 ± 0.3	23.2 ± 1.1	15.4 ± 1.7	3.5a	3.7 ± 1.2
Yagambi km5 (Resistant)	–	–	0 ± 0.0	4.3a	–
Butobe (AAA-EA) Susceptible	4.9 ± 0.1	12.9 ± 0.2	15.0 ± 6.0	9.3b	5.4 ± 0.8

LSD (0.05) = 1.23.
Source Nowakunda et al. (2015), Kubiriba et al. (2016).

Since the first set of crosses were introduced in 1994, several waves of crosses have followed and resulted in more lines of improved hybrids and male diploids which are still at various levels of evaluation for pest and disease resistance and consumer acceptability. From the most advanced wave, two hybrids better than the earlier released have been preliminarily selected also for final release to farmers (Kubiriba et al., 2016) (Table 5).

8.2 Molecular breeding

By 2000, it was realized that the most farmer-preferred cultivars were not going to be improved and would finally disappear from production. It was also not going to be possible to get hybrids that taste as good as land races because of sourcing resistance from wild banana diploids that are astringent. Furthermore, it was not going to be possible to develop hybrids resistant to banana Xanthomonas wilt that destroys all banana cultivars and their wild relatives. NARO, backed by the Government of Uganda, decided to apply molecular breeding (also called precision breeding by some scientists) to supplement the conventional breeding efforts.

The first step involved developing embryogenic cell suspensions of East African highland bananas (*Musa* AAA-EAHB) that would serve as a medium of transformation. This is the most limiting step in molecular breeding of highland bananas because the frequency of success is so low that nearly all cultivars are considered recalcitrant. The first success was realized with cultivar 'Nakyetengu' (*Musa* AAA-EAHB) after several years of effort. An embryogenic callus with friable tissue and embryos was obtained and a cell suspension that continued growing was initiated (Namanya et al., 2014). Since then, embryogenic cell suspensions have been obtained from other highland banana cultivars including 'Nakinyika', 'Mpologoma', 'Nakasabira', 'Nakiembe' and hybrid KABANA 6H (M9).

The next step involved transforming the obtained banana embryogenic cells with genes targeting specific traits using *Agrobacterium tumefaciens*. Transforming highland banana cells also required a lot of optimization experiments to make the technique work. This was also achieved after a long period of trying. The NARO research team has since 2005 successfully inserted the resistance genes for bacterial Xanthomonas wilt, weevils and nematodes; and those for enhancing provitamin A into bananas. The genes inserted are:

- Bacterial wilt resistance: Plant ferredoxin-like protein (*Pflp*) and hypersensitivity response assisting protein [*Hrap* genes cloned from sweet pepper and provided to NARO by Academia Sinica through African Agricultural Technology Foundation (AATF) and IITA] (Namukwaya et al., 2011; Tripathi et al., 2010).
- Provitamin A enhancement: Phytoene synthase (*a-Psy*) cloned from Asupina (a type of banana found in Asia) and provided by Queensland University of Technology.

Transgenic banana lines expressing genes described earlier in cultivar apple banana (model cultivar) were generated, characterized and screened in greenhouses. Selected lines showing strong response were evaluated in confined field trials. Ten lines showing 100% resistance to banana *Xanthomonas* wilt (Tripathi et al., 2014) and 50 lines with over four times more provitamin A were selected demonstrating effectiveness of the technologies. NARO has now embarked on the phase of product development and is on course to release the first transgenic banana by 2021. This could as well be the first transgenic crop produced in sub-Saharan Africa by African scientists.

8.3 Tissue culture for development of improved bananas

Three tissue and cell culture techniques, that is, embryo rescue, cell culture and shoot tip culture, are an invaluable support to the development of improved bananas (De Langhe and Vuylsteke, 1985). In conventional breeding, hybrid seed generated from pollinated bananas is the starting material for hybrid generation. When planted directly in soil, 97% of hybrid banana seeds do not germinate. However, with *in vitro* embryo rescue techniques, 30% of hybrid seed germinates into normal shoots within 7–14 days. This technique nurtures the immature or weak embryo, thus allowing it the chance to survive. Embryos are manually excised and placed immediately onto a culture medium that provides the proper nutrients to support survival and growth. The main factors that influence success are: the time of culture, composition of the medium, temperature and light.

Banana lines selected are progressed as single independent genetic lines. Through shoot tip culture techniques, such selected lines are mass propagated into multiple clones to enable further replicated evaluations in screen houses, field trials or for delivery to uptake pathways. Development of tissue culture technology has been the foundation of high-quality, disease-free planting material production at a mass scale, particularly in vegetatively propagated crops. Micropropagation facilitates production of large number of plantlets/unit time, thus helping in rapid introduction and dissemination of new varieties. Using tissue culture, it is possible to develop planting material which is free from sucker-borne diseases and pests and facilitates safe movement and easy handling of germplasm between laboratories within and across countries.

8.4 Bananas in non-banana-growing areas of northern and eastern Uganda

Northern Uganda sometimes experiences food insecurity situations. The number of food-insecure people in Uganda is estimated at about 1.1 million people with 81% of this population located in northern Uganda. The typical farming system in northern Uganda is dependent on mixed and annual cropping with soybeans, groundnuts, sorghum, millet, maize, beans, pigeon peas, cassava, sweet potatoes, rice and cow peas, all of which are highly susceptible to climate variability. With the advent of the varying and unpredictable climatic conditions, farmers who depend on these annual crops are always vulnerable to famine and loss of household incomes. The continuous production of bananas all year round offers a significant opportunity for more sustainable food supply and income generation than traditional cash crops for many such smallholder farmers.

The National Agricultural Research Organisation's Banana Research Program has been promoting the production of the recently released banana hybrids for improved income and food security in northern Uganda since 2012. Recent survey data show that 74% of the host farmers consider banana as one of the important crops for food and household income security. More than 70% of the farmers in the northern region were expanding their banana plots and giving out suckers to fellow famers, implying increased interest.

9 Marketing as a constraint to banana production in Uganda

It is estimated that 35–40% of the banana produced in Uganda is sold on the local market, mainly the urban centres. The rest is consumed as a carbohydrate source on-farm where smallholder subsistence farmers produce it. The highland banana marketing chain is made up of producers, collectors (brokers and bicycle traders), wholesalers (lorry traders), exporters (though very few), retailers and consumers. It involves movement of bananas to lorry traders by either the producers or brokers. These traders transport and resell the bananas to retailers who in turn sell to consumers. The marketing system presents to farmers a serious problem of low farm-gate prices and consequently low incomes that in turn severely impact productivity of the crop.

This marketing problem is considered one of the topmost constraints to productivity of the crop because of their influence on the amount of money peasant farmers earn and can possibly reinvest. Low incomes earned limit abilities of farmers to re-invest in replenishing soil nutrients, managing moisture stress (usually by mulching) and managing pests and diseases in the banana crop. The low farm gate prices are attributed to a disorganized producer-unfavourable marketing system with the following features (Fehr, 2010):

1. Farmers who sell their produce as individuals relying on personal negotiating power to get the best price from skilled traders bent on using misinformation techniques to give the lowest possible price. This denies farmers the higher bargaining power they would have if they were organized in marketing groups/associations.
2. A stronger relationship between brokers, who receive bananas from farmers, and wholesalers (lorry traders) than with farmers. This tends to diminish farmers' chances of getting good prices from their produce.
3. Low produce volumes leading to loss of interest by wholesalers.
4. Limited marketing skills and high overhead costs.
5. High costs of transport and poor infrastructure.

The local banana market has been growing with increasing urbanization since the 1970s. Bananas are produced in remote areas away from the major consuming urban centres. Southwestern Uganda supplies most of the bananas sold in the markets of Kampala (76.1%), Entebbe (98.9%) and Jinja (39.4%), the three biggest cities in the country (Mugisha and Ngambeki, 1994). The market increase was also assisted by the deterioration of the marketing mechanisms for traditional cash crops (coffee and cotton) in the 1970–80s (Mugisha and Ngambeki, 1994).

The export market in Uganda is underdeveloped. There are few firms involved in banana export which mainly export apple bananas, bogoya (Gros Michel) and some highland cooking bananas to a few European countries (UK, Denmark and Netherlands), Middle East (Saudi Arabia) and neighbouring countries (Rwanda, Kenya and Sudan) (Embrechts et al., 1996; Fehr, 2010). The Embrechts study attributed the low banana exports to high costs of transport and packaging, bulkiness, quick perishability, lack of a steady and reliable supply of needed varieties and poor infrastructure.

9.1 Intervention targeting better prices for farmers

A private company, Gourmet Gardens Ltd, has tested an innovative strategy for linking subsistence banana farmers to export markets with remarkable success (Fehr, 2010). The company, serving as an honest broker, was able to bring together the interests of farmers and consumers in order to exploit the market of dried bananas in Europe. The strategy has the following elements:

1. Assisting farmers to form producer organizations and to pool their production.
2. Centralizing processing facilities to minimize the challenges of product quality and overhead costs.
3. Fair trade certification and trademarking for the purpose of branding and adding value to the product.
4. Provision of company-owned extension services to ensure quality standards demanded by European consumers.

The strong partnership of farmers, the company and consumers has ensured trust and loyalty that have in turn led to the success achieved. Farmers are able to earn more significantly from their bananas and to reinvest in order to sustain high productivity. They have transformed into commercial farmers. This success has been attributed to the NARO, which is adopting the approach to marketing fresh bananas on the local and regional markets. We are at the stage of organizing farmers into banana producer and marketing associations.

10 Future trends and conclusion

This chapter concludes that *Musa* cropping systems are diverse in Africa. Plantains dominate the Western and Central African coastal and tropical forest ecologies. East African highland bananas dominate the East African highlands, mainly in the GLRA. The cropping systems vary and are characterized by different levels of management inputs and therefore productivity. Average productivity of less than 13 t/ha is extremely low, far below its production potential of 60–100 t/ha. *Musa* spp has the potential to trigger a green revolution for sub-Saharan Africa if the constraints hampering productivity can be addressed. This will entail reorienting subsistence banana farmers into commercial farmers, which in turn will require mobilizing and organizing farmers to negotiate better farm-gate prices that take into account the cost of inputs. Currently, subsistence banana farmers are ripped off on the surplus sold after meeting family needs.

The yield gaps can be considerably reduced by practices that intensify banana production as currently used by commercial and semi-commercial farmers. New technologies including plantain and highland banana hybrid deployment will cover the other yield gaps that will remain after adopting the existing technologies.

The cropping systems across African banana and plantain-growing areas are majorly characterized by low levels of management inputs and complexity and therefore low productivity. It is possible to significantly increase productivity from the current levels through intensive production. Research efforts should shift from pilot studies on individual constraints to system analysis so that the interactive driving synergistic forces of the production system are accurately estimated for optimum management and greater impact. Additionally, data generated in future should support intensive and sustainable production such as judicial use of water for production.

The future research and development effort should be driven by commercialization of the dominant subsistence banana production. Subsistence farmers will need to shift from the goal of producing for home consumption to producing for known markets. Only then will they be able to afford inputs that will increasingly be required to sustain or increase productivity. This will require investing in organizing farmers to produce and market their produce in groups. Strategies that are supportive of the organization of up-end value chain actors will be critical. For instance banana unions where the farmers, traders, processors and policy makers are brought into a partnership to the greater benefit of producers and consumers could be promoted. Organizing farmers will lead to increased farm gate price of bananas, which in turn will enable farmers to invest more in inputs needed to achieve sustainable intensive production.

Future *Musa* improvement activities should exploit *Musa* over 36 000 genes in the sequenced *Musa* genome to broaden the trait base beyond black Sigatoka resistance and shortening the breeding cycle. The breeding programme should include introgressing weevil, nematode and drought tolerance. Additional traits that *Musa* lack such as resistance to Xanthomonas wilt should be introduced using molecular breeding. Subsequently, hybrids with multiple traits together with soil inputs should improve sustainability and productivity of small EAHB holdings.

EAHB breeding will also need to integrate innovations in biology and genetics to enhance improvement. For instance, molecular markers that are coinherited with a trait can be used to introgress that trait more efficiently in marker-assisted selection. Additionally, since development of transgenic products is becoming expensive due to over regulation, efforts should be devoted to developing systems that give similar products such as cis-genics and gene edits that do not require the rigorous deregulatory systems. Such efforts will shorten the time it takes to release improved varieties.

11 Where to look for further information

More information can be accessed from the following sources:

- International Society for Horticultural Science; Proceedings for Conferences held every two years. The latest was held in Brisbane, Australia, in 2014. Proceedings are available on http://www.ishs.org/.
- Promusa – This is a network of scientists of the global fraternity working on bananas (http://www.musalit.org/). It is coordinated by Bioversity International.
- *African Crop Science Journal* – Vol 14, No 2 (2006). This was a special issue devoted for the technologies and information generated for BXW control in Uganda.
- Root, tuber and banana (RTB)-CGIAR project, where most of the banana activities are implemented.
- *Agriculture in Uganda*, Vol. II, Crops, 2001: The book is a production of National Agricultural Research Organisation.

- Research partnerships between the National Agricultural Research Organisation, International Institute of Tropical Agriculture and Bioversity International; annual reports, publications and other information are available on their websites; www.banana.go.ug, www.iita.org and http:/www.musalit.org/.

12 References

Abera, A., Gold, C. S., Kyamanya, S. and Karamura, E. B. (2000), 'Banana weevil *Cosmopolites sordidus* Germar ovipositional preferences, timing of attack and larval survival survivorship in a mixed cultivar trial in Uganda', *Acta Horticulturae* 540: 487–96.

Akinyemi, S. O. S., Aiyelaagbe, I. O. O. and Akyeampong, E. (2010), 'Plantain (*Musa* spp.) cultivation in Nigeria: A review of its production, marketing and research in the last two decades', In: Dubois, T., Hauser, S., Staver, C. and Coyne, D. (Eds), *Proceedings of an International Conference on Banana & Plantain in Africa Harnessing International Partnerships to Increase Research Impact.* Acta Horticulturae, International Society for Horticultural Science, Leuven, Belgium, 879, pp. 211–18.

Bagamba, F. (1994), Resources allocation efficiency in banana based cropping systems in Uganda. M.Sc. Thesis, Makerere University, Kampala.

Baiyeri, K. P. (2002), 'Nitrogen fertilizer influenced harvest index of plantain (*Musa* spp. AAB cv. Agbagba) in sub-humid zone of southeastern Niger', *Journal of Sustainable Agriculture* 20: 95–102.

Bananuka, J. A. (2001), The effect of drought stress and cultivar on growth and yield of bananas. Ph.D. Thesis, Makerere University, Kampala.

Bananuka, J. A. and Rubaihayo, P. R. (1994), 'Banana management practices and performance in Uganda', In: Adipala, E., Bekunda, M. A., Tenywa, J. S., Ogenga-Latigo, M. W. and Mugah, J. O. (Eds), *African Crop Science Proceedings*, African crop science society, Kampala, Uganda, pp. 177–82.

Barekye, A., Kashaija, I. N., Tushemereirwe, W. K. and Adipala, E. (2000), 'Comparisons of damage levels Caused by *Radopholus similis* and *Helicotylenchus multicinctus* on banana in Uganda', *Annals of Applied Biology* 137: 273–8.

Brandes, E. W. (1919), 'Banana wilts', *Phytopathology* 9: 339–92.

Bridge, J. (1993), 'Worldwide distribution of major nematode parasites of bananas and plantains', In: Gold, C. S. and Gemmill, B. (Eds), *Biological and Integrated Control of Highland Banana and Plantains Pests and Diseases*, pp. 147–24. IITA, Ibadan.

Buddenhagen, I. W. and Elsasser, T. A. (1962), 'An insect spread bacterial wilt epiphytotic of Bluggoe banana', *Nature* 194: 164–5.

Buregyeya, H. 2010, Evaluation of the contribution of birds, bats and farm tools in the long distance transmission of banana bacterial wilt. Doctoral dissertation, Masters' thesis, Makerere University, Kampala Uganda.

Buregyeya, H., Kubiriba, J., Tusiime, G., Kityo, R., Ssekiwoko, F. and Tushemerierwe, W. K. (2014), 'Role of birds and bats in long distance transmission of banana bacterial wilt in Uganda', *International Journal of Agricultural Innovations and Research*, 2, 636–40.

Cabaleiro, C. and Segura, A. (1997), 'Field transmission of grapevine leafroll associated virus (GLRa-V-3) by mealybug, *Planococcus citri*', *Plant Disease* 81: 283–7.

Carrière, S. (2003), *Les orphelins de la forêt: pratiques paysannes et écologie forestière (Ntumu, Sud-Cameroun).* IRD, Paris.

Chamberlin, J. (2007). Defining Smallholders Agriculture in Ghana: Who are smallholders, what do they do and how are they linked with markets? http://www.ifpri.org/themes/gssp/gssp.htm.

Cutler, J. M. and Loomis, R. S. (1977), 'Role of changes in solute concentration in maintaining favorable water in field-grown cotton', *Agronomy Journal* 69: 773–9.

Damour, G. Ozier-Lafontaine. H. and Dorel, M. (2012), 'Simulation of the growth of the banana (*Musa* spp.) cultivated on cover crop with simplified indicators of soil water and nitrogen availability and integrated plant traits', *Field Crop Research* 130: 99–108.

De Langhe, E. and Vuylstek, D., (1985), 'Feasibility of *in vitro* propagation of banana and plantains', *Tropical Agriculture (Trinidad)*, 62, 323–8.

Drechsel, P., Graefe, S., Sonou, M. and Cofie, O. O. (2006), *Informal Irrigation in Urban West Africa: An Overview*, Volume 102, p. 34. International Water Management Institute, Colombo.

Duman, F. (2012), 'Uptake of mineral elements during abiotic stress', In: Ahmad, P. and Prasad, M. N. V. (Eds), *Abiotic Stress Responses in Plants: Metabolism, Productivity and Sustainability*. Springer, New York, pp. 267–81.

Dury, S., Bricas, N., Tchango-Tchango, J., Temple, L. and Bikoi, A. (2002), The determinants of urban plantain consumption in Cameroon. *Food Quality and Preference* 13(2): 81–8.

Edmeades, S., Smale, M., Kikulwe, E. M., Nkuba, J. M. and Byabachwezi, M. S. R. (2007), 'Characteristics of banana-growing households and banana cultivars in Uganda and Tanzania', In: Smale, M. and Tushemereirwe, W. K. (Eds), *An Economic Assessment of Banana Genetic Improvement and Innovation in the Lake Victoria Region of Uganda and Tanzania*. Research Report No. 155. International Institute of Tropical Agriculture, Washington, DC.

Ekesa, B., Poulaert, M., Davey, M. W., Kimiywe, J., Van den Bergh, I., Blomme, G. and Dhuique-Mayer, C. (2012). Bioaccessibility of provitamin A carotenoids in bananas (*Musa* spp.) and derived dishes in African countries. *Food Chemistry* 133(4): 1471–77.

Embrechts, J., Bananuka, J. A., Couveur, Y., Bagamba F. and Lallemand D. (1996), Banana Subsector review, Republic Uganda. Commissioned by the European Union (STABEX). Service contract No. SC/26/96. October 1996.

English, A. (2008), Determinants of Liberian farmgate cocoa prices. M.Sc. dissertation, University of Tennessee, Knoxville, 155p.

FAOSTAT. (1990), *Production Statistics*. Food and Agriculture organization of the United Nations, Rome, Italy. http://faostat.fao.org/.

FAOSTAT. (2012), *Production Statistics*. Food and Agriculture organization of the United Nations, Rome, Italy. http://faostat.fao.org/.

Feakin, S. D. (1971), *Pest Control in Bananas. PANS Manual 1*. PANS, London.

Fehr, C. (2010), 'Banana (*Musa* spp) growing for export: Innovative approaches in production and marketing- the need for partnership', *Acta Horticulturae* 879: 53–6.

Firth, D. J., Johns, G. G. and Whalley, R. B. D. (2003), 'Glass house and field studies on the effects of ground covers on banana and macadamia growth and water relations', *Australian Journal of Experimental Agriculture* 43: 1245–54

Gobin, A., Campling, P. and Feyen, J. (2001), 'Logistic modelling to identify and monitor local land management systems', *Agricultural Systems*, 67(1), 1–20.

Gold, C. S. (1995), 'Banana weevil: Ecology, pests status and prospects of integrated control with emphasis on East Africa', In: *Proceedings of a Symposium on Biological Control in Tropical Crop Habit*, 30 October – 4 November 1994. Third international Conference on Tropical Entomology, Nairobi, Kenya.

Gold, C. S. and Bagabe, M. (1994), 'Banana weevil (*Cosmopolites sordidus*) infection of cooking and beer banana in adjacent stands in East in Uganda', In: *African Crop Science Conference Proceedings* 1, African Crop Science Society, Kampala, Uganda, pp. 296–9.

Gold, C. S., Ogenga Latigo, M. W., Tushemereirwe, W. K., Kashaija, I. N. and Nankinga, C. (1993), 'Farmer perception of banana pest constraints in Uganda', Results from a rapid rural appraisal, In: Gold, C. S. and Gemmill, B. (Eds), *Proceedings of a Research Coordination Meeting for Biological and Integrated Control of Highland Banana and Plantain Pests and Diseases*, IITA, Cotonour, Benin, pp. 3–24.

Gold, C. S., Speijer, P. R., Rukazambuga, N. D. and Karamura, E. (1994), 'Assessment of banana weevil in East African highland banana systems and strategies for control', In: (Valmayor, R. V., Davide, R. G., Stanton, J. M., Treverrow, N. L. and Roa, V. N. (Eds), *Proceedings of a Workshop on Nematodes and Weevil Borer Affecting Asia and Pacific Serdang*, Selangor, Malaysia, INIBAP, Los Banos, pp. 170–90.

Gowen, S. R. (1995), 'Pests', In: Gowen, S. R. (Ed.), *Bananas and Plantains*, Chapman and Hall, London, pp. 382–402.

Gowen, S. and Quencherve, P. (1990), 'Nematode parasites of banana plantains and abaca', In: Luc, M., Sikora, R. A. and Bridge, J. (Eds), *Plant Nematodes in Subtropical and Tropical Agriculture*, CAB International, Wallingford, pp. 431–60.

Harper, G., Dahal, G., Thottappilly, G. and Hull, R., (1999), 'Detection of episomal banana streak badnavirus by IC-PCR', *Journal of Virological Methods* 79: 1–8.

Harper, G., Hart, D., Moult, S., Hull, R., Geering, A. and J. Thomas. (2005), 'The diversity of banana streak virus isolates in Uganda', *Archives of Virology* 150(12): 2407–20.

Harris, V. W. (1947), 'The banana borer', *The East African Agricultural Journal* 13: 15–18.

Hedge, D. M. (1988), 'Growth and yield analysis of 'Robusta' banana in relation to soil water potential and nitrogen fertilization', *Scientia Horticultura* 37: 145–55.

Ingram, W. R. (1970), 'Pests of bananas', In: Jameson, J. D. (Ed.), *Agriculture in Uganda*, 2nd Edition. African Books Collective, Oxford United Kingdom.

INIBAP. (1996), 'Major banana and plantain types cultivated around the world', pp. 23–5. In: *The Global Banana and Plantain Network: INIBAP Annual Report 1994*. International Network for Improvement of Bananas and Plantains (INIBAP), Montpellier, France.

Jalloh, A., Roy-Macauley, H. and Sereme, P. (2012), 'Major agroecosystems of West and Central Africa: Brief description, species richness, management, environmental limitations and concerns', *Agriculture, Ecosystems and Environment*, 157, 5–16.

James, A. P., Geijskes, R. J., Dale, J. L. and Harding, R. M. (2011), 'Development of a novel rolling-circle amplification technique to detect Banana streak virus that also discriminates between integrated and episomal virus sequences', *Plant Disease* 45: 57–62.

Jones, D. L. and Lockhart, B. E. L. (1993), *Banana Streak virus. Fact sheet* (No. 1). International Network for Improvement of Banana and Plantain, Montpellier, France.

Kangire, A. (1998), Fusarium wilt (Panama disease) of exotic bananas and wilt of East African highland bananas (Musa, AAA-EA) in Uganda. Doctoral dissertation, University of Reading, United Kingdom.

Kanmegne, J., Smaling, E. M., Brussaard, L., Gansop-Kouomegne, A. and Boukong, A. (2006), 'Nutrient flows in smallholder production systems in the humid forest zone of southern Cameroon', *Nutrient Cycling in Agroecosystems*, 76(2), 233–48.

Karamura, D. and Pickersgill, B. (1999), 'A classification of the clones of East African highland bananas (Musa) found in Uganda', *Plant Genetic Resources Newsletter (IPGRI/FAO) Bulletin des Ressources Phytogenetiques (IPGRI/FAO) Noticiario de Recursos Fitogeneticos (IPGRI/FAO)* 119: 1–6.

Karamura, D. A., Karamura, E. B. and Gold, C. S. (1996), 'Cultivar distribution in primary banana growing regions of Uganda', *MusAfrica* 9: 3–5.

Karamura, E., Frison, E., Karamura, D. A. and Sharrock, S. (1998), 'Banana production systems in eastern and Southern Africa', In: Picq, C., Fouré, E. and Frison, E. (Eds), *Banana and Food Security*, pp. 401–12. INIBAP, Montpellier. ISBN: 2-910810-36-4.

Kashaija, I. N. (1996), Factors influencing Nematodes population densities and rot damage on banana cultivars in Uganda. Ph.D. Thesis, University of Reading, UK.

Kashaija, I. N., Speijer, P. R., Gold, C. S., and Gowen, S. R. (1994), 'Occurrence, distribution and abundance of plant parasitic nematodes on bananas in Uganda. Preliminary results of the diagnostic survey', *African Crop Science Journal*, 2, 99–104.

Kashaija, I. N., McIntyre, B. D., Ssali, H. and Kizito, F. (2004), 'Spatial distribution of roots, nematode populations and root necrosis in highland bananas Uganda' *Nematology* 6(1): 7–12.

Kubiriba, J. and Tushemereirwe, W. K. (2014), 'Approaches for the control of banana Xanthomonas wilt in East and Central Africa', *African Journal of Plant Science*, 8(8), 398–404.

Kubiriba, J., Legg, J. P., Tushemereirwe, W. and Adipala, E., (2001), 'Disease spread patterns of banana streak virus in farmers, fields in Uganda', *Annals of Applied Biology*, 139(1), 31–6.

Kubiriba, J., Tushemereirwe, W. K., Kenyon, L. and Chancellor, T. C. B., 2013, 'Field spread of banana streak virus (BSV)', *African Journal of Agricultural Research*, 8, 1881–90.

Kubiriba, J., Muthomi, J., Ndungo, V., Kwach, J., Rockefeller, E., Rwomushana, I., Tushemereirwe, W. and Opio, F., 2014, 'Strategies for rehabilitation of banana fields infested with Xanthomonas campestris pv. musacrearum', *Journal of Crop Protection* 3(1): 21–9.

Kubiriba, J., Ssali, R. T., Barekye, A., Akanwasa, K., Tushemereirwe, W. K., Batte, M., Karamura, E. B. and Karamura, D. (2016), 'The performance of East African highland bananas release in farmers' fields and the need for further improvement', In: Smith, M., Van den Bergh, I. and Picq, C. (Eds), *Proceedings of IX International Symposium on Banana: ISHS-ProMusa Symposium on Unravelling the Banana's Genomic Potential*. Acta Horticulturae, 1114, ISHS, Leuven (Belgium), pp. 231–8. ISBN: 978-94-62611-08-5.

Leakey, A. L. C. (1970), 'Banana diseases', In: J. D Jameson (Ed.), *Agriculture in Uganda*, Second Edition, Oxford University Press, United Kingdom, pp. 143–5.

Lescot, T. and Ganry, J. (2010), 'Plantain (*Musa* spp.) cultivation in Africa: A brief summary of developments over the previous two decades', In: Dubois, T., Hauser, S., Staver, C. and Coyne, D. (Eds), *Proceedings of an International Conference on Banana & Plantain in Africa Harnessing International Partnerships to Increase Research Impact*. Acta Horticulturae 879, International Society for Horticultural Science, Leuven, Belgium, pp. 445–55.

Lockhart, B. E. L. (1986), 'Purification and serology of a bacilliform virus associated with banana streak disease', *Phytopathology* 76(10): 995–9.

Magnaye, L. V. (1994), Virus diseases of banana and current studies to eliminate the virus by tissue culture, pp. 38–43. In: Tangonan, N. G. (Ed.), *Towards Making Pest and Disease Management Relevant to Big and Small Banana Growers*. Proceedings of the first PPS-SMD National symposium on Pests and diseases in Philippines, 23–24 April 1993, Davao City, Philippines. Phytopathology Society Inc., Southern Mindanao Division.

Masanza, M. (2003), Effect of crop sanitation on banana weevil (*Cosmopolites sordidus*) populations and associated damage. Doctoral dissertation, PhD thesis, Wageningen University, Wageningen.

Mugisha, J. and Ngambeki, D. S (1994), 'Marketing system of the Uganda banana industry', In: *Africa Crop Science Proceedings* 1, pp. 384–7.

Murekezi, A. K. (2003), Profitability analysis and strategic planning of coffee processing and marketing in Rwanda: A case study of a coffee growers' association. Doctoral dissertation, University of Reading, United Kingdom.

Mutsaers, H. J. W., Mbouémboué. P. and Boyomo, M. (1981), 'Traditional food crop growing in the Yaoundé area (Cameroon) Part II.Crop associations, yields and fertility aspects', *Agro-Ecosystems* 6(4), 289–303.

Mwebaze, J. M., Tusiime, G., Tushemereirwe, W. K. and Kubiriba, J. (2006), 'The survival of *Xanthomonas campestris* pv. *musacearum* in soil and plant debris', *African Crop Science Journal* 14(2): 121–7.

Namaganda K., A., Stultiens, A. and Naluwooza, M. (2016), Promoting Genealogy and Local History Collections using Social media in Uganda. IFLA World Library and Information Congress, International Federation of Library Associations and Institutions, The Hague, the Netherlands.

Namanya, P. Mutumba, G. Magambo, S. M. and Tushemereirwe, W. (2014), 'Developing a cell suspension system for Musa AAA-EA cv 'Nakyetengu': A critical step for genetic improvement of matooke East African highland banana', *In Vitro Cellular & Developmental Biology – Plant* 50(4): 445–50. doi: 10.1007/s11627-014-9598-0 (available on line at link.springer.com).

Namukwaya, B., Tripathi, L., Tripathi, J., Arinaitwe, G., Mukasa, S. B. and Tushemereirwe, W. K. (2011), 'Transgenic banana expressing *Pflp* gene confers enhanced resistance to Xanthomonas wilt disease', *Transgenic Research*, 21(4): 855–65.

Nankinga, C. (1994), Potential of indigenous fungal pathogens for the biological control of the banana weevil, *Cosmopolites sordidus* (Germar) in Uganda. M.Sc. Thesis, Makerere University, Uganda.

Nankinga, C. M. and Ogenga-Latigo, M. W. (1996), 'Effect of method of application on the effectiveness of *Beauveria bassiona* against the banana weevil, *Cosmopolites sordidus*', *African Journal of Crop Protection* 6: 12–21.

Ngeleza, G. K., Owusua R., Jimah K. and Kolavalli, S. (2011), Cropping practices and labor requirements in field operations for major crops in Ghana. What needs to be mechanized? Discussion Paper 01074, IFPRI, Washington, p. 28.

Night, G., Gold, C. S. and Power, A. G. (2010), 'Survivorship and development of the banana weevil *Cosmopolites sordidus* (Coleoptera: Curculionidae) on different banana cultivars in Uganda', *International Journal of Tropical Insect Science* 30(04), 186–91.

Nowakunda, K., Barekye A., Ssali R. T., Namaganda J., Tushemereirwe W. K., Nabulya G., Erima R., Akankwasa K. and Hilman E. (2015), ''Kiwangazi' (syn KABANA 6H) black Sigatoka, nematode and banana weevil tolerant 'matooke' hybrid banana released in Uganda', *Horticultural Science* 50(4): 621–3.

Nweke, F. I., Njoku, J. E. and Wilson, G. F. (1988), 'Productivity and limitations of plantain (*Musa* spp. cv. AAB) production in compound gardens in Southeastern Nigeria', *Fruits* 43(3): 161–6.

Nyombi, K., Van Asten, P. J. A., Corbeels, M., Taulya, G., Leffelaar, P. A. and Giller, K. E. (2010), 'Mineral fertilizer response and mineral use efficiencies of East African Highland banana (*Musa* spp AAA-EAHB cv. Kisansa)', *Field Crops Research* 117: 38–50.

Okech, S. H. Van Asten, P. J. A. Gold, C. S. and Ssali, H. (2004), 'Effects of potassium deficiency, drought and weevil on banana yield and economic performance in Mbarara, Uganda', *Uganda Journal of Agricultural Sciences* 9: 511–19.

Pettigrew, W. T. (2008), 'Potassium influences on yield and quality production for maize, wheat, soya bean and cotton', *Physiologia Plantalum* 133: 670–81.

Robinson, J. C. (1996), *Bananas and Plantains*. CAB International, Wallingford, UK.

Rowe, P. R. (1984), 'Breeding bananas and plantains', *Plant Breeding Review* 2: 135–55.

Rukazambuga, N. (1996), *The Effect of Banana Weevil* (*Cosmopolites sordidus,* Germar) *on the Growth and Productivity of Bananas (Musa AAA-EA) and the Influence of Host Vigour on Attack*. University of Reading, UK.

Rukazambuga, N. D. T. M., Gold, C. S. and Gowen, S. R. (1998), 'Yield loss in East African Highland banana (*Musa* spp., AAA-EA group) caused by the banana weevil, *Cosmopolites sordidus* Germar', *Crop Protection* 17(7): 581–9.

Sharma, S. R. (1987), 'Banana bunchy top virus - a review', *International Journal of tropical plant diseases* 6: 19–41.

Singhi, D. N. (2004), 'Effect of graded dose of N on growth and yield attributes of ratoon banana, In: Kumar, A. (Ed.), *Environment and Health*. A.P.H. Publishing corporation, New Delhi, India, pp. 133–5.

Smithson, P. C. McIntyre, B. D. Gold, C. S., Ssali, H. and Kashaija, I. N. (2001), 'Nitrogen and potassium fertilizer vs. nematode and weevil effects on yield and foliar nutrient status of banana in Uganda', *Nutrient Cycling in Agroecosystems* 59: 239–50.

Sonwa, D. Ngongmeneck, B. A. Weise, S. W., Tchatat, M. Adesina, A. A. and Janssens, M. J. J. (2007), 'Diversity of plants in cocoa agroforests in the humid forest zone of Southern Cameroon', *Biodiversity and Conservation* 16(8), 2385–400.

Speijer, P. R. Gold, C. S., Karamura, E. B. and Kashaija, I. N. (1994a), 'Banana weevil and nematode distribution patterns in Highland banana systems in Uganda: Preliminary results from a diagnostic survey', In: Adipala, E., Bekunda, N. A., Tenywa, J. S., Ogenge-Latigo, M. W. and Mugah, J.O (Eds), *African Crop Science Conference Proceedings*, pp. 285–9. African crop Science Society, Kampala.

Speijer, P. R. Rukazambuga N. D. and Karamura, E. B. (1994b). Assessment of banana weevils in East Africa highland banana systems and strategies for control. In: Valmayor, R., Davide, R. G., Stanton, J. M., Treverrow, N. L., and Roa, V. N. (Eds), *Proceedings of Conference – Workshop on Nematodes and Weevil Borer Affecting Bananas in Asia and the Pacific Serdang*, Selangor, Malaysia 18–22 April 1994. INIBAP, Los Banos.

Ssekiwoko, F., Taligoola, H. K. and Tushemereirwe, W. K. (2006), '*Xanthomonas campestris* pv. *musacearum* host range in Uganda', *African Crop Science Journal* 14(2): 111–20.

Stover, R. H., (1962), 'Intercontinental spread of banana leaf sport (*Mycosphaerellamusicola* Leach', *Tropical Agriculture, Trinidad* 39: 327–38.

Stover, R. H. and Waite, B. H. (1960), 'Pathogenically and distribution of *Fusarium oxysporum Cubense* races 1 and 2', *Canadian Journal of Botany* 38: 51–61.

Su, H. J., Hwang, S. C. and Ko, W. H. (1986), Fusarial wilt of Cavendish bananas in Taiwan. *Plant Disease*, 70(9), 814–18.

Swennen, R. (1990), *Plantain Cultivation Under West African Conditions.* International Institute of Tropical Agriculture, Amarin Printing Group Co., Nigeria, p. 24.

Swennen, R. and Vuylsteke, D. (2001), 'Banana (Musa L.)', In: Raemaekers, R. (Ed.), *Crop Production in Tropical Africa*, pp. 530–52. DGIC (Directorate General for International Co-operation), Brussels, Belgium.

Taulya, G. (2015), Understanding productivity of East African Highland banana. Ph.D. Thesis, Wageningen University, Wageningen.

Thresh, J. M. and Owusu, G. K. (1986), 'The control of cocoa swollen shoot disease in Ghana; An evaluation of eradication procedures', *Crop Protection* 5: 41–52.

Tinzaara, W. Gold, C. S. Ssekiwoko, F. Tushemereirwe, W. Bandyopadhyay, R. Abera, A. and Eden-Green, S. (2006), 'Role of insects in the transmission of banana bacterial wilt', *African Crop Science Journal* 14 (2): 105–10.

Tripathi, L., Mwaka, H., Tripathi, J. N. and Tushemereirwe, W. K. 2010. Expression of sweet pepper Hrap gene in banana enhances resistance to *Xanthomonas campestris* pv. *musacearum. Molecular Plant Pathology* 11: 721–31.

Tripathi, L., Tripathi, J., Kiggundu, A., Korie, S., Shortkoski, F. and Tushemereirwe, W. K. (2014), 'Field trial of Xanthomonas wilt disease-resistant banana in East Africa', *Nature Biotechnology* 32(9): 868–70.

Tushemereirwe, W. (1996), Factors influencing expression of leaf spot diseases of highland bananas in Uganda. Ph.D. Thesis, University of Reading, Reading.

Tushemereirwe, W. and Waller, J. M. (1993), 'Black leaf streak (*Mycospharellafijensis)* and associated disease of bananas in Uganda', *Plant Pathology* 42: 471–2.

Tushemereirwe, W. K., Gold C. S., Speijer P. R. and Holderness, M. (1993), 'Foliar diseases of banana in Uganda, Results from a diagnostic survey', *African Crop Science Journal* 1(2): 145–9.

Tushemereirwe, W., Karamura, E. B. and Karyeija, R. (1996), 'Banana Streak Virus (BSV) and associated filamentous virus (unidentified) disease complex of highland bananas in Uganda', *Informusa* 5: 9–12.

Tushemereirwe, W. K., Kangire, A., Kubiriba, J. and Nowakunda, K. (2000), 'Fusarium wilt resistant bananas considered appropriate replacements for cultivars susceptible to the disease in Uganda', *Uganda Journal of Agricultural Sciences* 5(6)62–4.

Tushemereirwe, W. K., Karamura, D., Ssali, H., Bwamiki, D. Nankinga, C. Bagamba, F. Kangire, A. and Ssebuliba, R. (2001), '(Bananas *Musa* spp)', In: Mukiibi, J. K. (Ed.), *Agriculture in Uganda, Volume II, Crops*, (pp 281–319). Oxford University Press, United Kingdom.

Tushemereirwe, W. K., Kasheija, I. N., Tinzaara, W. Nankinga, C. and New, S. (2003a), Banana production manual: A guide to successful banana production in Uganda, p. 77. NARO and ADC IDEA Project. Kamala, Uganda.

Tushemereirwe, W. K., Kangire, A., Smith, J., Ssekiwoko, F., Nakyanzi, M., Kataama, D., Musiitwa, C. and Karyaija, R. (2003b), 'An outbreak of banana bacterial wilt in Uganda', *Informusa* 12: 6–8.

Tushemereirwe, W. Kangire, A. Ssekiwoko, F. Offord, L. C. Crozier J. Boa, E., Rutherford, M. and Smith, J. J. (2004), 'First report of *Xanthomonas campestris* pv. *musacearum* on banana in Uganda', *Plant Pathology* 53: 802.

Tushemereirwe, W. K., Kangire, A. and Tinzaara, W. (2006a), 'Banana bacterial wilt in Uganda: A disease that threatens livelihoods', *African Crop Science Journal* 14 (2): 73–183.

Tushemereirwe, W., Okaasai, O., Kubiriba, J., Nankinga, C., Muhangi, J., Odoi, N. and Opio, F. (2006b), 'Status of banana bacterial wilt in Uganda', *African Crop Science Journal* 14(2): 73–82.

Van Asten, P. J. A., Gold, C. S., Wendt, J. De, Waele, D., Okech, S. H. O., Ssali, H. and Tushemereirwe, W. K. (2005), 'The contribution of soil quality to yield and its relation with other banana yield loss factors in Uganda', In: Blomme, G., Gold, C. S. and Karamura, E. (Eds), *Proceedings of a Workshop Held on Farmer Participatory Testing of IPM Options for Sustainable Banana Production in Eastern Africa*, Seeta, Uganda, 8–9 December 2003, International Plant Genetic Resources Institute, Montpellier, pp. 100–15.

Van Asten, P. J. A., Fermont, A. M. and Taulya, G. (2011), 'Drought is a major yield loss factor in East African highland banana', *Agricultural Water Management* 98: 541–52.

Vuylsteke, D., Karamura, D., Ssebuliba, R. N. and Makumbi, D. (1996), Seed and pollen fertility in the East African highland bananas. Musa Africa, Plantain and banana Newsletter No. 10. International Institute of Tropical Agriculture, Ibadan, Nigeria.

Vuylsteke, D., Ortiz, R., Ferris, S. and Crouch, J. (1997), 'Plantain improvement', *Plant Breeding Reviews* 14: 267–320.

Wairegi, L. W. I. (2010), Management practices and opportunities in East African highland banana (*Musa* spp. AAA-EA) production in Uganda. Ph.D. Thesis, Wageningen University, the Netherlands.

Wairegi, L. W. I. and Van Asten, P. J. A. (2011), 'Norms for multivariate diagnosis of nutrient imbalance in East African Highland bananas (*Musa* spp AAA)', *Journal of Plant Nutrition* 34: 1453–72.

Wairegi., L. W. I. and Van Asten, P. J. A. (2010), 'The agronomic and economic benefits of fertilizer and mulch use in banana systems in Uganda', *Agricultural Systems* 103(8): 543–50.

Weerasinghe, P., Premalal, N. H. R. and Saranasinghe, S. N. K. (2004), 'Influence of nitrogen on crop performances and leaf nitrogen status of dense-planted banana', *Annals of Sri Lanka Department of Agriculture* 6: 217–26.

Yirgou, D. and Bradbury, J. F. (1968), 'Bacterial wilt of Enset (*Ensete ventricosum*) incited by *Xanthomonas musacearum*', *Phytopathology* 59: 111–12.

Yirgou D. and Bradbury, J. F. (1974), 'A note on wilt of banana caused by the Enset wilt organism, *Xanthomonas musacearum*', *East African Agricultural and Forestry Journal* 40: 111–14.

Chapter 4

Challenges and opportunities for smallholders in banana value chains

William Tinzaara, Bioversity International, Uganda; Dietmar Stoian, Bioversity International, France; Walter Ocimati, Enoch Kikulwe and Gloria Otieno, Bioversity International, Uganda; and Guy Blomme, Bioversity International, Ethiopia

1 Introduction
2 Challenges along the value chain
3 Enabling an environment for addressing the challenges
4 Future trends
5 Conclusion
6 Where to look for further information
7 References

1 Introduction

Bananas and plantains (*Musa* spp.) rank sixth on the list of staple crops in the world (FAO, 2017). With a global production of around 86 million tons per annum, about 18 million tons are exported, while the remainder is consumed domestically. Though exports, mainly comprising dessert bananas, account only for about 20% of global production, they account for about half of the value generated (about US$13 billion out of a total of US$26 billion). Cooking bananas and plantains account for the other half of the value generated and, given that they account for about 80% of global banana production, play a critical role in terms of global food security and in national and household economies.

The bulk of export bananas belongs to the Cavendish subgroup (AAA), which by the late 1950s emerged as principal dessert bananas due to their resistance to Panama disease (Fusarium wilt) after a global outbreak had wiped out the previously dominant Gros Michel subgroup (Dita et al., 2010). The varieties consumed in the producing countries, mainly cooking bananas and plantains, show a great deal of regional variation (Sharrock and Frison, 1998). True plantains (AAB), which need to be cooked before eating, are dominant not only in West Africa, but also in Central and South America. In Southeast Asia and the Americas, cooking bananas (ABB) and dessert bananas (AAB) are the most common, while cooking bananas (AAB) are popular in the Pacific. In East and Central Africa (ECA), three main banana types are grown: 1) cooking bananas, mainly East African Highland bananas (AAA), which are largely produced for home consumption with surplus sold to

http://dx.doi.org/10.19103/AS.2017.0020.10

the local market; 2) beer banana cultivars (AB, ABB and AAA-EA), which are a main source of household income when processed into banana beer and sold in local markets; and 3) dessert types, mainly comprising AAA and AB types.

In ECA alone, bananas and plantains provide food and income for more than 50 million smallholder farmers, with an annual production worth US$ 4.3 billion, equivalent to about 5% of the region's gross domestic product (FAOSTAT, 2014). The consumption of bananas in ECA is the highest in the world, providing 3–22% of total daily calorie consumption estimated at 147 kcal per person (FAOSTAT, 2014). Bananas and plantains have multiple uses. As food, they can be prepared through boiling, steaming, mashing, baking, drying and pounding (FAO, 1990). The dried leaves, leaf sheaths and petioles (i.e. the stalk attaching the leaf blade to the stem) are used as ropes, cover for fermenting cassava, nesting materials for egg-laying poultry, building materials for temporary shelters, sponges and roofing material (Akinyemi et al., 2010; Kamira et al., 2015). Banana leaves are also used for wrapping, packaging, marketing and serving of food (i.e. as plates) (Kamira et al., 2015; Karamura, 1993). The fruit peels are used as feed for livestock, while the dried peels are used in soap production (Akinyemi et al., 2010; Kamira et al., 2015).

With average yields often not exceeding 30% of the crop's potential of over 60–70 tonnes ha^{-1} yr^{-1} (Van Asten et al., 2005), banana productivity is generally low in the ECA region which comprises Uganda, Tanzania, Rwanda, Burundi, Democratic Republic of Congo (DRC) and Kenya. At the global level, banana production and yields are constrained by lack of production efficiency exacerbated by pests and diseases, such as black leaf streak, banana bunchy top disease (BBTD), Xanthomonas wilt of banana (BXW), Fusarium wilt, weevils and nematodes; soil degradation; limited application of sound farming practices; lack of fertilizers and other inputs; short shelf life of the fruit; limited availability of labour; post-harvest losses and poor market access (Akankwasa et al., 2008). These challenges need to be addressed by key stakeholders in the value chain, comprising smallholder farmers, buyers, processors and traders, in collaboration with service providers from outside of the chain, such as government agencies, non-governmental organizations (NGOs) and private businesses providing technical, business and financial services. Figure 1 shows a generic value chain of banana, with ramifications to domestic and international markets.

There have been diverse research efforts focused on the key challenges facing smallholder banana growers, but limited efforts have been made to develop integrated approaches that are adapted to the livelihoods of resource-poor smallholders. Available conventional technologies for the management of biotic and abiotic constraints of banana production include cultural control practices, clean planting material, and biological or chemical control, which are often costly, labour-intensive and/or ecologically unsustainable. In addition, it is evident that no single control method will provide a sustainable solution. Use of an integrated soil fertility and pest and disease management approach has therefore been recommended. Recent research provides insights into innovative pest and disease management practices which are ecologically sustainable. These can be combined with disease-resistant planting material emerging from conventional breeding and genetic engineering, as well as evidence-based recommendations for how to reduce post-harvest losses and improve the role of smallholders in the banana value chain.

This chapter summarizes key research findings in relation to challenges facing smallholders in domestic banana value chains, identifies opportunities for smallholders and other value chain actors and service providers for addressing these, and concludes with recommendations for future research.

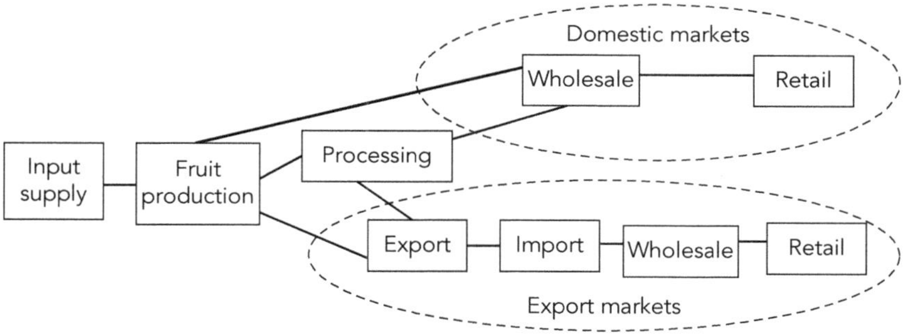

Figure 1 Generic banana value chain.

2 Challenges along the value chain

With regard to challenges in banana value chains, four phases can be distinguished: 1) pre-production, 2) production, 3) post-harvest management and 4) marketing. Table 1 provides an overview of these phases and the details are described below.

2.1 Pre-production challenges

2.1.1 Availability of labour for land preparation

Availability of labour is a key factor affecting production and yield in smallholder systems. In many of these systems, variability of labour is a principal challenge, particularly with regard

Table 1 Principal challenges along banana value chains

Pre-production	Production	Post-harvest	Marketing
• Limited availability of labour for land preparation • Limited use of and access to quality planting materials/seeds • Social constraints • Land tenure and fragmentation	• Limited availability of labour for plantation management • Poor access to fertilizers and pesticides • Poor soil fertility and water/moisture stress • Poor crop management practices • Pests (weevils, nematodes) • Bacterial, fungal and viral diseases • Climate variability and change	• Post-harvest losses • Value adding	• Asymmetric market information • Low prices for produce/seasonal fluctuation of prices • High tax burden, especially for transport agents • Poor road networks increasing transport costs • Unequal benefit sharing among value chain actors • Low quality of the products compromising prices

to land preparation (including digging of holes for planting) which is labour-intensive and, if reliant on hired labour, costly. Labour availability is often subject to seasonal fluctuation, as there may be competing claims in view of overlapping planting and harvesting seasons of other crops (Qaim, 1999). Considering the labour-intensive nature of land preparation, households with a higher number of members of economically active age are more likely to perform these tasks effectively and on time. For commercial production systems, however, land preparation is mechanized and labour requirements decrease accordingly.

2.1.2 Use of and access to quality planting material

Limited access to and use of improved seeds has been found to be a major barrier to banana production (Kikulwe et al., 2007). Farmers in ECA region are predominantly reliant on suckers obtained from their own or neighbours' fields (Ocimati et al., 2013). The use of suckers significantly increases the incidence and impact of banana pests and diseases (Fujardo, 2010).

The high reliance on suckers has thus led, over time, to a build-up of pests and diseases on smallholder farms. Access to clean seeds has partly been hampered by the low investment in micropropagation facilities. Poor road networks in many parts of sub-Saharan Africa make the distribution of new seed varieties in rural areas very costly (Carruthers et al., 2009). Where local multiplication of improved seed varieties is possible, farmer networks can play an important role in addressing this challenge.

2.1.3 Social constraints

Gender

Despite their prominent role in the management of banana plantations (Table 2), women are faced with several gender-based constraints, such as lack of control over land and limited share of income derived from bananas (Ajambo et al., 2017). In many mixed households, women farmers have less power to make decisions on how to use the land and banana income and are often less involved in the marketing of bananas.

Alliances and networks

Limited interactions among value chain actors has been identified as a constraint to production, processing and marketing of bananas (Ariho et al., 2015). Alliances between value chain actors and farmer networks can help overcome bottlenecks and facilitate innovation in the value chain through provision of information, capacity development, and mechanisms for sharing benefits and risks. Information may be disseminated through formal institutions, such as public sector extension services, NGOs and mass media; or through informal mechanisms such as farmers' organizations or networks of friends, relatives and acquaintances. The different means of information dissemination have different costs and, depending on the socio-economic characteristics and location, farmers may prefer different channels and ways to receive information.

Minimum wages

The national labour laws of producing countries are often not fully implemented, and low wages discourage work in banana plantations. The working environment can be improved by increasing worker participation in decision-making regarding production, productivity and profitability within the banana subsector. Wages can be increased by targeting

Table 2 Gendered involvement in banana production in East Africa

Banana management activity	Men	Women
De-sheathing/pruning	+	+++
Weeding	+	+++
De-suckering	++	++
Corm removal	+++	+
Mulching	+	+++
Earthing up	++	++
De-budding	+	+++
Pests and disease control (trapping + other cultural practices)	+	+++
Chemical control of pests and diseases	+++	+
Soil and water conservation practices	+++	+
Digging water trenches	++	++
Selecting materials and planting	++	++
Applying manure	++	++
Applying fertilizer	+++	+
Marketing	+++	+

Note: strong (+++), some (++) and low (+).

investments that require higher levels of skill and technical expertise, upgrading wage packages and indexing the minimum wages with inflation.

2.1.4 Land tenure and fragmentation

The size of landholdings and the type of tenure system often discourage investments in banana production in Africa. Most banana crops are produced on small and fragmented landholdings in ECA. For example, the area under banana cultivation varies from 0.28 to 1 acre in Uganda (UBOS, 2010). Land fragmentation exacerbates the effects of land degradation, hampering agricultural development due to the lack of economies of scale (Niroula and Thapa, 2005). Small, fragmented landholdings also limit mechanization of farms and production at commercial scales. In the ECA region, farmers typically rely on family labour and rudimentary farm tools, rendering banana production a costly and less productive endeavour.

Land fragmentation can be attributed to a combination of local customs and legal stipulations regarding land rights and ownership. Agricultural expansion is hindered by legal restriction in relation to land tenure and cultural habits linked to the inheritance system expansion (Akinyemi et al., 2010). In Nigeria, a large portion of the land is communally owned, and private land is allocated by community members only for establishing a residence. In this system, customary rules of tenure restrict smallholders' access to communal land sufficient to feed and support their families (Arua and Okorji, 1998). This has been fostered by the high population densities and rate of population increase. For example, the population densities of Burundi and Rwanda, some of the countries dominated by smallholder banana producers are respectively, estimated at 435 and 470 people/km^2 and high growth rates between 2.3 and 3.3% per annum, which are among the highest in the world (The World Bank Group, 2016). In addition, in most African countries policies related to land fragmentation are lacking or relaxed, thus limiting banana production to a subsistence level. The underdevelopment of industrial and service sectors also means that most people in several of the banana-producing countries are employed in the agricultural sector, perpetuating land fragmentation. Investments in industrialization and the service sector coupled with appropriate policies and strategies are needed to facilitate land consolidation in a sustainable manner.

2.2 Production challenges

At the production stage, there are several challenges that affect banana production. A combination of abiotic (land, soil nutrient deficiencies and drought stress) and biotic (black leaf streak, BXW, BBTD, Foc, weevils, nematodes) stresses and lack of information, a low genetic base, social constraints and climatic change varying in relative importance in space and time, constrain production in most banana-producing regions of the world (Fermont et al., 2009; Wairegi et al., 2010). Potential yield gains from managing some of the principal biotic and abiotic constraints are summarized in Table 3.

2.2.1 Availability of labour for plantation management

Family members constitute the highest proportion of total labour involved in banana plantation management in most African farming systems (Anselm et al., 2005). In addition to planting, other labour-intensive activities include corm removal, making soil bands and trenches, de-suckering, and manure preparation and application (Table 4). As labour shortage is usually experienced during the time of land preparation and planting, households that can afford it supplement their family labour with hired labour to meet labour requirements.

2.2.2 Access to fertilisers and pesticides

Growth in agricultural productivity, for example, in ECA, has been dependent on own farm inputs (mainly manure and crop residues), a method that only helps recycle nutrients within the farming system, and does not add to the stock of nutrients in the system. Improving banana and plantain yields thus requires external chemical inputs such as mulch, fertilizers, pesticides and herbicides (Fig. 2). As most banana-growing countries in Africa are net importers of these goods, production costs are high. However, though

Table 3 Estimated yield gain due to management of priority banana production constraints in ECA

Constraint	Yield gain (%)	Source
Black leaf streak (Black Sigatoka)	30–50	Tushemereirwe et al. (2004a)
Xanthomonas wilt of banana (BXW)	80–100	Kubiriba et al. (2014)
Banana weevil	60	Okech et al. (2004)
Nutrient deficiency (K and N)	28–68	Nyombi et al. (2010)
Drought stress	20–65	Van Asten et al. (2011)

Table 4 Ranking of labour intensity of various banana management practices

Banana management activity	Labour intensity	Rank of all activities: 9 = least labour-intensive; 1 = most labour-intensive
Planting	5	1
Corm removal and splitting stems	5	2
Making soil bands and making trenches	5	3
De-suckering	5	4
Making compost manure	3	5
Weeding	2	6
De-sheathing	2	7
Applying manure/fertilizer	2	8
Pests and disease control	1	9

the use of pesticides and other chemical inputs can increase yields more effectively than non-chemical alternatives, they can also cause negative environmental and health effects (Viljoen, 2010). Given the high cost, negative effects and unavailability of fertilizers/pesticides, manure and mulch use remain the only viable alternatives and hence agro-ecological strategies for ensuring their continuous supply to farmers need to be explored/fine-tuned, promoted and integrated into the extension packages.

There are studies suggesting that the use of inputs like pesticides, fertilizers, manure and mulch is substantially low especially in African banana cropping systems (Nyombi, 2013). The input use must be emphasized to improve banana productivity and realize more profits from banana production. Future research should promote investment in developing agribusinesses aimed at establishing links with banks and microfinance institutions to provide credit and loans to farmers to purchase high-quality seeds, fertilizer and other necessary inputs. Access to credit can also help farmers establish (low-cost) community-based mother gardens and micropropagation chambers to increase access to clean and relatively cheap planting materials.

2.2.3 Soil fertility and water/moisture stress

Soil nutrient deficiencies, poor/inefficient nutrient cycling and retention of adequate soil moisture have been identified as major obstacles to enhancing banana production (Van Asten et al., 2011). Sustainable increase in banana productivity should go hand in hand with meeting the crop nutrient requirements. However, the profitability of mineral fertilizer application reduces with increasing distance from the main markets (Wairegi et al., 2010). This calls for combined use of mineral and organic nutrient sources and soil water conservation measures for improving nutrient use efficiency since most banana production zones are distant from the markets. The interactions suggest that there are trade-offs to be optimized and synergies to be harnessed for management of these constraints. We postulate that poorly managed plant stands may affect the ability of the cropping systems to sustain their nutrient stocks. Banana is a shallow-rooted crop with high nitrogen and potassium requirements. Potassium and nitrogen are highly leachable nutrients and,

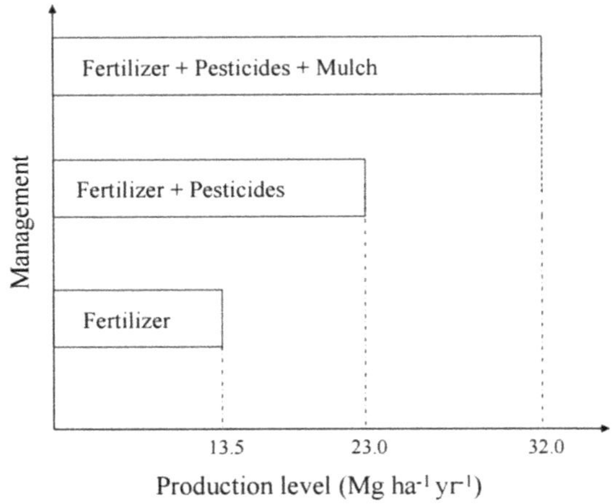

Figure 2 Response of East African Highland bananas to management options in Masaka, central Uganda. Source: Rubaihayo et al. (1994).

among cations, potassium is the most dependent on standing biomass to maintain its levels in the cropping system. Poor biomass can lower potassium stocks, which in turn exacerbates nutrient leaching to depths inaccessible to the banana plant/mat. Similarly, banana thrives under a thick organic mat/mulch layer.

Though the evidence is less clear, it is believed that high banana weevil and nematode pest status reflects poor management, declining soil fertility and the presence of already stressed plants. Poorly nourished banana plants may be less capable of defending themselves against (i.e. more susceptible to) weevil and nematode attack. Stressed plants are also subject to the high mat phenomenon, by which the corm comes out of the ground, increasing the plant's vulnerability to weevil attack (Gold et al., 2001).

2.2.4 Crop management practices

Small-scale banana producers often adhere to traditional practices of producing banana in their backyard or where it incidentally grows, generally without access to improved cultivars. Most of the recommended crop management practices are knowledge-intensive, labour-intensive and costly to apply and often poorly adopted by small-scale farmers (Ochola et al., 2013). Knowledge extension to farmers coupled with adaptation of technologies to local conditions is crucial for banana cultivation. This has, however, been hampered by poor and, in some instances, non-existent extension structures and/or services.

2.2.5 Pests (weevils and nematodes)

The banana weevil *Cosmopolites sordidus* (Germar) (Coleoptera: Curculionidae) and a complex of root parasitic nematodes (*Radopholus similis*, *Pratylenchus goodeyi* and *Helicotylenchus multicinctus*) are the major pests of bananas. They destroy roots and parts of the rhizome, thus reducing anchorage and interfering with water and nutrient uptake. These pests are most damaging at altitudes of <1500 m, can cause yield losses >50% and can reduce the effect of investment in soil fertility and water management (Gold et al., 2001). The weevil is a hydrophobic, soil-borne pest and therefore soil management (e.g. factors affecting soil moisture) may influence the pest's population densities, behaviour and damage levels. Both the weevil and nematodes are disseminated through the movement of infested planting material.

Research findings suggest that no single control strategy will provide complete control of banana pests. Therefore, a well-defined integrated pest management (IPM) approach might offer the best chance for successful pest control. For small-scale producers, the components of such a programme include cultural (e.g. residue management), botanical (use of neem and concoctions/plant extracts), conventional and non-conventional genetic improvement (host plant resistance) and (in some cases) chemical control. Several of these IPM components are being used by farmers in Uganda although lack of labour remains the main constraint on the use of these practices (Bagamba et al., 2007).

2.2.6 Bacterial, fungal and viral diseases

Both smallholder and large commercial banana production systems are limited by a range of diseases caused by fungi, viruses and bacteria. The fungal diseases include black leaf streak (commonly known as black Sigatoka, is a leaf spot disease caused by the fungus *Mycosphaerella fijiensis*), Fusarium wilt caused by *Fusarium oxysporum cubense* (Foc), BBTD caused by the banana bunchy top virus (BBTV) and banana streak caused by the *banana streak virus*, while the bacterial diseases include BXW caused by *Xanthomonas campestris* pv. *musacearum* and Moko/Bugtok and blood disease caused by *Ralstonia solanacearum* and the blood disease bacterium (*Ralstonia syzygii* subsp. *celebesensis*), respectively (Blomme et al., 2017).

The disease challenges are to a large extent caused by the fact that bananas are vegetatively propagated so that pathogens can move around in infected planting material. In addition, few acceptable resistant cultivars have been developed through conventional breeding due to the near sterility of commonly grown/accepted cultivars. Research centres are very active in developing methods for the control of banana diseases. Some of these efforts include the development of resistance towards Fusarium wilt, BXW and BBTD through genetic modification, development of diagnostics to support the dissemination of disease-free banana planting material and cultural control options.

Fusarium wilt

Fusarium wilt is a destructive disease and a big threat to the global banana production. It is a soil-borne fungus caused by *Fusarium oxysporum* f. sp *cubense* (Foc). It infects bananas through their roots and moves relatively slowly in the soil, though can be moved over long distances through infected planting material, irrigation and runoff water and in soil attached to machinery or footwear (Stover, 1962). Importantly, it can remain in soil (as ascospores) for decades even in the absence of bananas (Ploetz, 2015). Foc has several races, with each race infecting different cultivar groups. The most important races are Race 1 and (Tropical) Race 4 (TR4). Race 1 is ubiquitous in banana-growing regions and has historically had a major impact on commercial banana production. This fungus was the reason why the major multinational banana-producing companies changed from susceptible 'Gros Michel' to resistant 'Cavendish' cultivars. Both Cavendish and East African highland bananas are resistant to Foc Race 1 but 'Lady finger' in Australia and 'Sukali Ndizi' in Uganda are susceptible as are many other cultivars (Tushemereirwe et al., 2004a). Foc TR4 is very different. Previously limited to Darwin in Australia and Southeast Asia (Ploetz, 2006), it has recently been reported to be also present in Oman, Pakistan, Lebanon, Jordan and Mozambique (Ordonez et al., 2015). Compared with Foc TR1, Foc TR4 is a much more serious threat as it infects a wider range of cultivars including Cavendish. There are currently no validated methods available for effective control of Foc TR4 other than taking advantage of cultivar resistance and using genetic modification (including somaclonal selection) as a promising way for generating resistant bananas (Krishna et al., 2016).

Black leaf streak disease (commonly known as black Sigatoka)

Black leaf streak disease (commonly known as black Sigatoka) is a leaf spot disease of banana and plantain caused by the ascomycete fungus *Mycosphaerella fijiensis* (Morelet). The disease causes necrotic lesions on the leaves of many Musa cultivars, reducing their photosynthetic ability, prematurely ripening the fruit and leading to yield reductions of 30–50%, in some cases as high as 80% (Mobambo et al., 1993).

Control of black Sigatoka in commercial systems is mostly reliant on frequent (weekly) fungicide application (de Bellaire et al., 2010). Such frequent application of fungicides is costly and out of reach of small-scale farmers. Frequent fungicide application also has socio-economic implications that include human health and environmental hazards (Ploetz, 2001). Small-scale farmers, for example, in ECA prune old leaves to minimize infections in the younger leaves. Pruning also allows in solar radiation that reduces the survival and development of the fungal spores (Amponsah et al., 2010).

Banana bunch top disease

BBTD is regarded as the most serious viral disease of banana, causing significant yield losses. The disease is present in most of the banana-growing areas of Asia with notable exceptions of Thailand and Malaysia; in Central Africa (Burundi and DRC) and Zambia, Angola and Malawi in Southern Africa and Hawaii in the Americas (Kumar et al., 2011). The disease is caused by the BBTV, a circular ssDNA virus belonging to the genus *Babuvirus* in the family Nanoviridae. The virus is transmitted from plant to plant by the aphid *Pentalonia nigronervosa* that are widespread in Africa (Adegbolo et al., 2013) and through infected planting materials. Plants infected at early growth stages are severely dwarfed and do not bear fruit. Currently, management of BBTD is reliant on cultural practices such as timely rogueing of infected plants, killing *in situ* with herbicides the infected plants and use of clean planting materials. Methods for diagnosis are still poor in some of the affected regions, especially in sub-Saharan Africa. There are no known resistant cultivars. As such, breeding for transgenic virus resistance is now considered the most appropriate strategy for long-term BBTD control.

Banana Xanthomonas wilt (BXW) and Moko/Bugtok disease

Bacterial wilts that have become a major threat to banana production include banana Xanthomonas wilt, Moko/Bugtok and blood disease. The Xanthomonas wilt disease was first observed on enset in Ethiopia in the 1930s (Castellani, 1939) and was identified as Xanthomonas wilt in 1968 on enset (Yirgou and Bradbury, 1968). Its infection causes losses in banana production through early fruit ripening and rotting of fruits, and through leaf yellowing and wilting, and eventual death of plants. The disease was reported to be found in both Uganda and east DR Congo in 2001 and has now spread to all countries in ECA (Smith et al., 2008; Tushemereirwe et al., 2003, 2004b; Blomme et al., 2014). To date, although all types of bananas in the ECA have been found susceptible, the ABB beer banana cultivars that are susceptible to insect-mediated infections probably play a critical role in facilitating the spread of the disease (Tripathi and Tripathi, 2009).

The Moko disease is found in Latin America, the Caribbean and Southeast Asia (Buddenhagen, 1994). The wilt caused by the banana blood disease pathogen is affecting dessert bananas and local cooking banana cultivars in Indonesia and New Guinea (Eden-Green, 1994; Davis et al., 2001). All the bacterial wilts lead to plant leaf yellowing, dry rot of the flower, browning of the fruit pulp and appearance of bacterial ooze in cut plant tissues (mainly pseudostem and leaves) (Blomme et al., 2017), resulting in production collapse.

Transmission of these bacterial diseases occurs mainly through insect vectors (e.g. stingless bees) and the use of contaminated farm tools and infected seedlings. Spread of the *Ralstonia* spp. also occurs through soil, irrigation and runoff water. Management practices employed for the control of bacterial wilts include the elimination of infected plants, use of disease-free planting materials, disinfection of cutting instruments/tools with 5% chlorine or 90% ethanol, early removal of male flower buds and strict quarantine (Blomme et al., 2014).

2.2.7 Climate variability and change

Climate variability and change is another factor affecting banana production worldwide. In addition to the typical variation of temperature and precipitation regimes, there is an increasing need for farmers to address the effects of climate change. According to climate models, by 2070 the land area suitable for bananas will increase globally by 50% due to more favourable temperature and rainfall patterns in the subtropics and tropical highlands (German Calberto et al., 2015). At the same time, higher temperatures will raise water demand, which is projected to increase by 12–15% (Washington and Pearce, 2012). Higher temperatures also favour the spread and impact of pests and diseases. Colder dry seasons, on the other hand, will increase the time for ripening.

The impact of climate change on temperature and rainfall in any specific location is reasonably uncertain; thus policies will need to be flexible and adaptable, so as not to overly rely on a given strategy. By examining detailed maps of productivity trends in the light of climate change, it is possible to identify climate risk areas (areas with projected losses) and climate opportunity areas (those with future gains). Researchers will need to work together with farmers to develop new crop varieties along with supporting agronomic and husbandry methods, extension services and mechanisms for scaling up and out, for successful agricultural adaptation to climate change. Moreover, the more than 500 banana/plantain varieties around the world (German Calberto et al., 2015) present a key resource for climate change adaptation that needs to be further explored and used to improve yield. Adjusting agricultural practices for varieties grown in given areas can help maintain or increase production levels. Investment in breeding for climate-resilient banana cultivars should be considered a priority in research efforts. There is a need to initiate work with local farmers to promote education on pest and disease control as well as best practices for water-saving irrigation techniques.

2.3 Processing challenges

2.3.1 Post-harvest losses

Post-harvest losses are a major challenge in the value chains of cooking bananas and plantains. Generally, the primary factors causing post-harvest losses in fresh produce are of mechanical, physiological, pathological or environmental nature (Kader and Rolle, 2004).

Losses are incurred by secondary factors resulting mainly from inadequate technology applications and quality control. In Africa, high post-harvest losses caused by inadequate post-harvest handling are one of the major factors limiting the expansion of banana production (Olorunda, 2000). Similarly, lack of adequate infrastructure for pre-packaging, cold storage, packaging and distribution, post-harvest treatments and washing, along with production constraints, are reported to cause low productivity and post-harvest losses in most banana-producing countries in ECA (Gabre-Mariam, 1999). Considerable volumes of cooking bananas and plantains are damaged, or lost altogether, before reaching the target markets (Gabre-Mariam, 1999).

As a perishable crop, banana requires proper mechanisms for storage. A key limitation is the lack of infrastructure and technologies to store bananas, so as to use them during times of scarcity. In Uganda, for example, bananas are in high supply during the peak season in June and July, increasing the risk of waste and losses. In addition, traders offer low prices in times of oversupply, while farmers have few alternatives to immediate selling. In contrast, prices soar during periods of low production, particularly in the months of November and December, when only few farmers have bananas to sell. Recent studies in Uganda indicate that during low production (scarcity), on-farm post-harvest losses affect about 3.3% of the produced volume (physical losses) and 5.4% of the generated value (economic losses), increasing during the peak period to 9.6% and 8.1%, respectively (Kikulwe et al., 2017). Overall, post-harvest losses in the cooking banana value chain amount to 12%. Most losses are reported along the downstream segments of the value chain (Hailu et al., 2013). In general, economic losses often outweigh physical losses in banana value chains.

In the upstream segments of the value chain, particularly at farm level, post-harvest losses are often associated with theft (physical losses) and selling of immature bananas (economic losses). At the retail level, physical losses are also due to theft, while economic losses are due to bruising, finger-plucking and ripening. In addition, release of ethylene during bulk storage makes the fruits ripen faster and increases the risk of rotting before reaching their destination. Poor handling during ripening and transport can significantly reduce green life and shelf life which is 5–10 days after harvest. Strategies such as appropriate transportation and proper storage for limiting post-harvest losses are not readily available and, where they are, access might not be equitable in terms of gender and other factors of social differentiation (Hailu et al., 2013). Largely lacking are gender- and, overall, client-differentiated solutions for improved storage, transport and handling of bananas from the farm to the market.

2.3.2 Value adding

As a multi-use crop, banana holds the potential of significant value adding through product differentiation and processing into diverse food and other products. This potential is often only partially realized due to poor technology, low investments in agro-processing, lack of business and technical skills and weak orientation towards the needs of diverse consumer groups. Thus, bananas are often sold as fresh produce, in local markets, and with little differentiation. In addition, there is a massive underutilization of banana residuals which could serve as a raw material for value-added products, increasing agricultural productivity and reducing waste (Mohammadi, 2006). Opportunities exist for industrial use of diverse banana by-products, such as peels, leaves, pseudostem, stalk and inflorescence. This can lead to various food and non-food applications, including use as thickening agent, colouring and flavour agent; and as alternative sources of macro- and micronutrients, nutraceuticals, livestock feed, natural fibres and sources of natural bioactive compounds and bio-fertilizers (Padam et al., 2014). In addition, local processing into vinegar, juice, beer and other beverages is a form of value addition particularly suitable to women, either as individual entrepreneurs or organized into collective enterprises.

Further options include processing of bananas into various products like biscuits, sweets, crisps, cakes, jam, sauce and yoghurt that are more palatable (Adeniji et al., 2010; Akinyemi et al., 2010), and blending banana fruits and foliage into livestock feeds, especially in Africa and Central America where other resources are limited. However, some challenges need to be addressed to realize their potential for value addition, particularly as regards improvements of local infrastructure, transport, retail outlets, processing facilities and associated investments. This requires joint efforts by public and private sectors, including co-investments, credit lines and tax breaks. An enhanced enabling environment would also encompass public purchasing policies (e.g. banana products as part of school meals) and other policies that promote the production, processing and consumption of different banana products.

2.4 Marketing challenges

With a global value of international and national trade exceeding US$25 billion a year, banana markets constitute one of the principal commodity markets and provide employment and income to more than 50 million people worldwide. While export markets are largely controlled by a few multinational companies and major retailers in the importing countries, domestic markets involve a broad array of stakeholders including smallholders, transport agents, processors, petty traders and other retailers. Domestic markets for cooking bananas and plantains, in turn, are found across a broad spectrum of small- and medium-size towns in rural and peri-urban areas as well as larger urban agglomerations and cities. While these markets increase opportunities for farmers and other value chain actors to generate income, they also face several challenges, such as poor transportation which increases transport costs, limited infrastructure for storage and marketing, lack of incentives for quality banana and product differentiation and scant access to credit (Mulualem et al., 2015).

One of the smallholders' major impediments to commercial banana farming is limited access to reliable transport networks and viable markets (Zenebe et al., 2015). In addition to poor handling causing bruising and delays in transport due to poor roads and deficient transportation means, quality and price are compromised due to inadequate infrastructure for storing and marketing bananas in rural, peri-urban and urban market places. Limited flow of information from the markets to the farms and vice versa increases the likelihood of unnecessary shortages or peaks of supply, with price fluctuation affecting both farmers and consumers. During peak periods of production, supply may outweigh demand, leaving little incentive for farmers to market their produce in view of low prices. Fluctuation of prices is often aggravated by limited availability of technologies for processing and other forms for increasing shelf life. Income is further compromised by lack of product differentiation in terms of processing and quality grading. Smallholder farmers and processors lack guidance on quality standards and incentives for meeting them. Bananas are usually trucked to markets without specific protection, resulting in unnecessarily high losses through physical damage and quality deterioration. Many smallholder farmers are unorganized and rely on middlemen who pick up their produce at the farm gate.

Transactions are often based on asymmetric market information and rarely involve formal contractual commitments. Opportunities for processing banana, for example into juice, are hampered by limited information on markets for substitutes (juice from other fruits), and associated requirements in terms of quality and hygiene. Organizing into collective enterprises, such as cooperatives and farmers' associations, with a clear business model would allow smallholders to bulk production, increase their bargaining power, engage in processing and, overall, realize economies of scale. Still, close collaboration with value chain actors near the final markets is required to tap into their knowledge on consumer segmentation, capacity and willingness to pay for differentiated banana products, and price differentials across products and markets.

Cost competitiveness for marketing cooking bananas and plantains in international markets is severely limited in view of high domestic transportation costs, taxes for transport agents and shipping rates for sea freight, along with the need for a cold chain (cooled containers for fresh products) and relatively limited demand in importing countries as compared with that for dessert bananas. In domestic markets, the marketing of ripening varieties is more challenging than that of cooking varieties as post-harvest losses tend to increase during ripening. Such challenges compromise value chain efficiency and, hence, incomes of both traders and farmers. Post-harvest deficiencies compromising marketability are linked to bananas developing unacceptable qualities in terms of colour and texture due to overripening, inappropriate post-harvest handling and limitations in market infrastructure and organization. There is also a lack of grading and standard weight measurement, resulting in erratic price setting, affecting both farmers and consumers.

Addressing these challenges requires improving the information flows among value chain actors, also to reduce the likelihood of intermediaries taking advantage of poor rural producers who often lack market information (Ariho et al., 2015). Asymmetric market information is a general problem in most tropical countries where formal market information services like radio broadcasts are absent or lack local specificity. This is a clear intervention point for national and local governments. NGOs specialized in value chain development can also contribute to promote forward and backward linkages in the value chain that enable value chain actors to embrace market dynamism and improve their understanding of consumer preferences. In addition, stronger linkages with external service providers, such as input firms and business advisory services, can improve the efficiency of value chains, as do modern information and communication technologies. In Uganda, for example, banana traders use mobile phones to carry out transactions with producers in rural areas and buyers in urban markets (Kikulwe et al., 2014). Similarly, the use of mobile money systems allows banana-growing households to purchase significantly higher amounts of inputs – fertilizers, pesticides and hired labour – and to sell a larger proportion of their harvest in the market (ibid.). Such technologies help overcome cash limitations and enable transactions with buyers from outside their local communities (Sekabira and Qaim, 2017).

Looking forward, R&D efforts need to identify effective and efficient ways for improving information flows in the value chains, extending the shelf life of bananas, differentiating products in line with established, location-specific demand and broadening opportunities for processing and marketing of by-products. On the demand side, growing urban populations and middle-class, along with expanding supermarkets and other retail outlets, will provide quality and price signals to the farmers. The latter, in turn, could become organized into cooperatives and other forms of collective enterprises to strengthen their position in the value chain, enhance their advocacy vis-à-vis local and national governments and improve their access to credit. For example, farmer organizations such as High-ridge Banana Growers and Marketing Association in Kenya and Nakasongola District Farmers Association in Uganda provide various benefits to their members, including access to value-adding technologies, technical assistance and training, collective marketing, access to farming inputs and credit and women empowerment (Nyang et al., 2010). Complementary training may be provided by government agencies and NGOs, particularly as regards organizational and management training, timely production to meet market demand/periods of peak prices, maintaining product quality throughout the value chain and obtaining up-to-date market information, as well as basic business practices like recordkeeping and effective marketing. Finally, improvements in the marketing and general infrastructure requires state-borne investments in the road network and other basic infrastructure, while the private sector can be expected to provide co-funding for infrastructural improvements that directly enhance value chain efficiency (e.g. cold chain investments).

3 Enabling an environment for addressing the challenges

Enabling an environment for banana production, processing and marketing is invariably linked to a wide range of public policies and institutions, but also relates to private institutional arrangements, such as voluntary sustainability standards and the provision of technical, business and financial services by NGOs and private firms. In this sense, the enabling environment for value chains of cooking bananas and plantains also includes access to input markets, market information and crop production technologies to ensure productive, environmentally sound production oriented towards the demand of diverse consumers and, overall, positive implications for agricultural development and food security.

In a more classical sense, access to critical production factors such as land, labour and capital is a critical element of the enabling environment (Ariho et al, 2016). Improving access of smallholders to land, particularly women, is featuring prominently on the agendas of NGOs and advocacy organizations and, increasingly, also on those of political decision-makers. In Uganda, for example, the National Land Policy has recently been adjusted such that historical injustices are redressed to protect or improve the land rights of communities and women and, thus, their participation in banana production. Such legal adaptations are critical as lacking access to land is a major constraint on women's stronger participation in banana production as in the absence of land titles their access to credit is thwarted. In Uganda, women barely own 20% of land and yet they contribute over 65% of labour to the production, processing and marketing of bananas (Republic of Uganda, 2013). In addition to agricultural production, women play a key role in household reproductive activities. Efforts to improve productivity in banana and other crops therefore need to address gendered division of labour in smallholder households and the differentiated access of women and men to productive assets including land, machinery and equipment and other farming inputs.

Access to quality planting material is also a critical aspect of production. However, more than 90% of the planting material originates from informal exchange among neighbours, purchases in the local market and use of own materials (Otieno et al., 2016). This increases the vulnerability of banana plantations to diseases and pests. Provision of quality materials is dependent upon investments in local tissue culture laboratories and decentralization of governmental pest and disease control agencies. Such seed inspection units need to be equipped with more seed inspectors to reach out to broad numbers of farmers and sensitize them on the use of clean planting materials.

Addressing post-harvest constraints related to processing and marketing requires stronger engagement by both value chain actors and public sector entities for joint investments and action. As banana is a perishable crop, its processing and marketing are dependent on adequate infrastructure to facilitate transportation, storage and marketing. Improvements in the road network and warehouses are critical for reducing post-harvest losses, lowering transaction costs and overall marketing of bananas (Ariho et al., 2016). Private sector engagement in agro-processing and co-funding of infrastructure can be fostered through tax breaks, low-cost credit and other institutional incentives.

Linkages between the providers of technical, business and financial services can be strengthened by bringing them together with producers, processors and traders in business roundtables and encouraging complementary service offer. This can be paired with promoting cooperatives and other forms of collective enterprises as vehicles for farmers to access inputs and services and, overall, strengthening their role in the value chain. In Uganda, for example, the National Agricultural Advisory Services (NAADS) Act of 2001 provided the government extension institution with the mandate to support farmers by establishing functional organizations (structured and operational institutions) to ensure collective procurement and marketing, and collective access to financial services. The NAADS act was further mandated to aid farmers in creating options for financing and facilitating the participation of the private sector in funding agriculture and engaging in agro-processing (Republic of Uganda, 2001). However, the act was trimmed down in 2015 and has been replaced by a wealth creation programme led by the military who are unprepared to provide agricultural extension services, leading to a gap in extension and much needed farmer support.

Successful value addition and agro-processing not only require injection of financial capital, but also investment in human capital to develop productive and commercial means and skills of value chain actors including smallholders (Ariho et al., 2016; Spilsbury et al., 2004). Where extension services are absent or not up to their task, NGOs and private sectors need to provide the required services. These services need to be informed by solution-oriented research addressing the wide range of production, processing and marketing challenges indicated above. This, in turn, requires collaboration between national research institutions and universities and international research organizations.

Finally, strengthening the enabling environment for cross-border trade of cooking bananas and plantains in regional markets is a way forward. For example, there is great potential for fresh or value-added banana products in the East African Community (EAC), specifically in Kenya and Sudan (EAC, 2014). Recent advances in EAC integration, including the development of a customs union, have eliminated tariffs and reduced transaction costs related to customs, particularly clearing of goods and other administrative services. The EAC has also been working towards developing the East African industrialization policy and related strategies to create an integrated framework for industrialization through agro-processing and increased trade in value-added products among the EAC countries (EAC, 2014). It is expected that this will lead to growth of cottage industries and stimulate agro-processing and value addition for major crops in East Africa.

4 Future trends

With yields varying from as low as 5 t/ha on farmer's fields to as high as 70 t/ha on research stations (Fermont et al., 2009; Van Asten et al., 2005), yield gaps facing smallholder farmers are excessively high. As a result, research and development efforts focus on effective and efficient ways for increasing banana production and resulting food security and income through improvements at production, processing and marketing, levels for the benefit of smallholders and other resource-poor value chain actors. Table 5 provides an overview of key research areas, associated opportunities for increased production, processing and marketing, and suggestions for complementing current endeavours through additional efforts in the future.

Challenges associated with banana cultivation may primarily be solved through sustainable intensification of production, calling for urgent development and deployment of new validated technologies (Bagamba et al., 2007; Edmeades et al., 2007). Research with a focus on key biotic and abiotic constraints has well advanced as regards EAHB cropping systems. For example, norms for assessing nutrient imbalances applicable to EAHB using compositional nutrient deficiency diagnosis have been developed, as part of integrated systems and associated recommendations (Wairegi et al., 2010). A model for quantitative evaluation of fertility in tropical soils has been calibrated for EAHB (Nyombi et al., 2010) to quantify nutrient requirements for defined yield targets. A dynamic crop growth model – LINTUL-BANANA2 for water-constrained production – has been developed to quantify water requirements (Taulya, 2015). A range of pest management technologies already exists (see Masanza, 2003; Tinzaara et al., 2007). Significant progress has also been made for managing the expanding threat of banana diseases, such as BXW (Blomme et al., 2014; Smith et al., 2008; Tinzaara et al., 2016; Tripathi and Tripathi, 2009) Black Sigatoka (Tushemereirwe et al., 2004a), Foc TR1 and TR4 (Dita et al., 2013) and BBTD (Kumar et al., 2011, 2015).

Future research efforts should seek to unravel the interactions between abiotic and biotic constraints and to evaluate trade-offs and synergies between management options, with the aim of designing client-specific control packages for different types of smallholder farmers. The knowledge generated on biotic and abiotic interactions will be incorporated into a growth model for simulating EAHB production. This model, in turn, will inform the development of an aggregate bio-economic model for *ex ante* analysis of integrated soil fertility and pest management and good agricultural practices. Findings need to be translated into gender-differentiated capacity development initiatives to ensure adequate fit and utilization of the new technologies. Outreach to smallholders will be facilitated through the promotion of farmer cooperatives and associations as a vehicle for integrated approaches with a multiplier effect that focus on cultural practices, post-harvest losses, value adding, market orientation, leadership skills of women and savings and credit facilities (Bihunirwa et al., 2012).

Socially and environmentally sustainable intensification of banana production is also urgently needed to boost productivity for supplying growing populations. Recent research findings on key biotic and abiotic constraints in East African highland bananas (Nyombi et al., 2013) could be applied to sensitize smallholders as regards banana seed systems and appropriate technologies for managing soil, water and pests and diseases. There is also a need for appropriate gender-responsive decision support tools that can help promote intensification of cropping systems in banana-growing areas in line with the diverse realties and needs of male- and female-headed and mixed households. Strategies need to be developed that promote: i) access to clean planting material and ii) location- and

Table 5 Key research areas and opportunities for increasing banana production along the value chain

Key areas of research	Opportunities for increased production, processing and marketing	Current R&D efforts	Future R&D efforts
Pre-production • Labour • Land • Socio-institutional constraints • Planting material	Inputs for enhancing production are available in the markets (e.g. various highly productive genotypes/cultivars, organic and inorganic fertilizer formulations suitable for banana)	Understanding of banana seed systems and promotion of a smallholder-friendly seed system Increase knowledge on banana cultivar performance according to various agro-ecological zones	Explore strategies for land consolidation and mechanisation Explore strategies for improving farmers' access to inputs
Production • Labour • Soil and water • Pests and diseases • Inputs • Drought risks	Management technologies/ IPM package(s) available for key pests and most diseases Set of agronomic practices to boost production available	A model for quantifying nutrient requirements for defined yield targets Sustainable management of Xanthomonas wilt, *Fusarium* spp. and banana bunchy top disease that are key constraints to banana production	Development of gender-differentiated strategies for improved production, processing and marketing of banana Need to understand bottlenecks to adoption and rightful application of technologies Promote education on best practices for water-saving irrigation techniques Exploration of appropriate strategies for bridging the gap between research and extension
	Extensive studies on Musa germplasm diversity conducted, germplasm conserved *in situ* and *ex situ* in different collections (national and international) These materials also support breeding efforts Conventionally bred black Sigatoka-resistant plants. BXW-resistant genetically modified banana	Modelling BXW epidemiology Developing climate-smart, pest and disease-resistant banana germplasm by conventional breeding Mainstreaming of gender to boast production, incomes and nutrition in banana-based systems Developing and tailoring of bio-economic models for banana-based households/farming systems	Development of a banana growth model increase farmer awareness and utilization of appropriate technologies Understanding of weevil and nematode resistance mechanisms and genes that control these mechanisms Promoting GM bananas (commercializing and consumer acceptance)

Value adding • Shelf life • Handling • Storage • Transportation • Value addition • Product quality	Development of diversified banana products and by-products Different banana products and by-products Novel transport, storage and processing techniques	Understanding the causes of post-harvest losses and ways for reducing them Optimization of the banana handling and ripening process	Developing new differentiated products based on different banana varieties and banana by-products Capacity development for farmers to form, cooperatives and other types of enterprises
Marketing • Price levels and fluctuation • Tax regimes • Marketing infrastructure	Stimulation of demand for diversified banana products and by-products Improvements in tax regimes and facilities for transportation, storage and marketing	Banana value chain analysis	Market research to guide investment decisions and marketing strategies Policy research focusing on the enabling environment for production, processing and marketing of banana

client-specific agro-ecological intensification technologies, including options for locally adapted pest and disease control.

The development of resistant plants is a long-term strategy for pest and disease management, especially on small-scale farms, as the inclusion of such plants can be part of an integrated pest and disease management strategy. Genetic engineering is considered as one of the eco-friendly and safer methods to control banana pests/diseases and will continue to play a key role in the sustainable production of cooking bananas and plantains in Africa. This has recently been recognized by several African governments and international donors, reflected in higher financial support for research into banana and plantain biotechnology. The current efforts of national and international research centres to develop pest and disease-resistant varieties through conventional breeding based on careful selection of parents – particularly improved diploids with resistance – should continue to complement the biotechnology efforts. Such a strategy will increase the chances of developing, for example, resistance against the banana weevil. The development of rapid screening methods for testing weevil resistance in the field and laboratory is an essential step in developing resistant hybrids. A better understanding of resistance mechanisms and genes that control these mechanisms should form the focus of research priorities in developing successful resistance for the banana weevil and nematodes. Future research, therefore, should focus on long-term performance of genetically modified (GM) bananas under African field conditions, including the long-term response of weevils and effects to their natural predators. The major issues concerning GM banana plants that will need to be addressed before engaging on a broader scale in such an endeavour are their commercialization, affordability and consumer acceptance .

Limited findings are available on causes and control of post-harvest losses of ripening varieties, especially those due to diseases; ethylene physiology and ripe-fruit quality temperature management during handling; and modified-atmosphere storage and physiology. Research could help to optimize the ripening process and to reduce post-harvest losses. Further efforts are also needed to develop new products based on different banana varieties and banana by-products. Organoleptic properties, nutrient content and texture vary widely across varieties – an often-underexplored potential for new products. Similarly, peels, leaves, pseudostems, stalks and inflorescences are valuable raw materials that can be processed into diverse food, feed and other applications. Findings from such research need to be built into capacity development programmes for farmers, cooperatives and other types of enterprises which, in turn, requires the upgrading of the capacities of technical service providers who are usually specialized in production-focused topics. In addition, market research can guide investment decisions and inform the development of marketing strategies, based on consumer segmentation, analysis of capacity/willingness to pay and other critical success factors for linking supply and demand. Research findings can inform the programmes of business service providers who, in close collaboration with technical service providers, seek to strengthen the capacities of smallholder farmers, cooperatives and other forms of enterprises.

5 Conclusion

Production and trade in bananas, comprising dessert and cooking bananas and plantains, are pillars of global food security, providing about 400 million people worldwide with access to a staple food and more than 50 million people with a critical source of cash income. However, stages of pre-production, production, post-harvest management and marketing of banana face several challenges. At the pre-production stage, principal challenges relate to land, inputs and planting materials. At the production stage, appropriate technologies are needed for improving soil fertility, managing pests and diseases and optimizing water use. Post-harvest challenges are associated with transportation, storage and value addition through the development of products and by-products that capitalize on the genetic diversity of bananas. Finally, marketing challenges are linked to asymmetric access to market information, limited understanding of consumer preferences and related trends and deficiencies in the marketing infrastructure.

Arguably, banana production constitutes the largest set of challenges, particularly from the perspective of smallholders who are constrained in terms of land, labour, capital and other resources. These challenges include a complex of biotic (black leaf streak, BXW, BBTD, Foc, weevils, nematodes) and abiotic stresses (nutrient deficiencies, drought stress), low genetic diversity of the planting material used, social constraints and climate variability and change. Most of these challenges, especially the biotic ones, cause yield losses of 50% or more, part of which could be addressed through investments in soil fertility and water management. There have been significant efforts to identify viable solutions at the global level. Research and development efforts have led to improved pest and disease management technologies, soil and water management techniques and post-harvest handling. However, persistent yield gaps can be observed in most smallholder banana systems. In fact, most of the recommended crop management practices are knowledge-laden, labour-intensive and costly to apply and, hence, outside the reach of many smallholders.

In addition, the structure and functioning of domestic banana markets, the design of improved technologies and business support services and policies rarely do justice to the diverse realities and needs of smallholders. Smallholder farmers are typically constrained in terms of access to information on latest technologies and the innovations required to keep pace with the dynamic environment of farming, markets and policies. A significant portion of current research focuses on technologies that are environmentally sustainable, but not necessarily apt for the diverse conditions of resource-poor farmers. For example, results of conventional breeding and genetic engineering may yield pest-resistant and climate-resilient varieties which, if not developed with a clear focus on end user preferences, will not be adopted by farmers at large scale; or, if requiring a combined application of fertilizers and pesticides, will not be accessible to many smallholder farmers.

It is therefore critical that future research and development efforts take the diverse production and livelihood systems of smallholder farmers as a point of departure, accounting for varied needs and aspirations within and across households. Gender- and age-differentiated development of technologies, with a clear view on current and anticipated future market trends, will go a long way to make agricultural technology development relevant to those who are at the beginning of the value chain. Such research efforts will also need to sensitize service providers outside of the value chain, including government agencies, NGOs and private firms, of the diverse realities and technology needs of smallholder farmers. This calls for action research approaches with strong participatory and gender-responsive elements, which can be linked to multi-stakeholder platforms and other mechanisms that bring together partners from public and private sectors and civil society for joint action and learning. Research findings can thus feed into processes of joint reflection, the definition of complementary roles and the design of effective scaling mechanisms for increased livelihood benefits based on the production, processing and marketing of banana. Finally, public and private policies and complementary institutional frameworks can support multi-stakeholder action and learning through innovative institutional arrangements, public-private collaboration and investments and, overall, the adequate provision of technical, business and financial services needed by smallholders and other key stakeholders in banana value chains.

6 Where to look for further information

Bioversity International, a member of the CGIAR, headquartered in Rome, Italy, will provide assistance in case of further research about the information discussed in this chapter. Bioversity International works with partners around the world to identify and deliver innovative solutions to ensure biodiversity nourishes people and sustains the planet.

7 References

Adegbola, R. O., Ayodeji, O., Awosusi, O. O., Atiri, G. I. and Kumar, L. P. 2013. First report of banana bunchy top virus in banana and plantain (*Musa* spp.) in Nigeria. *Plant Disease*, 97(2): 290. https://doi.org/10.1094/PDIS-08-12-0745-PDN.

Adeniji, T. A., Tenkouano, A., Ezurike, J. N., Ariyo,C. O. and Vroh-Bi, I. 2010. Value-adding Post-harvest processing of cooking bananas (Musa spp. AAB and ABB genome groups). *African Journal of Biotechnology*, 9: 9135–41.

Ajambo, S., Mbabazi, E. G., Kikulwe, E., Nalunga, A. and Rietveld, A. 2017. Gender roles and constraints in the green cooking banana value chain: Evidence from South Western Uganda. *Enterprise Development and Microfinance Journal*, Submitted.

Akankwasa, K., Mugisha, J., Tushemereirwe, W. and Abele, S. 2008. Consumer willingness to pay for introduced dessert bananas in Uganda. *Paper Presented at the Conference 'Banana and Plantain in Africa: Harnessing International Partnerships to Increase Research Impact'*, 5–9 October, Mombasa, Kenya.

Akinyemi, S. O. S., Aiyelaagbe, I. O. O. and Akyeampong, E. 2010. Plantain (Musa spp.) cultivation in Nigeria: A review of its production, marketing and research in the last two decades. *Acta Horticulturae*, 879: 211–18.

Amponsah, N. T., Jones, E. E., Ridgway, H. J. and Jaspers, M. V. 2010. Effects of solar radiation and relative humidity on germination of Botryosphaeriaceae species conidia. *New Zealand Plant Protection*, 63: 28–32.

Anselm, A. E., Nwekeb, F. I., and Tollens, E. 2005. Hired labor use decisions in cassava-producing households of sub-Saharan Africa. *Agricultural Economics*, 33: 269–75.

Ariho, A., Makindara, J., Tumwesigye, G. and Sakira, A. 2015. Assessment of innovative market access options for banana value chain in Uganda. *Journal of Development and Agricultural Economics*, 7: 323–31.

Ariho, A., Makindara, J., Tumwesigye, G. and Sakira, A. 2016. Assessment of existing policy and legal framework for banana value chain development for Uganda. *International Journal of Asian Social Science*, 6: 146–65. Available at; http://www.aessweb.com/pdf-files/ijass-2016-6(2)-146-165.pdf.

Arua, M. and Okorji, O. 1998. *Access to Land and Agriculture in Nigeria*. University Press, Oxford.

Bagamba, F. 2007. *Market Access and Agricultural Production: The Case of Banana Production in Uganda*. PhD thesis submitted to Wageningen University, Netherlands.

Bihunirwa, A., Kinyua, H., Mugoya, M., Mohammed, S. and Rwakakamba, M. 2012. *Innovating to Compete: Smallholder Farmers' Agency and Markets in East Africa*. IIED/HIVOS/Mainumby, London/The Hague/La Paz, UK/The Netherlands/Bolivia.

Blomme, G., Dita, M., Jacobsen, K. S., Pérez Vicente, L., Molina, A., Ocimati, W., Poussier, S. and Prior, P. 2017. Bacterial diseases of bananas and enset: Current state of knowledge and integrated approaches toward sustainable management. *Frontiers in Plant Science*, 8: 1290. doi:10.3389/fpls.2017.01290.

Blomme, G., Jacobsen, K., Ocimati, W., Beed, F., Ntamwira, J., Sivirihauma, C., Ssekiwoko, F., Nakato, V., Kubiriba, J., Tripathi, L., Tinzaara, W., Mbolela, F., Lutete, L. and Karamura, E. 2014. Fine-tuning banana Xanthomonas wilt control options over the past decade in East and Central Africa. *European Journal of Plant Pathology*, 139: 265–81.

Buddenhagen, I. W. 1994. Banana diseases caused by bacteria. In: Ploetz, R. C., Zentmyer, G. A, Nishijima, W. T., Rohrbach, K. G. and Ohr, H. D. (Eds), *Compendium of Tropical Fruit Diseases*. American Phytopathological Society, Saint Paul, USA, pp. 15–17.

Carruthers, R., Krishnamani, R. R. and Murray, S. 2009. *Improving Connectivity: Investing in Transport Infrastructure in Sub-Saharan Africa*. Africa infrastructure country diagnostic Background Paper 7 (Phase II), World Bank, Washington, DC.

Castellani, E. 1939. Su un marciume dell' ensete. *L'Agric. Coloniale Firenze* 33, 297–300.

Davis, R. I., Moore, N. Y. and Fegan, M. 2001. Blood disease and Panama disease: Two newly introduced and grave threats to banana production on the island of New Guinea. In: Bourke, R. M., Allen, M. G. and Salisbury, J. G. (Eds), *Proceedings of the Papua New Guinea Food and Nutrition conference 2000*. ACIAR, Canberra, Australia, pp. 26–30.

de Bellaire, L. D. L., Fouré, E., Abadie, C. and Carlier, J. 2010. Black leaf streak disease is challenging the banana industry. *Fruits*, 65: 327–42.

Dita, M., Waalwijk, C., Buddenhagen, I. W., Souza, M. T. and Kema, G. H. J. 2010. A molecular diagnostic for tropical race 4 of the banana fusarium wilt pathogen. *Plant Pathology*, 59: 348–57.

Dita, M. A., Garming, H., Bergh Van den, I., Staver, C. and Lescot, T. 2013. Banana in Latin America and the Caribbean: Current state, challenges and perspectives. In: Van den Bergh, I., Edson, A. and Vincent, J. (Eds), *Proc. Int. ISHS ProMusa Symp. on Bananas and Plantains: Towards Sustainable Global Production and Improved Uses*. Acta Hort. 986, ISHS 2013, pp. 365–80.

East African Community, 2014. *The EAC Industrialization Policy Brief*. Available at: http://eacgermany.org/wp-content/uploads/2014/10/EAC-Industrialization-Policy-Brief.pdf.

Eden-Green, S. J. 1994. *Banana Blood Disease*. INIBAP Musa Disease Fact Sheet No. 3, INIBAP, Montpellier, France.

Edmeades, S., Smale, M., Kikulwe, E. M., Nkuba, J. M. and Byabachwezi, M. S. R. 2007. Characteristics of banana-growing households and banana cultivars in Uganda and Tanzania. In: Smale, M. and Tushemereirwe, W. K. (Eds), *An Economic Assessment of Banana Genetic Improvement and Innovation in the Lake Victoria Region of Uganda and Tanzania*. Research Report No. 155, International Institute of Tropical Agriculture, Washington, DC.

FAO, 1990. *Roots, Tubers, Plantains, and Bananas in Human Nutrition*. FAO Food and Nutrition Series, FAO, Rome. http://www.fao.org/docrep/t0207e/T0207E00.htm#Contents.

FAO, 2017. *Data: Production Quantity*. FAO, Rome Italy. http://www.fao.org/faostat/en/#data/QC (Accessed on 12 April 2017).

FAOSTAT, 2014. *Production Statistics*. FAO Rome, Italy. http://faostat.fao.org/.

Fermont, A. M., Van Asten, P. J. A., Tittonell, P., van Wijk, M. T. and Giller, K. E. 2009. Closing the banana yield gap: An analysis from small-holder farms in East Africa. *Field Crops Research* 112: 24–36.

Fujardo, J. 2010. *Quality Declared Planting Material: Protocols and Standards for Vegetatively Propagated Crops*. FAO, Rome, Italy. http://www.fao.org/docrep/013/i1195e/i1195e00.pdf.

Gabre-Mariam, S. 1999. *Banana Production and Utilization in Ethiopia*. Ethiopian Agricultural Research Organization, Addis Ababa, Ethiopia. http://www.eiar.gov.et.

German Calberto, G., Staver, C. and Siles. P. 2015. An assessment of global banana production and suitability under climate change scenarios. In: Elbehri, A. (Ed), *Climate Change and Food Systems: Global Assessments and Implications for Food Security and Trade*. Food Agriculture Organization of the United Nations (FAO), Rome.

Gold, C. S., Pena, J. E. and Karamura, E. B. 2001. Biology and integrated pest management for the banana weevil, *Cosmopolites sordidus* (Germar) (Coleoptera: Curculionidae). *Integrated Pest Management Reviews*, 6: 79–155. http://www.banana2008.com/cms/details/acta/879_31.pdf

Hailu, M., Workneh, T. S. and Belew, D. 2013. Review on post-harvest technology of banana fruit. *African Journal of Biotechnology*, 12: 635–47.

Kader, A. A. and Rolle, R. S. 2004. *The Role of Post-harvest Management in Assuring the Quality and Safety Horticultural Crops*. Agricultural Services Bulletin 152, Food and Agriculture Organization, 52p.

Kamira, M., Sivirihauma, C., Ntamwira, J., Ocimati, W., Katungu, M. G., Bigabwa, J. B., Vutseme, L. and Blomme, G., 2015. Household uses of the banana plant in eastern Democratic Republic of Congo. *Journal of Applied Biosciences*, 95: 8915–29.

Karamura, E. 1993. Strategic importance of banana/plantains in Uganda. In: Gold, C. S and Gemmill, B. (Eds), *Proceedings of a Research Coordination Meeting for Biological and Integrated Control of Highland Banana Pests and Diseases*, Contonou, Benin, 12–14 November 1991, IITA, Ibadan, Nigeria, pp 384–7.

Kikulwe, E. M., Fischer, E. and Qaim, M. 2014. Mobile money, smallholder farmers, and household welfare in Kenya. *PLoS ONE* 9(10): e109804.

Kikulwe, E. M., Nowakunda, K., Byabachwezi, M. S. R., Nkuba, J. M., Namaganda, J., Talengera, D., Katungi, E. and Tushemereirwe, W. K., 2007. Development and dissemination of improved banana cultivars and management practices in Uganda and Tanzania. In: Smale, M. and Tushemereirwe, W. K. (Eds), *An Economic Assessment of Banana Genetic Improvement and Innovation in the Lake Victoria Region of Uganda and Tanzania*. Research Report no. 155, International Food Policy Research Institute, Washington, DC.

Kikulwe, E. M., Okurut, S., Ajambo, S., Nowakunda, K. and Naziri, D. 2017. Assessing the extent and determinants of post-harvest losses along the cooking banana value chain in Uganda, unpublished.

Krishna, H., Alizadeh, M., Singh, D., Singh, U., Chauhan, N., Eftekhari, M. and Sadh, R. K. 2016. Somaclonal variations and their applications in horticultural crops improvement. *Biotechnology*, 6(1): 54.

Kubiriba, J. and Tushemereirwe, W. K. 2014. Approaches for the control of banana Xanthomonas wilt in East and Central Africa. *African Journal of Plant Science*, 8: 398–404.

Kumar, P. L., Selvarajan, R., Iskra-Caruana, M. L., Chabannes, M. and Hanna, R. 2015. Biology, etiology, and control of virus diseases of banana and plantain. *Advances in Virus Research*, 91: 229–69.

Kumar, P. L., Hanna, R., Alabi, O. J., Soko, M. M., Oben, T. T., Vangu, G. H. and Naidu, R. A 2011. Banana bunchy top virus in sub-Saharan Africa: Investigations on virus distribution and diversity. *Virus Research*, 159: 171–82.

Masanza, M. 2003. *Effect of Crop Sanitation on Banana Weevil Cosmopolites Sordidus (Germar) Populations and Associated Damage*. Wageningen University and Research Centre, The Netherlands, p. 165, library.wur.nl/WebQuery/wurpubs/fulltext/121436.

Mobambo, K. N., Gauhl, F., Vuylsteke, D., Ortiz, R., Pasberg-Gauhl, C. and Swennen, R.1993. Yield loss on plantain from Black Sigatoka leaf spot and field performance of resistant hybrids. *Field crops Research* 35: 35–42.

Mohammadi, I. M. 2006. Agricultural waste management extension education (AWMEE) The ultimate need for intellectual productivity. *American Journal of Environmental Science*, 2: 10–14.

Mulualem, A. M., Jema, H., Kebede, W. and Amare, A. 2015. Determinants of post-harvest banana loss in the marketing chain of Central Ethiopia. *Food Science and Quality Management*, 37: 62–7.

Niroula, G. S. and Thapa, G. B. 2005. Impacts and causes of land fragmentation, and lessons learned from land consolidation in South Asia. *Land use Policy*, 22:358–72.

Nyang, C. N., Webo, C. and Roothaert, R. L. 2010. *The Power of Farmer Organisations in Smallholder Agriculture in East Africa: A Review of 5 project initiatives of the Maendeleo Agricultural Technology Fund*. FARM-Africa Working Paper No. 13, London, UK.
Nyombi, K. 2013. Towards sustainable highland banana production in Uganda: Opportunities and challenges. *African Journal of Food, Agriculture, Nutrition and Development*, 13: 7545–61.
Nyombi, K., Van Asten, P. J. A., Corbeels, M., Taulya, G., Leffelaar, P. A. and Giller, K. E. 2010. Mineral fertilizer response and nutrient use efficiencies of East African highland banana (Musa spp., AAA-EAHB, cv. Kisansa). *Field Crops Research*, 117: 38–50.
Ochola, D., Jogo, W., Karamura, E. B., Rietveld, A., Tinzaara, W., Ocimati, W. and Karamura, D. A. 2013. Farmers' awareness and perceived benefits of agro-ecological intensification practices in banana systems in Uganda. *African Journal of Biotechnology* 12: 4603–13.
Ocimati, W., Karamura, D., Rutikanga, A., Sivirihauma, C., Ndungo, V., Ntamwira, J., Kamira, M., Kanyaruguru, J. P. and Blomme, G., 2013. Agronomic practices for musa across different agro-ecological zones in Burundi, Eastern Democratic Republic of Congo and Rwanda. In: G. Blomme, P. Van Asten and B. Vanlauwe (Eds), *Banana Systems in the Humid Highlands of sub-Saharan Africa*, CAB International, UK, p. 175.
Okech, S. H., Van Asten, P. J. A., Gold, C. S. and Ssali, H. 2004. Effects of potassium deficiency, drought and weevils on banana yield and economic performance in Mbarara, Uganda. *Journal of Agriculture Science*, 9: 511–19.
Olorunda, A. O. 2000. Recent advances in post-harvest technologies of banana and plantain in Africa. *Acta Horticulturae*, 540: 517–27.
Ordonez, N., García-Bastidas, F., Laghari, H. B., Akkary, M. Y., Harfouche, E. N., Al Awar, B. N. and Kema, G. H. J. 2015. First report of *Fusarium oxysporum* f. sp. *cubense* tropical race 4 causing Panama disease in Cavendish bananas in Pakistan and Lebanon. *Plant Disease*, 100(1), 209.
Otieno, G., Lopez Noriega, I. and Reynolds, T. W. 2016. Smallholder access to quality and diverse seed in Uganda: Implications for Food Security. Policy brief published by Bioversity international and available at https://www.bioversityinternational.org/index.php?id=244&tx_news_pi1%5Bnews%5D=8431&cHash=3f77abfe225c72cafa4509a3d7a616de.
Padam, B. S., H. S. Tin, F. Y. Chye and M. I. Abdullah. 2014. Banana by-products: An under-utilized renewable food biomass with great potential. *Journal of Food Science Technology*, 51: 3527–45.
Ploetz, R. C. 2001. Black sigatoka of banana. *The Plant Health Instructor*. DOI: 10: 1094. PHI-I-2001-0126-01. Available at: http://www.apsnet.org/education/feature/banana/Top.html.
Ploetz, R. C. 2006. Fusarium wilt of banana is caused by several pathogens referred to as *Fusarium oxysporum* f. sp. *cubense*. *Phytopathology*, 96: 653–6.
Ploetz, R. C. 2015. Management of Fusarium wilt of banana: A review with special reference to tropical race 4. *Crop Protection*, 73: 7–15.
Qaim, M. 1999. *Assessing the Impact of Banana Biotechnology in Kenya*. ISAAA Briefs No. 10. ISAAA, Ithaca, New York, USA.
Republic of Uganda. 2001. The National Agricultural Advisory Services Act 2001. Available at: http://eacgermany.org/wp-content/uploads/2014/10/EAC-Industrialization-Policy-Brief.pdf.
Republic of Uganda. 2013. Uganda National land Policy. Available at: http://landportal.info/sites/default/files/the_uganda_national_land_policy-_february_2013.pdf.
Rubaihayo, P. R., Odongo, O. J. B. and Bananuka, J. A. 1994. Some highland banana production constraints in Masaka district of Central Uganda. In: *African Crop Science Conference Proceedings, Volume* 1. African Crop Science Society, Uganda, pp. 188–92.
Sekabira, H. and Qaim, M. 2017. Mobile money, agricultural marketing, and off-farm income in Uganda. *Agricultural Economics*, 48(5): 597–611.
Sharrock, S. and Frison, E. 1998. Musa production around the world-trends, varieties and regional importance. INIBAP annual report, pp. 42–7. http://www.musalit.org/pdf/IN020261_en.pdf
Smith, J. J., Jones, D. R., Karamura, E., Blomme, G. and Turyagyenda. F. L. 2008. *An Analysis of the Risk from Xanthomonas Campestris pv. Musacearum to Banana Cultivation in Eastern, Central and Southern Africa*. Bioversity International, Montpellier, France.

Spilsbury, J., Jagwe, K., Wanda, K., Nkuba, J. and Ferris, R. S. B. 2004. *Evaluating the Marketing Opportunities for Bananas and Its Products in the Principal Banana Growing Countries of ASARECA. ASARECA Monograph 7*, IITA, Ibadan, Nigeria.

Stover, R. H. 1962. *Fusariam Wilt (Panama Disease) of Bananas and Other Musa Species*. CMI, Kew, Surrey, UK.

Taulya, G. 2015. *Ky'osimba Onaarya: Understanding Productivity of East African Highland Banana*. PhD Thesis, Wageningen University, The Netherlands. 167p.

The World Bank Group, 2016. Population density (people per sq. km of land area). http://data.worldbank.org/indicator/EN.POP.DNST?end=2015&start=2015&view=map. Accessed on the13 April 2017.

Tinzaara, W., Karamura, E. B., Kubiriba, J., Ochola, D., Ocimati, W., Blomme, G. and Ssekiwoko, F. 2016. The banana Xanthomonas wilt epidemic in east and central Africa: Current research and development efforts. *Acta Horticulturae* 1114: 267–74

Tinzaara, W., P. Nemeye, C. S. Gold, E. B. Karamura, W. Tushemereirwe, P. E. Ragama and F. Bagamba, 2007. Enhancing banana production for improved livelihoods in the Eastern Africa Great Lakes Region: challenges for research. In: Njeru, R. W., Kagabo, D. M., Ndabamanye, T., Kayiranga, D. and Ragama, P. (Eds), *Sustainable Agriculture and Productivity for Improved Food Security and Livelihoods. Proceedings of the National Conference on Agricultural Research Outputs*, 26–27 March 2007, Kigali, Rwanda, Institut des Sciences Agronomiques du Rwanda, pp: 204–25.

Tripathi L. and Tripathi J. N. 2009. Relative susceptibility of banana cultivars to *Xanthomonas campestris* pv. *musacearum. African Journal of Biotechnology*, 8: 5343–50.

Tushemereirwe, W. K., Kangire, A. Smith, J. Ssekiwoko, F., Nakyanzi, M., Kataama, D., Musiitwa, C. and Karyeija, R. 2003. An outbreak of bacterial wilt on banana in Uganda. *InfoMusa*, 12:6–8

Tushemereirwe, W. K., Kangire, A., Ssekiwoko, F., Offord, L. C., Crozier, J., Boa, E., Rutherford, M. and Smith, J. J. 2004b. First report of *Xanthomonas campestris* pv. *musacearum* on banana in Uganda. *Plant Pathology*, 53: 802 p.

Tushemereirwe, W. K., Kangire, A., Kubiriba J., Nakyanzi, M. and Gold, C. S. 2004a. Diseases threatening banana biodiversity in Uganda. *African Crop Science Journal*, 12: 2004: 19–26

UBOS, 2010. *Uganda Census of Agriculture 2008/2009: Crop Area and Production Report*, volume IV. Uganda Bureau of Statistics, Kampala Uganda.

Van Asten, P. J. A., Fermont, A. M. and Taulya, G. 2011. Drought is a major yield loss factor for rainfed East African highland banana. *Agricultural Water Management*, 98: 541–52.

Van Asten, P. J. A., Gold, C. S., Wendt, J., De Waele, D., Okech, S. H. O., Ssali, H. and Tushemereirwe, W. K. 2005. The contribution of soil quality to yield and its relation with other banana yield loss factors in Uganda. In: Blomme, G., Gold, C. S. and Karamura, E. (Eds). *Proceedings of a Workshop Held on Farmer Participatory Testing of IPM Options for Sustainable Banana Production in Eastern Africa*, Seeta, Uganda, 8–9 December 2003. International Plant Genetic Resources Institute, Montpellier, pp. 100–15.

Viljoen, A. 2010. Protecting the African Banana (*Musa* spp.): Prospects and challenges. In: Dubois, T., Hauser, S., Staver, C. and Coyne, D. (Eds), *Proc. IC on Banana and Plantain in Africa*. Acta Horticulturae 879, ISHS 2010.

Wairegi, L. W. I., Van Astena, P. J. A, Tenywa, M. M. and Bekunda, M. A. 2010. Abiotic constraints override biotic constraints in East African. *Field Crops Research*, 117: 146–53.

Washington R. and Pearce. H., 2012. *Climate Change in East African Agriculture: Recent Trends, Current Projections, Crop-climate Suitability, and Prospects for Improved Climate Model Information*. CGIAR Research Program on Climate Change, Agriculture and Food Security (CCAFS), Copenhagen, Denmark. Available online at: www.ccafs.cgiar.org.

Yirgou, D. and Brudbury, J. F. 1968. Bacterial wilt of *Enset (Ensete ventricosa)* incited by *Xanthomonas musacearum. Phytopathology*, 58: 111–12.

Zenebe, W., Ali, M., Derbew, B., Zekarias, S. and Bekele, A., 2015. Assessment of banana production and marketing in Ethiopia. *International Journal of Sciences: Basic and Applied Research*, 24: 283–307.

Part 2

Improving practices across the banana value chain

Chapter 5

A functional approach to bunch formation in banana

D. W. Turner, School of Agriculture and Environment, The University of Western Australia, Australia; and D. J. Gibbs, Consultant, Australia

1 Introduction
2 Gross morphology and bunch form
3 Phases of development in banana
4 The floral phase: overview
5 Inflorescence
6 Formation of hands and fruit: independent but closely coordinated processes
7 Growth of the aerial true stem
8 Case study: photosynthate availability and bunch formation in banana
9 Conclusion
10 Where to look for further information
11 References

1 Introduction

Sustainable management of bananas, whether in smallholder farming systems or in production for export, requires a sound knowledge of development and its response to environmental and soil conditions. Central to plant development are the processes of flower formation, which in bananas lay the foundations of yield of fresh fruit. The main objective of growing bananas is to produce fruit. For the purposes of crop management, we need to decide 'critical' stages of development in the life cycle such as the production of lateral shoots (suckers) and fruit-forming flowers. A functional approach integrates bunch formation with genetic and environmental cues and points to underlying mechanisms that can be used to guide plant improvement. Such an approach has been successful in wheat (*Triticum aestivum*), a temperate monocotyledon (Kamran et al. 2014).

For flowering plants the basic developmental cycle is vegetative–reproductive–vegetative–reproductive and so on through the generations. The banana follows this fundamental model but in two developmental pathways, one that involves seed set and germination (sexual) and one that does not (vegetative). The wild bananas reproduce by

http://dx.doi.org/10.19103/AS.2017.0020.11

both sexual and vegetative means. Edible seedless bananas were derived from natural hybridisation of the wild species *Musa acuminata* and *Musa balbisiana* and these form the basis of banana cultivation and commerce (Simmonds 1962; De Langhe et al. 2009). They undergo vegetative reproduction by the growth of lateral shoots (suckers) and each shoot has a fruiting phase. The advantage for human use is that the fruit develop without the need for pollination and are largely sterile, hence no seeds. With considerable effort in pollination, enough seeds can be produced in many edible cultivars to keep plant breeders interested, although not very happy. Overlapping generations of suckers promote productivity in the edible bananas (Dorel et al. 2016). Sympodial growth of the rhizome is the morphological basis of the banana's perennial nature and an individual banana clump (mat) could live for many generations (centuries). In reality, many clumps of wild bananas survive only for a few generations because they are seral plants in disturbed areas of forest, which do not survive when the climax vegetation becomes established (Simmonds 1962). For the wild bananas, seed-bearing fruit are the pathway for long-term survival, whereas for the edible cultivars, seedless fruit are essential. In both cases, bunch formation is central to their life cycles.

Here, we analyse data from the literature to identify processes critical to the initiation and formation of the bunch. The three main questions are:

1. What is the sequence of events during the floral phase from the initiation of a bunch to its emergence at the top of the pseudostem (flowering)?
2. How does this sequence relate to observable changes in the plant?
3. What is the relationship between the number of hands that contain fruit and the number of fruit-forming flowers within those hands?

We show how manipulative experiments are needed to further our knowledge of bunch formation in banana.

2 Gross morphology and bunch form

The fruiting banana plant of a single generation consists of rhizome, root system, leaves, pseudostem, lateral buds, suckers, elongated aerial true stem, bunch (inflorescence) with fruit-forming and usually male flowers. Each component undergoes processes of initiation, differentiation, growth, maturity, senescence and finally death. These processes are influenced by controls that are internal (e.g. gene networks, hormones or carbon supply) or external to the plant (e.g. environmental and edaphic factors).

The fruiting axis or peduncle of the banana family (Musaceae) is very distinctive with a female section at the base and beyond this a male section. The main feature of the female section is flowers of different types, but which all produce fruit because they have functional carpels. These flowers may be fully hermaphrodite, hermaphrodite, functionally female or female, depending on the extent of development of the stamens or staminodes. Fully hermaphrodite (perfect) flowers have well-developed stamens and are capable of self-fertilisation. Hermaphrodite flowers have either less than a full complement of fully developed stamens or well-developed staminodes but no pollen. Functionally female flowers have less developed staminodes and no pollen of sound appearance. Female flowers have staminodes that are either poorly developed or absent. Different types of female flowers may appear on the same bunch. For example, *Musa acuminata* ssp. *banksii* has functionally female flowers in the first 1 to 2 basal hands, while the remainder of the flowers on the bunch that produce fruit are hermaphrodite (Argent 1976). A transitional node or hand often appears between the female and male sections of the peduncle. This hand contains one or more flower types that produce fruit, while the remaining flowers are male. In this context, the hand is not hermaphrodite, only flowers within the hand may be classified in this way.

3 Phases of development in banana

3.1 Phases and 'clocks'

For convenience, the life cycle of a shoot of the banana can be divided into several phases: juvenile, mid-vegetative, floral and fruiting. From the juvenile to the start of the floral phase it is worthwhile thinking of two developmental clocks operating (Turner et al. 2016a). One is a 'leaf clock', reflected in the sequence of leaves produced at the vegetative apex and which subsequently emerge at the top of the pseudostem. The leaf clock is driven mainly by temperature and to a lesser extent the stage of plant development. In juvenile plants the phyllochron (time between the appearance of successive leaves at the top of the pseudostem) is about 4 days, but in plants approaching bunch emergence, this may increase to 10 days. The second is a 'bunch clock' that parallels the leaf clock in the juvenile phase, but which operates separately from it in the mid-vegetative phase. These two clocks interact in such a way that the total number of leaves produced before flowering on any individual shoot cannot be predicted.

In the juvenile phase of development, as defined by Fortescue et al. (2011), the banana is not able to flower. Its development is driven by temperature that can be quantified by using the accumulation of degree-days. However, cool temperatures and soil-water balance, being the amount of water in the soil, modify this temperature response. Cool temperatures, independent of their effect in slowing the rate of appearance of new leaves, can shorten the juvenile phase (Turner et al. 2016b). At 24°C 15 leaves may be produced in the juvenile phase, while at 16°C only 7 leaves are produced. Soil-water deficits or excess may extend the juvenile phase (Fortescue et al. 2011). The appearance of lateral shoots (suckers) above the ground is correlated with the end of the juvenile phase.

During the mid-vegetative phase, the rate of development is again driven by temperature, but this effect is modified by an internal plant component, photoperiod, and soil-water balance (Fortescue et al. 2011). The plant component ensures that the bunch clock slows to about half the rate of the leaf clock during the mid-vegetative phase. Long photoperiods hasten the bunch clock. The banana has a long-day quantitative response to photoperiod, similar to that of wheat and *Arabidopsis* (Fortescue et al. 2011; Turner et al. 2016a). The quantitative response differs from an obligate response, where flowering is initiated when a 'critical' photoperiod is perceived by the plant. In a long-day quantitative response, the initiation of flowering is hastened by long days, but long days are not essential for flower initiation to occur. As in the juvenile phase, soil-water balance can delay the transition from vegetative to floral phases. The interaction between seasonal changes in temperature, photoperiod and soil-water balance will influence the timing of the transition from vegetative to reproductive phases in banana, contributing to seasonal variation in flowering and bunch harvest.

The photoperiod response of banana is quite sensitive, two to three times more so than wheat, for example (Fortescue et al. 2011). High sensitivity is useful for a plant that grows close to the equator, where only small changes in photoperiod occur. It is tempting to speculate that *Musa* spp. may have developed sensitivity to photoperiod early in their evolution, as at their ancestral area of origin (northern Indo-Burma, 23° to 30°N), the photoperiod varied seasonally by at least 2 h and up to 4 h across the year (Janssens et al. 2016). As the members of the Musaceae moved towards the equator, selection was for increased photoperiod sensitivity, rather than against it.

3.2 A few words about methodology

In the vegetative and floral phases of development, it is necessary to keep in mind the relationship between the production of individual leaves at the apex and when those same leaves appear at the top of the pseudostem (Fig. 1). Counting leaves as they emerge in sequence from the pseudostem provides the base for describing developmental events. However, its value for timing events such as bunch initiation, for example, is limited since the total number of leaves that will be produced by the apex *(N)* is unknown. In addition,

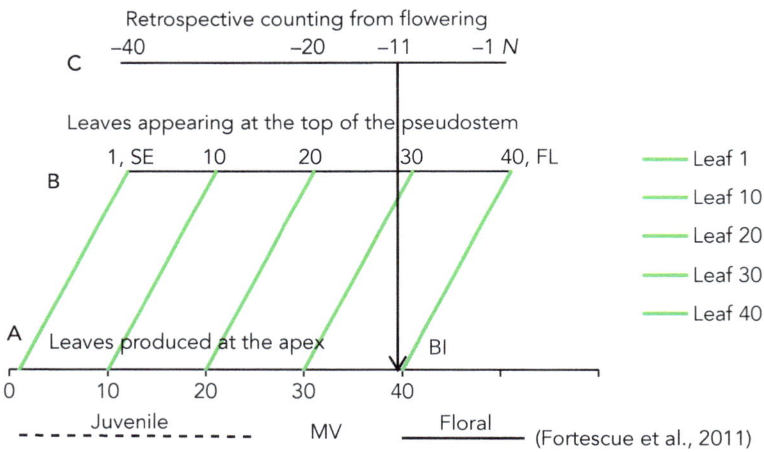

Figure 1 A diagram of a sequence of 40 leaves produced at the apex of a shoot (horizontal line A) leading to bunch initiation (BI). Those same leaves appear, in sequence, at the top of the pseudostem (intersection of sloping green lines with horizontal line B) with leaf 1 emerging from the soil (SE) and leaf 40 appearing at the top of the pseudostem immediately before flowering (FL). The slope of the green lines assumes that 11 unemerged leaves are growing within the pseudostem up until bunch initiation (Summerville 1944; Barker and Steward 1962a; Israeli and Blumenfeld 1986). Horizontal line C illustrates retrospective counting once all leaves (N) have emerged. Then, with the flowering of lines B and C aligned, the vertical arrow starting at leaf −11 in C coincided, in this example, with the emergence of the 29th leaf at the top of the pseudostem (B) and with bunch initiation at the apex (A). The juvenile, mid-vegetative (MV) and floral growth stages for a plant crop, according to Fortescue et al. (2011), are shown for illustration.

events such as the formation of fruit-bearing hands, which occur during the floral phase within the pseudostem, are not directly visible. In the field and in experiments the timing of bunch initiation can be determined retrospectively by counting back 11 leaves once flowering has occurred because there are about 11 developing leaves within the pseudostem of banana up to bunch initiation. To determine the sequence of events after bunch initiation, it is necessary to link these events with the number of leaves remaining (<11) to emerge (Fig. 1, see leaf −11). In this way a reliable sequence of events during bunch formation can be determined. That there are about 11 leaves in the pseudostem at floral initiation has been found in 'Gros Michel' (AAA) and in members of the Cavendish subgroup (AAA), and this leaf number has been successfully used to study development in plantains (AAB). If cultivars are found to have greater or fewer than 11 leaves at bunch initiation, then there will be a shift in the relationship between leaf number and process, but the principles remain the same.

4 The floral phase: overview

4.1 Introduction

Molecular changes at the apex initiate the transition from vegetative to floral. A major reorganisation of meristematic activity around the apex takes place with associated impacts on the base of those leaves yet to emerge from the pseudostem. For these leaves the formation of leaf-opposed lateral buds is inhibited and, for those leaves emerging during floral transition, the activity of the intercalary growth zone that contributes to their rate of leaf emergence appears to be reduced. At the apex itself, the main events are the formation of the inflorescence consisting of nodes (hands) of fruit-forming flowers forming the female peduncle, the slow elongation of the aerial true stem and peduncle, and then the formation of nodes of male flowers.

The appearance of nodes of male flowers at the apical meristem coincides with the start of the rapid elongation of internodes of mainly the aerial true stem and to a lesser extent the peduncle. The aerial true stem supports the last 11 or so leaves to emerge before bunch emergence. The peduncle is a continuation of the true stem, but it supports bracts and flowers rather than leaves. It has three sections – *transitional*, with usually two sterile nodes each with a bract that abscises at bunch emergence and which contain no flowers; *female* peduncle – that supports the hands of fruit and *male* peduncle with male flowers. The internodes of the aerial true stem elongate until their associated leaf emerges at the top of the pseudostem. When the inflorescence nears the top of the pseudostem, the contribution of the internodes of the elongating aerial true stem to the upward movement of the bunch is seamlessly taken over by elongation of the internodes of the transitional peduncle, and to a lesser extent, the female and male peduncle. As the bunch moves clear of the top of the pseudostem, elongation of the aerial true stem ceases. The elongating peduncle then extends the bunch.

4.2 The timing of events

The floral phase deals with the formation of the bunch, and the timing of the sequence of events is important for understanding the effects of genotypes and environment. The most appropriate measure of development, by proxy in this instance, is the association between the sequence of new leaves appearing at the top of the pseudostem and events at the apex within. Several studies, based on dissection, including those on plant crops in the subtropics (Summerville 1944) and ratoon crops in the tropics (Ganry 1977; Stover 1979), support the view that, on average, 11 to 12 leaves are present within the pseudostem when the apex transforms from the mid-vegetative to the floral phase (Table 1). For any individual shoot, the total number of leaves it produced can only be determined retrospectively at flowering, but use of the 11 to 12 last leaves to indicate the timing of bunch initiation and the following events at the apex is useful nonetheless. Only by dissection can the link between the formation of the inflorescence and the emerging leaves be established.

Table 1 The relationship between the last 11 leaves to appear at the top of the pseudostem leading up to flowering and the corresponding events at the apex between bunch initiation at leaf N-11 and flowering at leaf N

Number of leaves emerged at the top of the pseudostem	Events at the apex and development of the aerial true stem
N-11	Transformation from mid-vegetative phase to floral phase. Bunch initiation, redistribution of meristematic activity at the apical meristem, slowing of the rate of extrusion of one or more leaves as they emerge (Ganry 1977; Stover 1979). Formation of leaf-opposed buds suppressed (Barker and Steward 1962b).
N-10	
N-9	Nine to eleven leaves are present within the pseudostem when the first floral organs can be detected (Barker and Steward 1962b). The internodes of the last 11 leaves become the aerial true stem and together with the fledgling peduncle elongate slowly during the differentiation of the nodes of fruit-forming flowers (Ganry 1977).
N-8	If dissected, the structure of the inflorescence can be seen with a hand lens and is about 30 cm above ground level in 'Grand Nain' (Cavendish, AAA) plants 2.54 m high (Stover 1979).
N-7	If dissected, the inflorescence is large enough to be seen by the eye (Ganry 1977).
N-6	
N-5	Rapid elongation of the internodes of the aerial true stem, and to a lesser extent the peduncle, begins and coincides with the production of male flowers that can be discerned (Ganry 1977). Also Skutch (1927) and see Fig. 8.
N-4	
N-3	
N-2	Rapid elongation of the aerial true stem and peduncle to push the inflorescence to the top of the pseudostem.
N-1	
N	Flowering (bunch emergence). Elongation of the aerial true stem ceases.

A number of studies provide the basis of our knowledge about reproductive events and their timing in the development of the inflorescence. Of particular interest is the initiation of the bunch, the differentiation of fruit-forming flowers and subsequently male flowers. Also of interest are correlations between dimensions of the bunch and the leaves that are developing at the same time, both of which may be subject to simultaneous changes in the allocation of carbon resources among the growing tissues of the aerial true stem, leaves, peduncle and the bunch (Turner 1980).

In the vegetative phases, tens of nodes are produced by the apical meristem, but in the floral and fruiting phases, hundreds of nodes may be produced (White 1928). Overall, the rate of node production in the reproductive phases is six times greater than in the vegetative phase (Turner 1981). This implies that six hands would be produced in the time that it previously took to produce one leaf. This represents the formation of a hand each 1–2 days under tropical conditions of about 20°C to 25°C. The formation of a fruit initial within a node, if produced in sequence, would therefore take place about once each 2–3 h. The transition from the vegetative rate of node production to the reproductive may be gradual, rather than a sudden change to six times the rate (Turner 1981). In some genomic subgroups, such as 'Cavendish' (AAA), a gradual change is likely to affect the rate of node production more than the rate of formation of flower initials within each node. This may account for the high number of flower initials in the axils of the more basal hands (nodes) of the inflorescence compared with those that are distal (Alexandrowicz 1955).

5 Inflorescence

5.1 Changes at the vegetative apex and external indications

The vegetative apex has five zones of meristematic cells and these reduce to two extensive zones as the apex becomes reproductive (Fahn et al. 1963). Despite the reduction in the number of zones, these changes represent a considerable reorganisation of activity at the apex as the plant moves from a vegetative to floral phase. The outer zone of the floral apex is the point of origin of the bracts and fruit of the inflorescence. The reorganisation of meristematic activity at the apex is correlated with other events nearby and remotely, and indeed there may be direct connections. Nearby, the formation of leaf-opposed lateral buds at the bases of the leaves that are yet to emerge is suppressed (Barker and Steward1962b; Fisher 1978).

A more remote indicator of rearrangement of meristematic activity is a reduction of about 20–30% in the rate of emergence of the leaf appearing concurrently at the top of the pseudostem, and some subsequent leaves (Fig. 2, Ganry 1977; Stover 1979). In the study of Ganry (1977), this reduction in the rate of emergence of leaves affected all leaves that emerged after the transition from vegetative to floral. Stover (1979) observed reduced rates of emergence, but these effects lasted only for one to four subsequent leaves. This reduction in the rate of extrusion may be caused by redirection of meristematic activity away from the intercalary growth region at the base of each leaf as it emerges so that the contribution made by the intercalary growth zone to the process of leaf extrusion either ceases or is attenuated. The temporary nature of the slowing of leaf emergence in Stover's study implies that other processes also affect the rate of leaf extrusion. Perhaps the rate

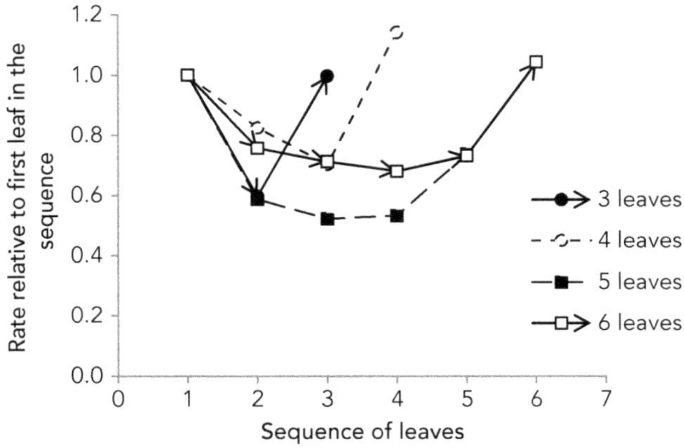

Figure 2 The temporary slowing of the rate of extrusion of successive leaves associated with the change in organisation of meristematic activity at the apex as it changes from vegetative to reproductive. Original data from Stover (1979) on 'Grand Nain' (Cavendish, AAA). Values for 33 ratoon suckers across three dates of appearance of suckers above the soil in Honduras. Data for rate of extrusion were standardised to the first leaf in each sequence of 3, 4, 5 or 6 leaves. Arrows indicate sequence for series with 3 or 6 leaves. Values of extrusion rate <0.85 are significantly different from 1.0 at $P = 0.05$.

of extrusion increases as the rapid elongation of the internodes basal to the inflorescence commences. Studies are needed to separate these effects.

Division and expansion of the cells along the whole of these developing leaves, outside of the intercalary growth region, combined with the elongation of the associated internodes, then become the main drivers for the extrusion of each leaf from the top of the pseudostem. Elongation of the internodes of the aerial true stem begins to push the pre-formed leaves and the new inflorescence upwards and this creates the aerial true stem seen at the centre of the pseudostem. Growth of the aerial true stem continues throughout flowering to form the peduncle that can be seen emerging from the pseudostem and to which the sterile transitional nodes (leaves to bracts), fruit-forming and male hands are attached (Ram et al. 1962; Argent 1976).

In the vegetative phase, leaf bases surround the apex situated on the rhizome and the lateral buds associated with each leaf are initiated in a leaf-opposed position. Leaves of the aerial true stem do not develop lateral buds (an exception can occur among plantains, AAB; Devos 1985) and the first one or two bracts produced on the transitional peduncle are also sterile. However, axial bud initials (flower initials) appear in subsequent floral bracts and there can be many buds within each axil (Fahn 1953; Ram et al. 1962). The clusters of flowers within these nodes are subtended by a bract and, in seedless bananas, they form edible fruits.

5.2 Thyrse with cincinni

The botanical term for the banana inflorescence is a thyrse (Kirchoff 1992), the same as for the heliconias. A thyrse is an inflorescence in which the main axis continues to grow and the lateral branches, in this case nodes of flowers, have determinate growth (Endress 2010).

The central axis of the thyrse of banana is the peduncle with a transitional, female and male section. The apex of the peduncle terminates some weeks or months after flowering. All nodes of the peduncle have a bract (modified leaf) that in the female and male sections subtends an axillary cushion of tissue from which the flower initials arise. Fahn (1953), Moncur (1988) and Kirchoff (2017) observed a sequence in the formation of flower initials within each hand. Fahn (1953) concluded that the flower initials represented a very compressed cincinnus, while Kirchoff (2017) interpreted the hand of *Musa velutina* as a highly modified cincinnus. Cincinni have a false main axis from which individual flowers arise alternatively in sequence (Buys and Hilger 2003). Fahn (1953) observed cincinni in the seeded diploids *Musa acuminata* and *Musa balbisiana* as well as in the cultivated triploid 'Dwarf Cavendish' (AAA, Cavendish subgroup). Moncur (1988) examined cultivars of the Cavendish subgroup. The sequence of flower formation in the cincinni moved from right to left (when viewing the inflorescence from the side with the apex pointing upwards) and in the study of Fahn (1953) alternated between the double rows of flowers. This sequence may vary across the genus (Kirchoff 2017). For a hanging bunch in the field, the sequence moves from left to right and can best be seen in the transitional hand that precedes the male phase of the inflorescence. The cincinnus in this hand has a mixture of fruit-forming and male flowers with the male flowers being at the extremity of the cincinnus. Subsequent cincinni produce only male flowers.

Because the cincinni in banana are highly modified, Kirchoff (1992) states that it is difficult to define the main axis. Cincinni are part of inflorescence development in *Heliconia* sp., a group closely related to the Musaceae (Kirchoff et al. 2009). Cincinni could also apply to single rows of flowers within Musaceae (Simmonds 1962; Häkkinen et al. 2008) as they approximate the boragoid form of a cincinnus observed from a lateral rather than a dorsal view, as described by Buys and Hilger (2003).

Knowledge of the mechanism of flower formation is important because the fruit-forming flowers become the fruit of commerce. The cincinnus interpretation offers a mechanism for coordination between two independent processes: the production of nodes with a subtending bract and the production of flower initials in the axils of the bracts. Within a cincinnus, flower initials are produced in a sequence and so the number of initials within a bract may be prescribed in part by genetic controls and in part by allocation of resources (Guitián et al. 2004). The number of flower initials developed in the axil of a bract will depend on the supply of sufficient carbon resources, the later-formed initials being affected first. Fahn (1953) noted that the sequence of formation of vascular tissues connecting the peduncle to the developing flowers followed the sequence in which the flower initials were formed except that connection began at the second initial rather than the first. The transitional node between the female and male sections of the peduncle is consistent with the thyrse/cincinnus interpretation. For example, in the monocotyledon *Commelina caroliniana*, the cincinni produce hermaphrodite flowers at first and then male flowers (Kaul and Koul 2008). This reflects distribution of resources within a cincinnus. In bananas its impact on the gender of flowers is demonstrated in the hand that separates the female from the male section of the peduncle (Alexandrowicz 1955). There is also a need to distribute resources, even if they are limited, to the elongating axis of the thyrse. Kaul and Koul (2008) argue the case that, when limited resources flow to the more distal sections of the inflorescence in *Commelina caroliniana*, then cincinni in the distal parts of the inflorescence form male flowers. Perhaps more importantly the thyrse/cincinni interpretation means that the final number of hands of fruit on a bunch that is of commercial significance is not determined by hand production *per se*. There are plenty of hands that make up the female and male sections of the peduncle. Instead, the cincinni switch from the production of fruit-forming flowers to male flowers. However, the switch is not 'clean' because the transitional hand has a mix of fruit-forming and male flowers within it, with more fruit-forming flowers to one side than the other (Alexandrowicz 1955).

In their study of 'Gros Michel' (AAA), Ram et al. (1962) concluded that the flower initials within a hand arose simultaneously within a crescent-shaped, meristematic body of cushion-like appearance. It is not clear from this model how the number of bud initials within a hand might be determined, unless it is restricted by the physical dimensions of the meristematic cushion.

5.3 Flower types: fruit-forming and non-fruit-forming

The classic appearance of a growing bunch of bananas or inflorescence is of a sterile section without flowers, then fruit at the basal (proximal) nodes and male flowers at the distal nodes, the remains of which may have fallen to the ground. These features separate the peduncle into transitional, female and male sections (Argent 1976). Within the female section of the peduncle, a number of different flower types can be identified: 'hermaphrodite' (e.g. *Musa boman, M. schizocarpa, M. ingens*) or female (e.g. *Musa balbisiana, M. maclayi, M. bukensis, M. peekelii, M. lolodensis*) (Argent 1976). The difference between hermaphrodite and female lies not in their ovaries but in the status of the stamens that vary from well developed in 'fully hermaphrodite' flowers to staminodes that are either absent or if present, poorly developed in 'female' flowers. What all these flower types have in common is an ovary capable of forming a fruit, which is not the case for male flowers. Thus, the female peduncle contains flowers that form fruit that may arise from a range of flower types, but all of which have functional carpels.

There is genotypic variation in how many nodes of fruit-forming flowers are produced in an inflorescence. Among the plantains (AAB) there is a range of genotypic expressions of the female and male sections of the inflorescence (De Langhe 1964). The female peduncle can have flowers that are abundant ('Okere'), minimal (one hand, 'Ifelete') or absent ('Bokangasoku'). Likewise, the male section of the peduncle can have flowers that are either abundant ('Okere'), or minimal ('Libanga likale') or absent ('Lokusu', 'Igbulu').

Of importance in commerce are the genetic and agronomic factors that affect the balance, especially increasing the number of flowers that produce fruit within a cincinnus, the number of nodes where the cincinni produce fruit-forming flowers and how the balance between these can best be manipulated.

6 Formation of hands and fruit: independent but closely coordinated processes

Within a bunch, a number of different patterns of female hand number (female cincinni per thyrse) and fruit per hand (fruit-formimg flowers per cincinnus) are associated with different cultivars and with different seasonal conditions for the same cultivar. Some of these patterns are similar to those found in the monocotyledon ornamental ginger (*Hedychium* spp., Zingiberaceae) (Sakhanokho and Rajasekaran 2010). Indeed, in the classification of *Musa*, the number of flowers per hand at one stage had taxonomic importance, at least to the resolution of 'many' or 'few', in distinguishing between sections of *Musa* (Cheesman 1947, sect. *Eumusa* – 'many'; sect. *Rhodochlamys* – 'few'). These two sections are now merged, based on molecular evidence (Häkkinen 2013), and as a consequence, a recent

Table 2 Scenarios of the number of flowers per hand on the female and male peduncle, and the relationship between the number of female hands per bunch and the number of fruit per hand within those bunches. (G and E) indicate main influences: G = genotype, E = environment

Scenario (Influence)	Features
1 (G)	The greatest number of flower initials in the basal hands of the female peduncle falls into three groups, few (5–10), several (11–15) or many (16–25+).
2 (G)	A decrease in the number of flowers per hand from basal to distal hands on the female peduncle, irrespective of whether the basal hands had few, several or many flowers.
3 (G)	The number of flowers per hand is constant from basal to distal hands and across the female and male sections of the peduncle.
4 (G)	As in scenario 2, but the number of flowers per hand in the male section of the peduncle is relatively constant.
5 (G)	As in scenario 2, but the male peduncle is absent.
6 (G + E)	There is a positive association between the number of female hands in an inflorescence and the mean number of fruit per hand (fruit per bunch divided by hands per bunch).
7 (E)	Across seasons, female hand number may be more stable than the number of fruit per hand.

revision of the sections of *Musa* no longer uses this morphological feature. However, these data are significant descriptors for wild as well as cultivated bananas (Häkkinen et al. 2008). For the discussion here, the number of flowers per hand reflects the activity of the cincinnus within the node and our interest lies in determining why some genotypes have 'many' while others have 'few' and others lie somewhere in between. Separating the likely effects of genotype and environment on the number of flowers per hand has commercial significance.

The different patterns of female hand number and fruit per hand are best accounted for if the processes of formation of hands (node formation on the thyrse) and flower initials (activity of the cincinnus within each node) are considered independent. Then, a number of scenarios are possible (Table 2). Some scenarios are characteristic of differences between cultivars and others are features of differences between seasonal conditions but within the same cultivar.

6.1 Scenarios characteristic of differences between genotypes

There are numerous examples of scenario 1 (Fig. 3). In the experiment of Swennen and Vuylsteke (1987), 'Agbagba' False Horn plantain, AAB, had few (7–8) female flowers in the basal hands; 'Mbang Okon', French Horn plantain, AAB, had several (13) female flowers in the basal hands; and in the experiment of Lassoudière (1978), 'Poyo' Cavendish, AAA, had many (21–24) female flowers in the basal hands (Fig. 3).

A decrease in the number of fruit-forming flowers per hand from the basal to the distal hands is a feature of scenario 2 and occurs, for example, in 'Poyo' and the plantains 'Mbang Okon', 'Ihitisim' (Horn, AAB) and 'Agbagba' (Fig. 3).

'Obino L'Ewai', a French Horn plantain, AAB, has several fruit-forming flowers in the basal hands, and this number of flowers is maintained from the basal to the distal hands and from the female to the male sections of the peduncle. In this cultivar the number of flowers per hand is similar in the most basal hand (hand 1) to that in hand 70, even though bunches had only 6 hands of fruit-forming flowers. It is an example of scenario 3 (Fig. 3).

In scenario 4 the number of fruit-forming flowers in the hands of the female section of the peduncle decreases from basal to distal hands, but in the male section of the bunch, the number of male flowers per hand stabilises and remains similar to that in the most distal female hand. This is a feature of Cavendish cultivars, AAA (Alexandrowicz 1955).

The Horn plantain (AAB) subgroup only has hands of fruit-forming flowers and these are often few in number. 'Ihitisim' is an example of scenario 5 (Fig. 3).

Genotype contributes to the relationship between the number of hands of fruit-forming flowers per bunch (female hands per bunch) and the number of fruit-forming flowers within those hands (fruit per hand) along the female peduncle (Table 2, scenario 6 and Fig. 4). Taking account of data from the tropics (Swennen and Vuylsteke 1987) and subtropics (Turner and Hunt 1984), the overall data are variable, but groups within the data sets show a positive relationship between the number of female hands per bunch and the number of fruit per hand (Fig. 4). This implies a greater level of available resources, whether under genetic or environmental control. Broadly, the cultivars fell into the 'few', 'several' and 'many' categories used to group the number of fruit per hand in scenario 1 (Table 2). The range of female hand number across cultivars within these groupings is similar: 5 to 11 and so useful comparisons can be made. The twofold range in female hand number per bunch in this set of data contrasts with the four-fold range in the number of fruit per hand. This suggests that the number of fruit-forming flowers in each cincinnus is more amenable

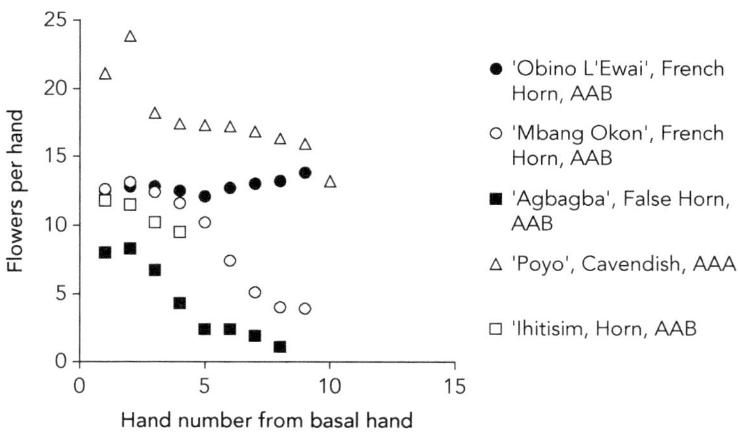

Figure 3 The relationship between the node (hand) position on the bunch (hand 1 is the basal node of the inflorescence) and the number of flowers in each female or male hand for a number of cultivars. 'Obino L'Ewai' had 6 hands of female flowers (fruit), 'Mbang Okon' 8 hands, 'Agbagba' 7 hands, 'Poyo' 10 hands and 'Ihitisim' 4 hands. Data from Swennen and Vuylsteke (1987) in Nigeria and Lassoudière (1978) in Ivory Coast.

to change than the number of female hands. A somewhat different picture emerges when groups of cultivars within the data set are considered.

Cultivars with only A genome (AAA, AAAA) had the highest number of fruit per hand ('many' category) and within that group those cultivars with more female hands had more fruit per hand. For each extra hand, the number of fruit per hand increased by 0.59 across the bunch. Cultivars containing the B genome (AAB, ABB) had less fruit per hand than the A genome genotypes whether grown in the tropics (Nigeria) or subtropics (New South Wales). Within the AAB groups those cultivars with more female hands per bunch tended to have more fruit per hand. For the AAB and ABB genotypes in NSW, the increase was 0.62 fruit per hand for each extra female hand per bunch, and for the False and French Horn plantains grown in Nigeria, the increase across cultivars was 0.58 fruit per hand for each extra female hand. The slopes of the fitted lines for these three categories were very similar (0.59, 0.62 and 0.58). The data for French plantains in Nigeria showed an increase of 1.04 fruit per hand for each extra female hand and fall well above the False and French Horn groups (Fig. 4). An exception was the Horn group from Nigeria which had very low female hand numbers per bunch (2–4) but several to many fruit per hand (Fig. 4).

For the two separate processes, hand production and the production of fruit initials within a hand, there are strong associations with genotype. The allocation of resources to these independent but coordinated processes produces a range in the number of female hands per bunch and fruit per hand, according to genotype. The number of female hands

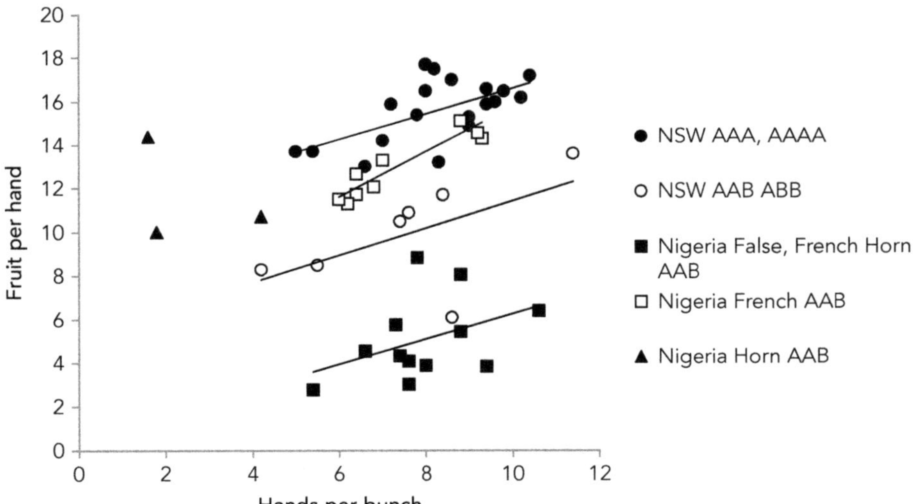

Figure 4 The relationship between the number of female hands on a bunch and the mean number of female flowers in those hands for plant crop bananas and plantains. 'NSW' are data from Turner and Hunt (1984) for AAA, AAAA, AAB and ABB cultivars grown at Alstonville, NSW (Lat 29°S). 'Nigeria' are data from Swennen and Vuylsteke (1987) for plantain subgroup (AAB) cultivars grown at Onne, Nigeria (Lat 5°N). The NSW data separated naturally into genomic groups (AAA, AAAA vs AAB, ABB) and the data from Nigeria separated naturally into French, French Horn, False Horn and Horn plantains within the AAB group.

per bunch and fruit per hand are important components of bunch weight, of significance to growers.

Scenario 6 is the positive association between the number of hands of female flowers on a bunch and the number of fruit-forming flowers per hand. This correlation is associated with differences among genotypes (Fig. 4), the main difference across genotypes being in the number of fruit per hand rather than the female hand number per bunch.

6.2 Scenarios characteristic of differences in seasonal and edaphic conditions

Scenario 6 also describes the effects of nutrition and salinity, for example, on bunch characteristics of cultivars of the Cavendish subgroup (AAA). Less favourable nutrition leads to reduced female hands per bunch and less fruit per hand (Summerville 1944; Alexandrowicz 1955; Champion 1967; Lassoudière 1978b; Turner 1981) as does increased salinity (Israeli et al. 1986). For salinity there may be an accumulative effect as salinity did not affect female hands per bunch or fruit per hand in the first cycle, but rather reduced yield by reducing fruit size. In the second cycle, salinity reduced hands per bunch and fruit per hand in equal proportions, in addition to reducing fruit size (Israeli et al. 1986).

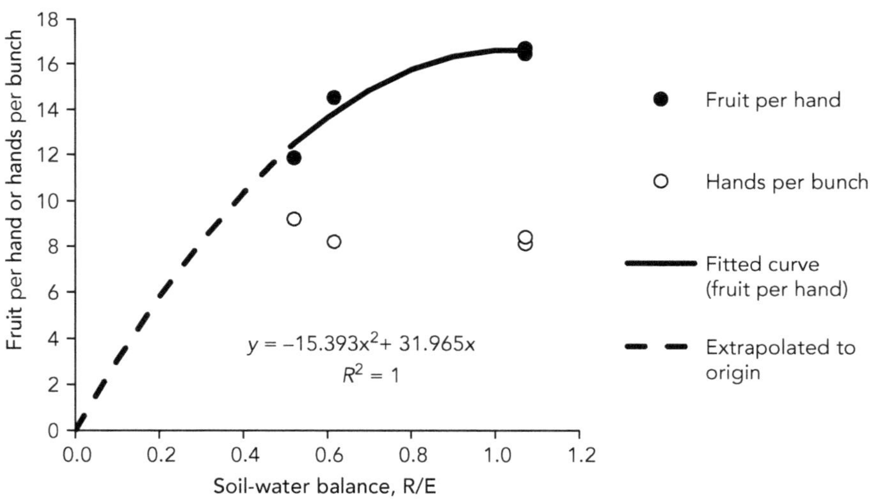

Figure 5 The relationship between an index of soil-water balance (rainfall/evaporation, Fortescue et al. 2011) when the bunch is forming (equivalent to the time taken for 4 new leaves to appear at the top of the pseudostem after the appearance of the 11th last leaf) and fruit number per hand, and hand number per bunch of 'Enyeru' (AAA-EAH) bananas. Data are for suckers emerging in each quarter of the year at Nyakyera, Uganda. The point of origin is assumed. Female hand numbers per bunch (8.5 ± 0.2) were not significantly different (P = 0.05) between quarters. Bunch and rainfall (R) data from Birabwa et al. (2010); evaporation (E) calculations based on temperature and equations from Linacre (1977).

For scenario 7, we explore two situations in the field: one in the tropics in Uganda where the temperature was stable throughout the year but water supply varied and the other in the subtropics of NSW in an irrigated plantation where water supply was adequate, but temperature and leaf area index varied throughout the year. In each case data on the female hand number per bunch and fruit number per hand were available, or could be calculated.

For scenario 7, the number of female hands per bunch may remain relatively constant across a year, but the number of flowers within the hands changes. An example of this is the seasonal variation in bunch components of the East African Highland banana 'Enyeru' (AAA-EAH). When grown at Nyakyera, Uganda, variation throughout the year in number of female hands per bunch (8.5 ± 0.2) was not significant ($P = 0.05$, Birabwa et al. 2010), but the number of flowers within hands differed significantly between seasons ($P = 0.05$) (Fig. 5). We found the number of fruit-forming flowers within those hands was highly correlated with the soil-water balance during bunch formation, where bunch formation was estimated retrospectively as the time taken for 4 new leaves to appear at the top of the pseudostem immediately after the 11th last leaf had appeared (Fig. 5). We assumed that a soil-water balance of 0.0 would mean no fruit initials were formed, acknowledging that this is an extrapolation of the data. In this case, an environmental factor has influenced the production of fruit-forming flowers in the cincinni, but not the number of nodes that become female hands.

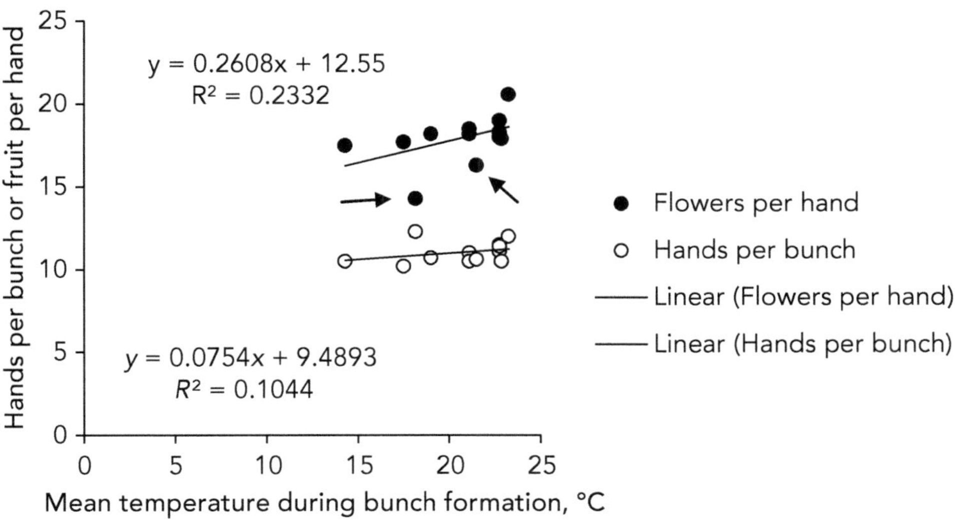

Figure 6 Relationship between temperature during bunch formation (the time it took for four leaves to appear at the top of the pseudostem after the 11th last leaf had appeared) and the number of female hands per bunch and the number of fruit per female hand for bunches emerging in each month of the year at Alstonville, NSW. Data from Turner (1981) on 'Williams' (AAA, Cavendish subgroup) at Alstonville, NSW (Lat 29°S). Arrows indicate bunches formed in spring when leaf area index was low and that subsequently emerged in summer (December–January).

In a well-watered situation in the subtropics, the monthly variation in female hand number per bunch and fruit number per hand shows separation when plotted against the temperature they experienced when the bunch was forming (Fig. 6). While female hand number per bunch shows no consistent trend with temperature, as bunches form under warmer weather the number of fruit per hand increased. This reflects scenario 7. There were two exceptions and these bunches (indicated by arrows, Fig. 6) have a lower number of fruit-forming flowers per hand than might be expected from their temperature alone. They formed during the spring when the leaf area index is about 2.0. Maximum leaf area index (4.5 to 5.0) occurs during late summer–autumn in subtropical NSW and this intercepts 90% of the incoming solar radiation (Turner 1972; 1990). A leaf area index of 2.0 could be expected to intercept 50% of the incoming solar radiation (Turner 1981) and so the reduced number of fruit-forming flowers in the hands at this time may be caused by reduced supply of carbohydrates to the developing bunch. The main factors contributing to seasonal change of leaf area index are the rate of appearance of new leaves (maximum in summer, minimum in winter) and the effect of cold weather on leaf senescence. The displacement between these two factors gives the spring minimum and late summer–autumn maximum in leaf area index.

7 Growth of the aerial true stem

The aerial true stem grows at the same time as the bunch is forming and is responsible for its elevation to the top of the pseudostem, resulting in bunch emergence. There is a need for the plant to balance resource allocation between bunch formation and then the need to move the bunch to the exterior of the plant to allow flowering and fruit filling to take place. In addition, the leaves within the pseudostem need to grow and emerge, parallel with the elongation of the aerial true stem. With this coordination of processes, perceptions of competition for resources between the developing bunch, growth of emerging leaves and elongation of the aerial true stem can easily arise.

The increasing rate of growth of the internodes of the aerial true stem within the pseudostem results in the point of attachment to the true stem of those leaves that formed before the transition to the floral phase occurring at ever-increasing height intervals (Fig. 7). The last 5–7 leaves are attached to the aerial true stem and are anatomically closer to the developing bunch than the older leaves. While the older leaves play a role in determining the number of fruit-forming flowers, the younger, last 5–7 leaves, play a significant role in the subsequent growth and development of the fruit and the male section of the inflorescence.

An examination of the data in Fig. 7 gives the appearance of an ever-increasing, or exponential, rate of movement of the aerial true stem up through the pseudostem leading to bunch emergence and flowering. Ganry (1977) carefully dissected the forming inflorescence and noted an increase in the rate of growth of the aerial true stem linked to the change from producing fruit-forming flowers to the formation of male flowers. In addition, the apex of the true stem changed from a wide to a narrow angle at the same time. This is reflected in the internode distance of successive leaves that provide a developmental measure of the change (Fig. 8, Skutch 1927).

If the data of Skutch (1927) in Fig. 7 are transformed to logs (Fig. 8), then two separate exponential phases of elongation of the aerial true stem can be discerned, akin to those

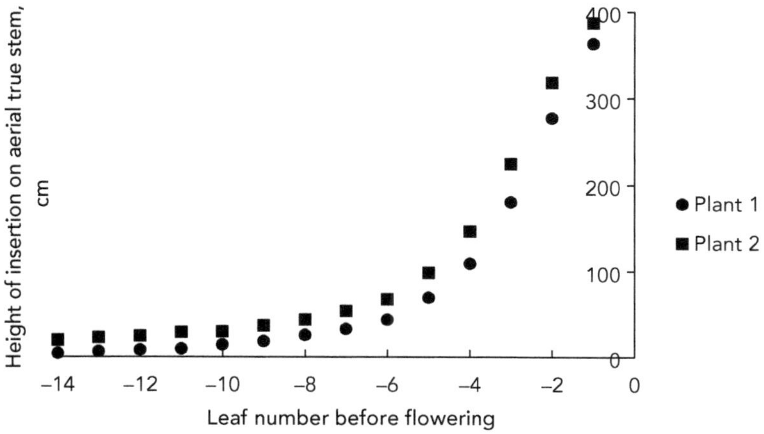

Figure 7 The relationship between height along the aerial true stem within the pseudostem and the points of insertion of the last 14 leaves produced before flowering (bunch emergence) on two plants of 'Gros Michel' (AAA) after the bunch has emerged. Data from Skutch (1927).

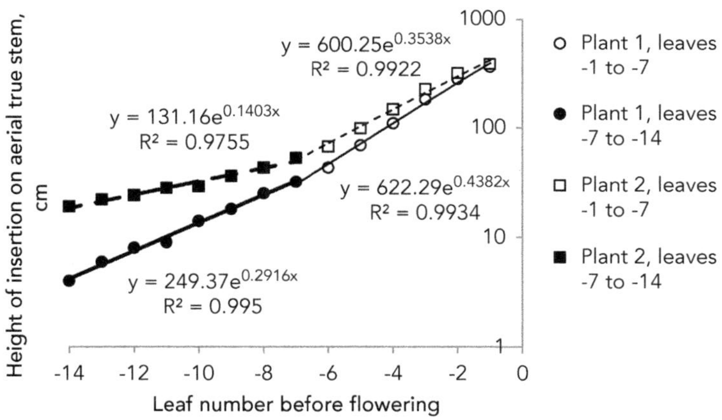

Figure 8 The relationship between the height of insertion of the last 14 leaves before flowering in two plants of 'Gros Michel' (AAA). Height data transformed to log to show exponential nature of growth and two identifiable phases. Data from Skutch (1927). See also Fig. 7.

reported by Ganry (1977). In plants 1 and 2 (Fig. 8) the relative growth rate (cm/cm/leaf, or $leaf^{-1}$) is 0.29 for plant 1 and 0.14 for plant 2. At about 7 or 8 leaves before bunch emergence, the relative rate increases by about twofold to 0.44 in plant 1 and 0.35 in plant 2. Ganry (1977) noted these changes at a similar stage, 8 leaves before bunch emergence in his study of the growing apex. Ganry (1977) also noted that, if the plant was not vigorous, then there was a delay in the start of the rapid rise of the aerial true stem, even though the number of fruit-bearing nodes was reduced.

The separation of the elongation of the aerial true stem into two distinct phases implies a change of resource allocation among the growing organs of the young inflorescence, the aerial true stem and the growing leaves that are yet to appear at the top of the pseudostem. Overall, this strategy may minimise competition for resources between stem elongation and the formation of the bunch, at least in terms of the number of hands of fruit and the number of fruit-forming flowers within each hand, these events being completed before the beginning of the rapid elongation phase. However, the particular event that causes the switch, whether genetic or resources or their combined effect, needs to be determined.

After bunch emergence, the true stem that forms the peduncle continues to grow, except for some Horn plantain (AAB) cultivars (De Langhe 1964). It is not clear how the apex of the male peduncle finishes once stem growth is complete – either by the formation of a final flower or by senescence and death. White (1928) observed a compound flower at the end of the male peduncle of a seeded banana that produced only six male bracts. How common this might be across the Musaceae is not known.

8 Case study: photosynthate availability and bunch formation in banana

8.1 Introduction

The number of female hands per bunch and fruit-forming flowers per hand are important components of bunch weight and therefore are of significance to banana producers. The production of hands per bunch (the thyrse) and flowers per hand (cincinni) are independent but coordinated processes, governed by genotype and allocation of resources to the developing bunch (see Section 6). We hypothesized that these components of bunch weight would be linked to availability of resources, in particular to the availability of photosynthate, at the time of bunch initiation and differentiation.

The availability of photosynthate within the banana plant is related to the leaf area of the plant and the amount of light it intercepts. Leaf area, and hence photosynthate availability, can be manipulated by defoliation of the plant.

To test our hypothesis we used data collected by Turner and Hunt (1987) in their defoliation experiments carried out in subtropical northern NSW, and examined the impact on floral formation of two levels of defoliation at different developmental stages in the banana.

In this case study we focused on the effect of defoliation at different leaf numbers before flowering (bunch emergence) on female hands per bunch.

8.2 Materials and methods

Experimental design, planting dates and treatments are described in detail in Turner and Hunt (1987). For this case study we used data from the three planting dates in spring (November), summer (January) and autumn (March) for the plant crop only. Defoliation was applied at four growth stages: when the plants had produced 5, 15, 25 or 35 leaves. Defoliation treatments were control (no leaves removed), half of each live leaf removed and all leaves removed.

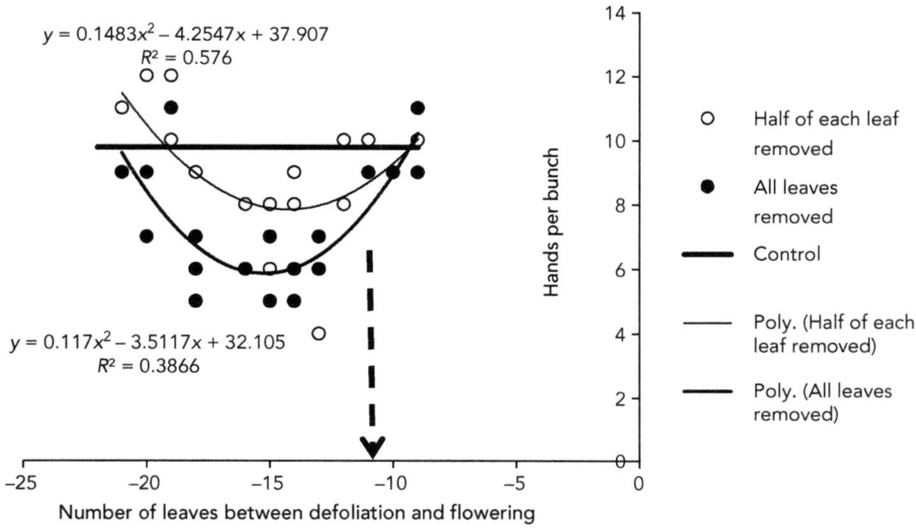

Figure 9 The number of leaves between defoliation and flowering in relation to the number of female hands per bunch. Flowering is at leaf 0 and negative values indicate leaves before flowering. Vertical dashed arrow indicates bunch initiation at −11 leaves before bunch emergence (flowering). Data are for those treatments where from 21 to 9 leaves emerged from the pseudostem between defoliation and flowering. Data points are for individual plants. Based on Student's *t*-tests, differences in number of hands per bunch >1.0 are significant at $P = 0.05$.

Using data from all three planting dates, we examined the relationship between the number of leaves emerging between defoliation and flowering and the number of female hands per bunch for the three treatments. On the basis of preliminary analysis, we focused on those treatments in which from 21 to 9 leaves emerged from the pseudostem between defoliation and flowering. For a hand to be counted as female, it had to have at least half of its flowers forming fruit.

Curve-fitting, quadratic equations, coefficients of determination (r^2) and Student's *t*-tests were generated using Microsoft Excel Version 14.6.6.

8.3 Results

Full-leaf defoliation at between 20 and 12 leaves before flowering (leaf −20 to leaf −12) reduced the number of female hands per bunch (Fig. 9). A similar but less pronounced pattern was observed for half-leaf defoliation, with defoliation at between 18 and 13 leaves (leaf −18 to leaf −13) reducing the number of female hands per bunch. Beyond the range from leaf –20 to leaf –9, defoliation had no effect on hand number per bunch. In both defoliation treatments, the minimum number of female hands per bunch occurred with defoliation at around 14–15 leaves before flowering (leaf −14 to leaf −15; Table 3). In the full-leaf defoliation treatment, the absolute reduction from the control in female hand number was about double that of the half-leaf defoliation treatment (Table 3).

Table 3 Features of the fitted curves in Fig. 9 illustrating changes in female hands per bunch as a consequence of defoliation from 21 to 9 leaves before flowering in banana

Features of fitted curves	Half of each leaf removed	All leaves removed
Minimum female hand number per bunch	7.4	5.8
Leaf number before flowering at minimum hand number per bunch	−14.3	−15.0
Absolute reduction in female hand number from control	2.4	4.0
Proportional reduction in female hand number from control (%)	24.4	41.1
Minimum leaf number before flowering where curve meets controls	−10.3	−9.1
Maximum leaf number before flowering when curve meets controls	−18.4	−20.9
Female hand number per bunch of controls	9.8	9.8

8.4 Discussion

In banana, at the time of bunch initiation, about 11 leaves were yet to emerge from the pseudostem (leaf −11, Fig. 9). At leaf −9, the formation at the apex of the female (fruit-bearing) hands is observable microscopically, with this process completed by about leaf −7 or leaf −6 (Section 4.2 and Table 1). We focused on those treatments in which from 21 to 9 leaves emerged from the pseudostem between defoliation and flowering (leaf −21 to leaf −9, Fig. 9), a range covering the mid-vegetative stage and the transition of the apex from the mid-vegetative to the floral stage of development.

Full-leaf defoliation reduced the hand number compared with the control (Fig. 9, Table 3) when it was applied from 20 to 12 leaves before flowering (bunch emergence), with a similar pattern observed for the half-leaf defoliation treatment. While this is consistent with our hypothesis that photosynthate supply influences bunch formation in banana, the impact of defoliation was much earlier in plant development than expected. Twenty to twelve leaves before flowering was at the mid-vegetative stage at the growing apex and before bunch initiation had occurred. Defoliation during bunch formation (from 11 to 7 leaves before flowering), surprisingly, had little effect on female hand number per bunch.

It appears that photosynthate availability can limit bunch formation in banana. Reducing the supply of current photosynthate by defoliation reduces female hand number per bunch, but only when defoliation occurs in the few weeks or months before bunch initiation occurs. For this to happen, it is likely that the carbon fixed by these leaves is stored in the rhizome and is subsequently used for bunch differentiation. This mechanism may buffer the early development of the bunch from sudden variations in carbon supply associated with daily and weekly changes in weather and edaphic conditions. A consequence of this is that stored carbon may be reduced by an extended drop in photosynthate supply caused by low leaf area index in the late winter and spring of subtropical regions (e.g. see Turner 1990).

Questions still remain. For example:

- Why is the number of female hands per bunch at a minimum when plants are defoliated at 14–15 leaves from flowering, that is some 3–4 leaves before bunch initiation, and irrespective of whether half of each leaf or all leaves are removed?
- How much might carbohydrates from the suckers, or indeed reallocation of resources to the suckers, contribute to the response?

Knowledge of the processes of bunch formation, particularly the independent but coordinated processes forming the female hands and fruit, are vital to improving yield. Manipulation of the system to test hypotheses provides a strong tool to allow us to understand better the relationships between processes involved in bunch formation, and to manage their interactions with environment, agronomic techniques and cultivars.

9 Conclusion

Efforts to make banana production sustainable have focused on dealing with pests and diseases that often require chemical intervention to manage them. New hybrid genotypes from banana breeding schemes have variable levels of disease and pest resistance, but are often less successful for yield and fruit quality, compared with standard cultivars (Dorel et al. 2016). Knowledge of the processes of bunch formation can provide a basis for genetic improvement in productivity, independent of those efforts focused on creating genotypes resistant to pests and diseases.

Apart from manipulating sucker set (Dorel et al. 2016), there is scope to manipulate, at a genetic level, the responsiveness of the juvenile phase to cool temperatures and the sensitivity of the mid-vegetative phase to photoperiod. Thus, there is room to manipulate the banana life cycle and match it better with the environment.

The sequence of events during bunch formation is best observed with reference to retrospective counting of leaves, starting at bunch emergence. Then, bunch formation begins at leaf −11 (11 leaves before bunch emergence), and formation of the fruit-bearing hands can be seen microscopically at leaf −9 and is completed at about leaf −7 or −6. Thus, about 4 leaves may be produced during the formation of those parts of the bunch of interest to commerce. Rapid elongation of the aerial true stem, and to a lesser extent the peduncle then begins, and male flowers are formed within the nodes on the peduncle. It is unclear which comes first, the elongation of the aerial stem, resulting in reduced resources for flower formation, or the change from fruit-forming flowers to non-fruit-forming flowers.

The process of hand formation (the thyrse) and flower formation within those hands (cincinnus) are independent, but coordinated processes. Plenty of hands are produced by the thyrse, but it is the activity of the cincinnus within each hand that drives the number of fruit-forming flowers in each female hand, and indeed, the number of female hands. Genotype and environment affect this coordination. There is evidence that environment affects cincinnus activity within hands more than between female hands (see Section 6.2, Figs. 5 and 6).

As the bunch forms, there is a considerable change in meristem activity within the apex and its surrounds. There is evidence that this slows the rate of appearance of leaves emerging from the top of the pseudostem at that time. However, the most useful external indicator is the number of leaves yet to appear before bunching. Frustrating as this is, since the number of leaves produced can only be known in retrospect, experience with cultivars and conditions can provide a workable estimate for deciding on management options.

Most of the work so far conducted on bunch formation has been descriptive. This has given us a firm foundation for our knowledge. In the future, manipulative experiments (see Section 8) are needed to explore the various processes involved in bunch formation and quantify relationships between environment, agronomic techniques and cultivars. Experiments reporting yield in bananas should include data on female hands per bunch and the total number of fruit per bunch to provide base data for future evaluation of thyrse–cincinnus relationships.

10 Where to look for further information

For further information about bunch development in banana, please see the following references: Fahn (1953), Barker and Steward (1962b), Ram et al. (1962) and Ganry (1977). For information about the evolution and domestication of bananas, please see Simmonds (1962), De Langhe et al. (2009) and Janssens et al. (2016). For recent information about the impact of environment on the early phases of banana development, please see Fortescue et al. (2011) and Turner et al. (2016b).

11 References

Alexandrowicz, L. (1955). Étude du developpement de l'inflorescence du bananier. *IFAC Annales* 9, 1–32.

Argent, G. C. G. (1976). The wild bananas of Papua New Guinea. *Royal Botanical Garden Edinburgh Notes* 35, 77–114.

Barker, W. G. and Steward, F. C. (1962a). Growth and development of the banana plant. 1. The growing regions of the vegetative shoot. *Annals of Botany* 26, 389–411.

Barker, W. G. and Steward, F. C. (1962b). Growth and development of the banana plant. 2. The transition from the vegetative to the floral shoot in *Musa acuminata* cv Gros Michel. *Annals of Botany* 26, 413–23.

Birabwa, R., van Asten, P. J. A., Alou, I. N. and Taulya, G. (2010). Got matooke (*Musa* spp.) for Christmas? *Acta Horticulturae* 879, 113–21.

Buys, M. H. and Hilger, H. H. (2003). Boraginaceae cymes are exclusively scorpioid and not helicoid. *Taxon* 52, 719–24.

Champion, J. (1967). *Notes et Documents sur les Bananiers et Leur Cultures. 1. Botanique et Genetique des Bananiers*. IFAC, Paris.

Cheesman, E. E. (1947). Classification of the bananas. II. The Genus *Musa* L. *Kew Bulletin* 1947 (2), 106–17.

De Langhe, E. (1964). Influence de la parthénocarpie sur la dégénérescence florale chez le bananier 3. La signification de la relation parthénocarpie/dégénérescence. *Fruits* 19, 311–22.

De Langhe, E., Vrydaghs, L., de Maret, P., Perrier, X. and Denham, T. P. (2009). Why bananas matter: an introduction to the history of banana domestication. *Ethnobotany Research and Applications* 7, 165–77.

Devos, P. (1985). The location of lateral buds of banana (*Musa* sp.) clarified by the discovery of two new types of adventitious buds within the AAB plantain group. *Fruits* 40, 471–4.

Dorel, M., Damour, G., Leclerc, N., Lakhia, S., Ricci, S., Vingadassalon, F. and Salmon, F. (2016). Parent plant vs sucker – how can competition for photoassimilate allocation and light acquisition be managed in new banana hybrids? *Field Crops Research* 198, 70–9.

Endress, P. K. (2010). Disentangling confusions in inflorescence morphology: Patterns and diversity of reproductive shoot ramification in angiosperms. *Journal of Systematics and Evolution* 48, 225–39.

Fahn, A. (1953). The origin of the banana inflorescence. *Kew Bulletin* 1953 (3), 299–306.

Fahn, A., Stoler, S. and First, T. (1963). Vegetative shoot apex in banana and zonal changes as it becomes reproductive. *Botanical Gazette* 124, 246–50.

Fisher, J. B. (1978). Leaf-opposed buds in *Musa*: their development and a comparison with allied monocotyledons. *American Journal of Botany* 65, 784–91.

Fortescue, J. A., Turner, D. W. and Romero, R. (2011). Evidence that banana (*Musa* spp.), a tropical monocotyledon, has a facultative long-day response to photoperiod. *Functional Plant Biology* 38, 867–78.

Ganry, J. (1977). Détermination '*in situ*' du stade de transition entre la phase vegetative et la phase florale chez le bananier, utilisant le '*coefficient de vitesse de croissance des feuilles*'. Essai d'interpretation de quelques processus de développement durant la période florale. *Fruits* 32, 373–86.

Guitián, J., Medrano, M. and Oti, J. E. (2004). Variation in floral sex allocation in *Polygonatum odoratum* (Liliaceae). *Annals of Botany* 94, 433–40.

Häkkinen, M. (2013). Reappraisal of sectional taxonomy in *Musa* (*Musaceae*). *Taxon* 62, 809–13.

Häkkinen, M., Hong, W. and Ge, X-J. (2008). *Musa itinerans* (Musaceae) and its intraspecific taxa in China. *Novon* 18, 50–60.

Israeli, Y. and Blumenfeld, A. (1986). *Musa* In Y. Halevy (ed.), *Handbook of Flowering*. Vol. 3, pp. 390–409. CRC Press, Boca Raton.

Israeli, Y., Lahav, E. and Nameri, N. (1986). The effect of salinity and sodium adsorption ratio in the irrigation water, on growth and productivity of bananas under drip irrigation conditions. *Fruits* 41, 297–302.

Janssens, S. B., Vandelook, F., De Langhe, E., Verstraete, B., Smets, E., Vandenhouwe, I. and Swennen, R. (2016). Evolutionary dynamics and biogeography of Musaceae reveal a correlation between the diversification of the banana family and the geological and climatic history of Southeast Asia. *New Phytologist* 210, 1453–65.

Kamran, A., Iqbal, M. and Spaner, D. (2014). Flowering time in wheat (*Triticum aestivum* L.): a key factor for global adaptability. *Euphytica* 197, 1–26.

Kaul, V. and Koul, A. K. (2008). Floral phenology in relation to pollination and reproductive output in *Commelina caroliniana* (Commelinaceae). *Australian Journal of Botany* 56, 59–66.

Kirchoff, B. K. (1992). Ovary structure and anatomy in the Heliconiaceae and Musaceae (Zingiberales). *Canadian Journal of Botany* 70, 2490–508.

Kirchoff, B. K. (2017). Inflorescence and flower development in *Musa velutina* H. Wendl. & Drude (Musaceae), with a consideration of developmental variability, restricted phyllotactic direction, and hand initiation. *International Journal of Plant Sciences* 178, 259–72.

Kirchoff, B. K., Lagomarsino, L. P., Newman, W. H., Bartlett, M. E. and Specht, C. D. (2009). Early floral development of *Heliconia latispatha* (Heliconiacae), a key taxon for understanding the evolution of flower development in the Zingiberales. *American Journal of Botany* 96, 580–93.

Lassoudière, A. (1978). Quelques aspects de la croissance et du développement du bananier Poyo en Côte d'Ivoire. 4. L'inflorescence. *Fruits* 33, 457–91.

Linacre, E. T. (1977). A simple formula for estimating evaporation rates, using temperature data alone. *Agricultural Meteorology* 18, 409–24.

Moncur, M. W. (1988). *Floral Development of Tropical and Subtropical Fruit and Nut Species. An atlas of scanning electron micrographs*. Natural Resources Series No 8. Division of Water and Land Resources. CSIRO, Melbourne. 181pp.

Ram, H. Y. M., Ram, M. and Steward, F. C. (1962). Growth and development of the banana plant. 3a. The origin of the inflorescence and the development of the flowers. 3b. The structure and development of the fruit. *Annals of Botany* 26, 657–73.

Sakhanokho, H. F. and Rajasekaran, K. (2010). Pollen biology of ornamental ginger (*Hedychium* spp. J, Koenig). *Scientia Horticulturae* 125, 129–35.

Simmonds, N. W. (1962). *The Evolution of the Bananas*. Longmans, London.

Skutch, A. F. (1927). Anatomy of leaf of banana *Musa sapientum* L var. Hort. Gros Michel. *Botanical Gazette* 84, 337–91.

Stover, R. H. (1979). Pseudostem growth, leaf production and flower initiation in the 'Grand Nain' banana. Bulletin 8, Tropical Agriculture Research Services, La Lima, Honduras.

Summerville, W. A. T. (1944). Studies on nutrition as qualified by development in *Musa cavendishii* Lamb. *The Queensland Journal of Agricultural Science* 1, 1–127.

Swennen, R. and Vuylsteke, D. (1987). Morphological taxonomy of plantain (*Musa* cultivars AAB) in West Africa. In G. J. Persley and E. A. De Langhe (eds), *Banana and Plantain Breeding Strategies: Proceedings of an International Workshop Held at Cairns, Australia*. ACIAR Proceedings No 21. Canberra, pp. 165–71.

Turner, D. W. (1972). Banana plant growth. 2. Dry matter production, leaf area and growth analysis. *Australian Journal of Experimental Agriculture and Animal Husbandry* 12, 216–24.

Turner, D. W. (1980). Some factors related to yield components of bananas in relation to sampling to assess nutrient status. *Fruits* 35, 19–23.

Turner, D. W. (1981). Crop physiology of bananas – *Quo vadis*? *Madras Agricultural Journal* 68, 78–84.

Turner, D. W. (1990). Modelling demand for nitrogen in the banana. *Acta Horticulturae* 275, 497–503.

Turner, D. W., Fortescue, J. A. and Daniells, J. W. (2016a). Banana agronomy – can unravelling the *Musa* genome help? *Acta Horticulturae* 1114, 245–60.

Turner, D. W., Fortescue, J. A., Ocimati, W. and Blomme, G. (2016b). Plantain cultivars (*Musa* spp. AAB) grown at different altitudes demonstrate cool temperature and photoperiod responses relevant to genetic improvement. *Field Crops Research* 194, 103–11.

Turner, D. W. and Hunt, N. (1984). *Growth, Yield and Leaf Nutrient Composition of 30 Banana Varieties in Subtropical New South Wales*. Technical Bulletin 31. Department of Agriculture, NSW, Sydney.

Turner, D. W. and Hunt, N. (1987). Planting date and defoliation influence the time of harvest of bananas. *Scientia Horticulturae* 32, 233–48.

White, P. R. (1928). Studies on banana. An investigation of the floral morphology and cytology of certain types of the genus *Musa* L. *Zeitschrift. Fuer. Zellforschung und Mikroskopische Anatomie Bd*. 7, 673–733.

Chapter 6

Banana plant propagation methods

Sharon D. Hamill, Department of Agriculture and Fisheries, Australia

1 Introduction
2 Background to banana evolution
3 Banana morphology and vegetative propagation
4 Bits and suckers
5 Propagation using conventional planting material
6 Banana tissue culture
7 Liquid culture systems
8 Banana off-types and how to reduce them
9 Future trends and conclusion
10 Where to look for further information
11 References

1 Introduction

The banana and plantain we eat today are sterile cultivars that evolved from the wild seeded diploid parents thousands of years ago (De Langhe et al., 2009; Perrier et al., 2011). Banana has been cultivated by mankind in early agricultural practices due in part to the fact that banana is vegetatively propagated, with evidence of domestic cultivation 7000 years ago in Papua New Guinea (Denham et al., 2003). When a superior selection was identified the plant was easily moved and propagated clonally, simply by taking and planting a sucker. For centuries banana and plantains have been one of the most widely consumed fruits, both as a staple food and as one of the main fruit crops of world trade.

The ease of banana propagation by taking a sucker has allowed farmers to multiply their banana selections and share them. The banana sucker is a resilient form of planting material as the growing point is protected within the sucker which contains a corm with a starchy store that allows the plant to readily re-establish. Banana suckers have been easily moved around the world to spread the crop but have also been the vector carrying banana pests and diseases with it (Kumar et al., 2015; Ploetz, 2015).

The way that the banana crop is propagated depends on how the crop is to be used and the technical tools available to the grower. Traditionally small shareholders and subsistence farmers who grow the crop for their own use will select suckers to plant in their farms.

http://dx.doi.org/10.19103/AS.2017.0020.12
Published by Burleigh Dodds Science Publishing Limited, 2018.

Larger agri-businesses that require best economic production practices and uniformity for farm efficiency and pest- and disease-free planting material will use tissue culture plants.

This chapter briefly describes the basic anatomy of the banana plant required to understand the part of the plant used for propagation. Also described are the various practical aspects of banana propagation using either conventional planting material or tissue culture plants and the benefits and disadvantages of each method.

2 Background to banana evolution

The origin of bananas and plantains is Southeastern Asia and the Pacific regions with one of the earliest origins of domestication reported to be Papua New Guinea (Denham et al., 2003; Sardos et al., 2016). There are hundreds of different cultivars as a result of natural crossing from the wild diploid parents *Musa acuminata;* and, later human domestication and movement of banana cultivars contributed further to hybridisation when man moved plants to different areas and crosses occurred between *M. acuminata* and *M. balbisiana* (Kennedy, 2009; Perrier et al., 2009, 2011). In the wild cultivars the fruit contains hard viable seeds that are readily dispersed over large distances by animals such as rodents, birds and bats that eat the fruit and their seeds (Meng et al., 2012). In this way progeny from wild seeded parents are spread over large areas. Some individual plants from the natural crosses were found to be seedless as they were parthenocarpic, female sterile and triploid. These edible banana cultivars were found by man and propagated clonally using suckers and in this way the selected cultivars could be maintained in a stable way and moved across regions (Kennedy, 2009; Perrier et al., 2011; Blomme et al., 2013). Fertility in banana is used in some banana breeding programmes and the fertile seed is collected and germinated usually by embryo culture (Pillay and Tenkouano, 2011). However, in this chapter the focus of discussion is restricted to the sterile edible banana cultivars that are vegetatively propagated.

3 Banana morphology and vegetative propagation

Banana and plantain belong to the genus *Musa*; they are monocots and include some of the world's largest herbaceous evergreen perennial plants. Banana plants have what is commonly called an underground corm, which is a true stem from which suckers are produced. The banana corm is technically a tuberous rhizome and the developing shoots initially grow horizontally from the rhizome before turning upwards to emerge from the soil. The banana plant has a trunk called a pseudostem which is made of overlapping leaf sheaths with successive leaves emerging in a spiral development arising from the corm to form the cylindrical trunk. There is a central meristem within the leaf sheaths that matures to initiate an inflorescence. At the time of maturity the meristem stops producing leaves and elongates to move through the pseudostem where it emerges as the bell which develops into the banana bunch (Stover and Simmonds 1987). When the bunch matures it is harvested and the mother plant dies and the suckers produce the next cycle of fruit.

The banana rhizome remains completely underground and numerous suckers are produced from the rhizome (corm) as lateral shoots during the life of the parent plant in sequential order (De Langhe 1961a). Each individual growing bud arising from the rhizome develops into a sucker and contains a single terminal meristem. The suckers develop from buds formed on the corm and each sucker also has buds on its corm, which will develop into suckers when they grow into plants. These buds may be called 'eyes' or 'bits' and can be removed and used to grow a banana plant. The banana plant propagates itself by suckers, which in commercial plantings are removed so that only a daughter plant is left to grow as the main plant for the next cycle.

4 Bits and suckers

There are two morphological types of suckers produced from the rhizome, known as 'sword' and 'water' suckers. The 'sword' suckers form from axillary buds located deeper in the soil on the parent rhizome and have narrow, upright and unformed leaves and a strong connection to the mother rhizome, which is important for initial sucker development, due to which they have well-developed rhizomes (Eckstein and Robinson, 1995) (Fig. 1).

The 'water' suckers originate from buds towards the top of the parent rhizome near the soil surface (Fig. 2) and do not have a strong connection to the parent rhizome and consequently are easy to dig out from the soil surface as they do not have a well-developed rhizome of their own. They can be distinguished by the broad and soft leaves that they have developed as they need to compensate for the lack of rhizome support from the parent (Robinson and Galan Sáuco, 2010). 'Water' suckers are often produced in the first two months of the banana plant's life; they are considered to be inferior and should not be used in banana propagation as they do not develop into a strong robust plant.

Suckers grow around the mother plant and, to manage the plantation and achieve a better crop yield, are removed. Usually when the main plant is nearing maturity one to two suckers are left attached to the mother plant, and are called 'followers', to form the next crop. Usually a strong sword sucker (daughter sucker) is left in the best position that roughly falls in line with the current row in the field and this sucker is usually left when the mother plant is close to bunching. Suckers should be selected to have roughly the same size and on the same side of the mother plants to maintain the row line and crop uniformity. The removal of unwanted suckers is called 'desuckering' and helps in the growth of remaining suckers and the developing bunch by reducing competition for water and nutrients. The first planted crop is called the 'plant' crop and the next crop that follows established from suckers produced by the 'plant crop' are called 'ratoon' crops.

Small growers, subsistence farmers and some organic growers use bits and suckers to establish and further propagate their banana crops. Bits and suckers are called 'conventional' planting material. Most countries with large commercial banana production now rely exclusively on tissue culture plants. However, in Australia larger growers use either conventional planting material or tissue culture plants or a combination of both. Bits and suckers present a high risk of introducing pests and diseases, while tissue culture plants, provided they are produced in a controlled way that prevents entry of pests and disease, provide a source of disease-free planting material. Planting material should never be collected from plants that have viruses or from land that has Panama disease (also known as Fusarium wilt) caused by *Fusarium oxysporum* f.sp. *cubense* (FOC) of which there are 4 races. Once Panama disease enters the soil the fungus will continue to survive for decades. Use of infected suckers or movement of infected soil is the main way in which this and other banana diseases have been rapidly spreading across the world (Bloome et al., 2013; Kumar et al., 2015).

Figure 1 Sword sucker (photo by Roger Goebel).

4.1 Advantages of using bits and suckers

Bits and suckers are readily available to all growers. Growers only need to plan for harvesting bits and suckers and can access the planting material all year round. The carbohydrate store in well-developed rhizome makes bits and suckers a durable planting material that can be planted without irrigation. Bits and suckers are suitable for smallholders and farmers who wish to use their own banana/plantain selections, and can use their own farm labour to dig the suckers that have a very low level of off-types and can be taken from mother plants shown to be of good quality. Bits and suckers can be trimmed and stored in the short term without the need for maintenance.

Published by Burleigh Dodds Science Publishing Limited, 2018.

Figure 2 Water sucker (photo by Roger Goebel).

4.2 Disadvantages of using bits and suckers

Bits and suckers are the main ways through which pests and diseases spread (Bloome et al., 2013). Pests such as nematodes and the banana weevil borer are carried in the plant tissue, and while nematode levels can be reduced by paring back the surface of the rhizome combined with hot water treatment, the nematode population is reduced and not entirely eliminated. Banana weevil borers tunnel deep into the rhizome and sometimes may not be easily detected and cannot be removed from the infested rhizome. Viruses such as Banana bunchy top virus (BBTV) and persistent soil borne diseases such as FOC are moved in infected plants and in contaminated soil adhering to the bits and suckers. Panama disease commonly spreads from farm to farm in infected or contaminated bits and suckers and on soil adhering to farm workers' footwear or vehicles. Banana diseases are

Published by Burleigh Dodds Science Publishing Limited, 2018.

the biggest threat to ongoing banana production, and the use of bits and suckers in areas with diseases such as Panama should be avoided.

Bits and suckers are bulky and heavy, which means that it is sometimes hard for the smallholder to transport large quantities of planting material which should be planted shortly after harvest. They may rot or fail to grow if planted into wet ground and this is influenced by the season. It is important to source bits and suckers that are as uniform as possible as they all do not shoot at the same time and some will fail to grow. This means that bits and suckers are not going to produce the uniform plants needed for efficient management in large commercial banana plantations. Although lack of uniformity means varied field management practices and different bunching times this may be suitable for a smallholder.

5 Propagation using conventional planting material

There are several ways to produce conventional planting material. Often the bits and suckers will be sourced from an older banana block that is being removed and suckers can be allowed to grow for several months before harvest. It is easier to dig up the bits and suckers from plants that have recently bunched. Digging the bits and suckers is easier in moist soil and they can be harvested using crowbars, mattocks and shovels. Backhoe tractors with special rippers or cutter bars can also be used. To prepare the planting pieces, as much soil as possible needs to be removed. A medium/high-pressure hose is often used to remove soil as long as there is no damage to the bud. When clean the suckers are separated from the butt using sharp de-suckering tools. Approximately 12–15 pieces can be produced per plant and they usually weigh between 0.5 and 1.5 kg each and each must have a prominent bud. Planting material is pared back to remove all roots and leaves and soil from the corm (Fig. 3). Planting material taken from an old block may have nematodes and banana weevil borers and must be inspected to ensure it is free from serious diseases and be discarded if infected. Bits and suckers taken from nematode-infested blocks are traditionally treated using a hot water bath where pieces need to be trimmed to a minimum size of 1 kg and immersed in water at 53–55 °C for 20 min (Colbran, 1964). Coyne et al. (2010) also described a simple method of treating nematode and banana weevil-infected planting material by placing trimmed bits and suckers into boiling water for 20–30 sec. Pared bits and suckers can be dipped in a nematicide registered for banana, but chemicals used to control nematodes are expensive and toxic and must be handled carefully. Stirling and Pattison (2008) described integrated pest management of nematodes in banana that includes crop rotation with non-host crop, nematode monitoring, using pest- and disease-free planting material and if required, limited use of nematicides.

Harvested and trimmed bits and suckers need to be grouped into uniform size to ensure uniform plants in the field and, after the cut surfaces have suberised (usually 1–2 d) they should be planted as soon as possible into moist soil at a depth of approximately 30 cm. Bits and suckers do not require irrigation at planting and depending on rainfall will only require irrigation after the plants have emerged. This provides benefits to growers who do not have access to constant water supply. Irrigation water should be clean to prevent the introduction of pests and diseases.

Field blocks specifically set up for the purpose of supplying good-quality bits and suckers can produce cleaner planting material than material sourced from farms.

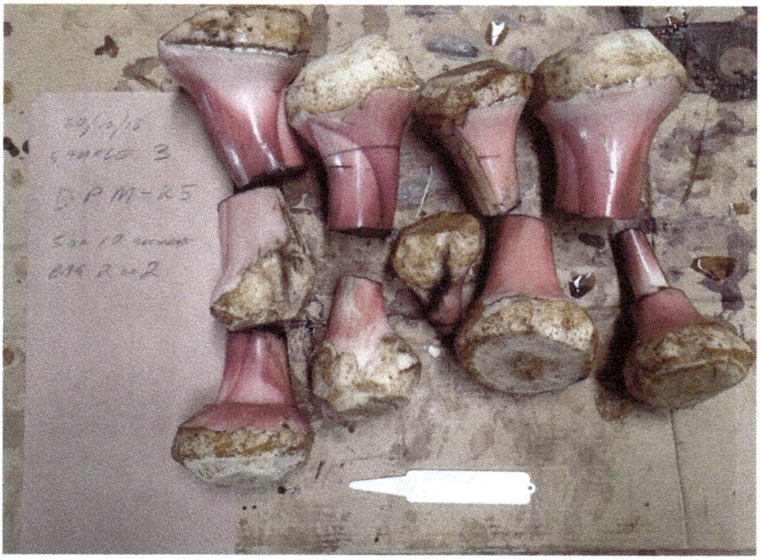

Figure 3 Suckers recently dug out and prepared and left to suberise prior to packing and sending (photo by Roger Goebel).

The field nursery should be established on ground that has not previously grown banana and does not receive soil and water run-off from other banana farms. The site must not have a history of nematode or Panama disease. Initial planting should be with tissue culture plants produced in an accredited scheme so that no pests or diseases are introduced. When using tissue culture the plants should be grown to bunching to ensure any inferior plants are removed before harvesting the suckers. Field nursery production uses different practices than those used for crop production. Since the aim is to produce many pieces of planting material the planting density can be higher and there is no need to remove suckers. To encourage good-quality sucker production the central growing point of the parent plants can be removed by cutting off the emerging flower to direct more reserves to the corm and pseudostem to feed the suckers and to open the canopy to provide more light (Robinson and Galan Sáuco, 2010). There have been a range of methods described to increase sucker production such as removing apical dominance; this is done by allowing the mother plants to grow until there is sufficient corm size and energy store and then destroying the central growing point. Other methods used to increase sucker production include cutting back after bunching, with various ways of cutting the parent pseudostem (Daniells and O'Farrell, 1987; Rodriguez et al., 2006) and decapitation of the parent at various ages and heights combined with elimination of the central growing point (De Langhe, 1961b). *In vivo* macropropagation (Njukwe et al., 2005) has the potential to produce 50–60 shoots per corm in 4–5 months and has been shown to be enhanced by application of cytokinins such as Benzylaminopurine or Thidiazuron (TDZ) (Msogoya and Mwakisitu, 2014; Thiemele et al., 2015). Bhende and Kurien (2015) provided a useful summary of some of the cultural practices developed for sucker production and the ProMusa website also provides additional information on planting material.

Published by Burleigh Dodds Science Publishing Limited, 2018.

Plants growing under optimal conditions will produce the best-quality suckers, and there are many other factors that influence sucker production. Factors such as the size of mother plant corm, amount of water and nutrients, cultivar being propagated, temperature, time of year planted, light, damage during harvest, pests or diseases and plant spacing will all influence sucker growth. These aspects will be discussed in following chapters.

6 Banana tissue culture

Larger commercial banana plantations more commonly use tissue culture plantlets produced using micropropagation. Micropropagation is a tissue culture method that directly induces explants to produce many shoots. Banana micropropagation was first described for 'Cavendish' by Ma and Shii (1972), and the first large plantations were established using tissue culture banana plants in Taiwan and Jamaica in the early 1980s (Ogelsby and Griffiths, 1986). Now there are arguably more banana plants being commercially micropropagated than any other fruit crop. Micropropagated banana plants have been readily adopted in modern farming systems as they provide multiple advantages compared to conventional material. While tissue culture banana plants are predominantly used by larger commercial banana farms the use of tissue culture plants by smallholder farms has also been shown to improve household income and food security (Kabunga et al., 2014).

Banana produced by tissue culture from virus-free plants provides an opportunity to start a plantation using pest- and disease-free plants. It is important that the mother plants used to initiate banana tissue cultures are free from viruses because viruses (such as BBTV) can be readily transmitted during micropropagation to produce virus-infected tissue culture plants (Thomas et al., 1995). In Australia, tissue culture banana plants may only be approved for use when produced in an accredited Quality Banana Approved Nursery (QBAN) tissue culture laboratory and must be initiated from suckers that have been sourced from a Panama disease and virus-free area and tested and found to be virus free. These banana tissue culture plants must only be grown in a (QBAN) nursery in an area free from banana pests and diseases. Tissue culture plants produced from banana plants tested to be virus-free is an important criterion required to safely move banana germplasm globally as described in *MusaNet Technical Guidelines for the Safe Movement of Musa Germplasm* (Thomas, 2015).

The world banana trade is dominated by Cavendish-type cultivars which are triploid AAA genome. Banana cultivars that contain *M. balbisiana* B genome, such as plantain, and cultivars that have been bred using parents with the B genome may contain several endogenous sequences of Banana streak virus (BSV) which are plant bacilliform pararetroviruses. It has been demonstrated that tissue culture triggers the activation of the BSV due to abiotic stress during the process (Côte et al., 2010).

Banana tissue culture plants provide a very uniform source of planting material and in the plant crop outperform conventional plants with regard to yield and cycle time, which will continue into the first ratoon (Eckstein and Robinson, 1995; Robinson and Galán Saúco 2010). Uniformity of plants produces improved farm efficiencies and allows for timing of bunch production. Tissue culture plants may produce a larger number of suckers in the field (Drew and Smith 1990; Israeli et al., 1995) which is useful in a nursery block but adds to labour costs in large-scale farms.

6.1 Advantages of banana tissue culture

Tissue culture banana plantlets are pest and disease free, if initiated from suckers that are also free from pests and diseases including viruses. Tissue culture plants are small, and large numbers of them can be cheaply transported over large distances and moved safely in enclosed containers across quarantine zones. Banana tissue culture plants are superior to conventional planting material as they can produce a larger bunch faster in the plant crop with increased productivity that may follow into first ratoon. Banana tissue culture plants can be extremely uniform if sorted in the nursery, which provides a high level of farm efficiency allowing farmers to manage and harvest all plants at the same time. The bunch date can be targeted using banana tissue culture plants. The plants have a well-established root system and with irrigation will have 100% survival in the field. Superior selections of banana can be rapidly multiplied using tissue culture for fast adoption by growers. Banana tissue culture provides a way to establish new farms free from banana pests and diseases, provided the farm is established on land that has not previously grown banana and does not already have banana pests or pathogens. Banana tissue culture is used in germplasm storage to safely maintain cultivars free from pests and diseases and adverse weather events.

6.2 Disadvantages of banana tissue culture

Banana tissue culture requires expensive equipment in production and requires skilled staff. The plants can be expensive to purchase and must be ordered a year in advance, so planning is more of a requirement with tissue culture bananas than when sourcing suckers. Tissue culture banana plants can carry viruses if the plants are not produced in a way to eliminate them. Clean tissue culture banana have been shown to be more susceptible to Panama disease when planted in infected soil (Smith et al., 1998). Extra care and management is required in banana tissue culture plants which must be irrigated at planting. Plants must be planted deeper than bits and suckers, or they will grow out of the ground becoming less stable. Banana tissue culture plants may require additional de-suckering. The process of banana tissue culture causes off-types which can be high in number if not managed. Banana tissue culture will also activate BSV in cultivars with the B genome.

6.3 Banana micropropagation

A wide range of banana cultivars have now been multiplied in tissue culture using similar methods which are described in several useful reviews (Israeli et al., 1995; Vuylsteke, 1998; Strosse et al., 2004; Smith et al., 2005). Banana tissue culture is widely accepted for rapidly producing disease-free plants with several advantages over conventional suckers and are routinely produced commercially (Hamill et al., 2009; Hamill, 2016). The apical meristems found in each sucker are commonly used to initiate tissue cultures although terminal floral axes have also been used (Cronauer and Krikorian, 1985). Suckers should be sourced from healthy superior mother plants growing in areas not known to have serious banana diseases such as Panama disease or BBTV.

The process of multiplying banana in tissue culture is predominantly by micropropagation, which is an organogenesis method where shoots develop directly from each plantlet, induced using growth regulators, and the shoots are removed and subcultured onto fresh media and the cycle repeated until the target number of plants is obtained. There are other methods that can multiply banana including cell suspension and somatic embryogenesis; however, these methods are used in research such as for plant transformation rather than commercial production as there is a concern over increased somaclonal variation (off-types) produced using those methods (Ghasemali et al., 2015; Moradi et al., 2017).

6.4 Case study of banana tissue culture

Micropropagated banana plants are grown in a sterile environment under artificially controlled conditions of temperature and light. The plants are usually grown in clear glass or plastic containers containing appropriate semi-solidified culture media that support each stage of the propagation cycle (i.e. initiation, multiplication, elongation and rooting). Generally banana plants are cultured between 25 and 28 °C and air conditioning is needed to control temperature. The culture rooms should be fitted with high-temperature cut-off thermostats so that the air conditioning system will turn off before causing fatal heat damage to the plants if they fail. The lights should be linked to this high-temperature cut-off so that they also turn off and prevent additional heat load.

Banana plantlets grow under white fluorescent lighting of intensity between 3000 and 10 000 lux with photoperiods between 12 and 16 h of light. Plants can be grown under 24 h light, and this will lessen some condensation within culture vessels caused as they heat and cool; however, electricity cost is increased. The intensity and type of lights vary between laboratories and a wide range of fluorescent lights with daylight spectrum are suitable. More recently LED lights are used as they offer improved light spectrum for better growth in banana (Nhut et al., 2002; Vieira et al., 2015), provide output that does not diminish over time and have less energy consumption than conventional lights and less heat output which reduces the heat transmission to shelves and saves air conditioning costs.

Sometimes banana micropropagation may fail due to ongoing problems with contamination of cultures. Attention to hygiene is extremely important in a tissue culture room especially in tropical areas with high humidity. Periodically, the whole room (floors, walls, shelves and ceilings) must be cleared and wiped down with an antiseptic solution, such as 80% ethanol or dilute bleach (0.5–1% a.i. of sodium hypochlorite). This will remove dust and destroy any mould spores that may be present. Air filters must be frequently cleaned. The area around the air conditioner must be kept free from vegetation and as mould will often grow near the air conditioner intake/outflow due to condensation and must be cleaned frequently to reduce spores entering the laboratory. The laboratory should be set up so that washing up areas and rough laboratories are kept separate from the clean areas of media preparation, culture dissection/subculture and culture rooms.

Murashige and Skoog (1962) (MS) medium at full strength is used for all stages of banana micropropagation, comprising initiation, multiplication, elongation and rooting phases. Some organic supplements (such as myo-inositol at 100 mgL^{-1}) are routinely added. Domestic grocery sugar is usually used as the sucrose supplement at 20–30 g/L. pH is adjusted to 6.0 prior to autoclaving, but should be checked after autoclaving as the pH level can drop significantly especially if the medium was not been mixed uniformly when it was dispensed.

Banana micropropagation is most commonly performed using semi-solid media. The type of solidifying agent, whether it is agar, agargel (mixture of agar and phytagel) or gelrite/phytagel can affect the performance of some cultivars. Agar in particular may

Published by Burleigh Dodds Science Publishing Limited, 2018.

vary from batch to batch and brand to brand. Agar must be melted at high temperature until fully dissolved, while phytagel/gelrite is dissolved cold, and the action of cations in the media allows gelling to occur. Gelrite or phytagel produce a very clear medium that will better allow for observation of bacterial contamination; agar is cheaper but makes an opaque medium.

6.5 Selection of material for culture initiation

Good selection of the source material is one of the most important steps to producing high-quality tissue culture plants. The mother plant to be used as a source of growing point should be grown using good management practices in an area free from diseases. The plant characteristics need to be identified as a superior, true-to-type example for that cultivar. Healthy sword suckers need to be carefully dug from the mother plants so that they are not damaged in the process. Twisting suckers may cause internal cracks and damage the growing point (Fig. 4). Careless digging may cause deep cuts damaging the growing point (Fig. 5). Sometimes the base of the sucker is cut in half, which damages the growing point (Fig. 8). Suckers should be carefully inspected for banana weevil borer tunnel holes since the insects usually tunnel through the rhizome into the growing point, destroying the sucker.

Suckers must be free from soil and have a surface-trimmed back and have roots removed. The cut surface should be allowed to suberise before sending suckers. The suckers should be wrapped in paper and placed in paper bags and taken or sent to the laboratory for culture initiation. Suckers should not be refrigerated as this can kill the growing point and they should not be placed in plastic bags if it will take several days before the suckers can be used since the moisture that accumulates in plastic encourages mould.

Once in the laboratory suckers should immediately be assigned a unique code so that the information of the sucker can always be traced back (Figs. 6 and 7). The unique code is then used in all tissue culture production processes. Traceability of tissue culture accessions is an important and often overlooked step and can be made sophisticated by

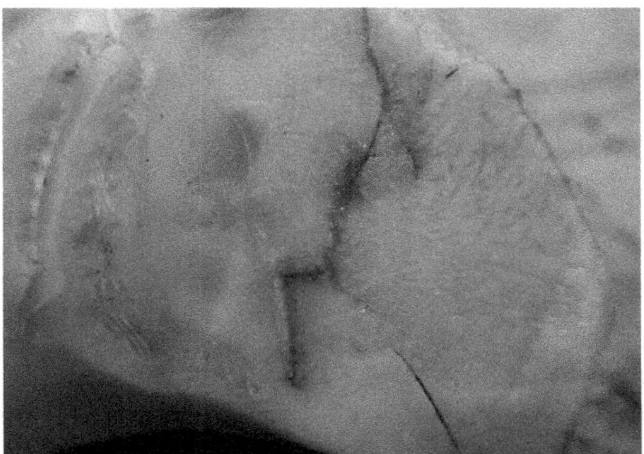

Figure 4 Sucker cracked internally (photo by Roger Goebel).

Published by Burleigh Dodds Science Publishing Limited, 2018.

Figure 5 Sucker damage (photo by Roger Goebel).

using a purpose-designed barcode scanning system or by simply assigning number codes that link to information in a spreadsheet or work book. Many suckers from multiple mother plants of each cultivar should be used to initiate tissue culture for each accession as this is a way to manage off-types and maintain quality, which is discussed later in this chapter.

A sharp knife and clean cutting board are used to cut back the suckers to extract the central block of tissue containing the growing point. Eighty per cent ethanol or other safe sterilising agent should be used to wipe down the cutting board and clean the knife

Figure 6 Suckers received in the laboratory, unpacked and labelled prior to initiation (photo by Sharon Hamill).

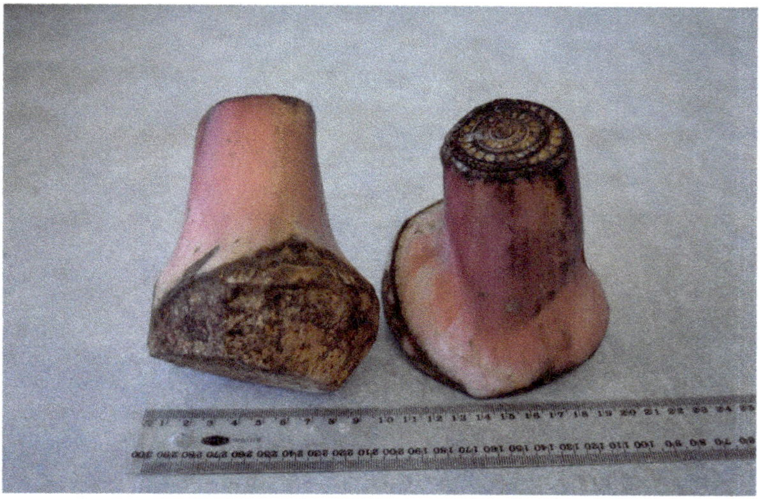

Figure 7 Sucker size (photo by Sharon Hamill).

between suckers. The sucker should be cut back and observed for vascular discoloration that may indicate disease and suckers with internal discoloration should be discarded. Extracting the block of tissue containing the meristem is undertaken in a clean laboratory, while the remaining steps in the initiation process should be undertaken using a sterile technique in a laminar flow cabinet. There are two main methods used to initiate banana tissue cultures: shoot-tip culture and meristem culture. Banana tissue culture is most

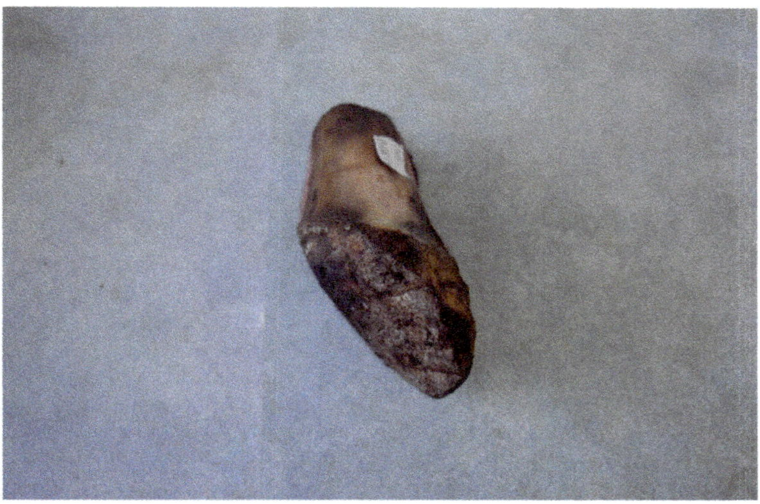

Figure 8 Sucker cut too close to growing point (photo by Sharon Hamill).

commonly initiated using shoot-tip culture using approximately a 5 × 7-mm block of sterile tissue containing the growing point extracted from each sucker (Hamill et al., 1993; Vuylsteke, 1998). Meristem culture, which requires dissection of the meristem with the aid of a microscope, extracts a piece of tissue less than 2 mm in diameter containing the apical dome covered with approximately three ensheathing leaf bases. Meristem culture is used to establish cultures with reduced levels of endophytic bacteria (Van den houwe et al., 1998) needed in research, cryopreservation and germplasm collections. The apical meristem in the bell can also be used in a similar way to the meristem in suckers to establish banana tissue culture, but commercially it is the suckers that are used to source the growing point for culture establishment. A bud on the side of the sucker or on a bit of the corm can also be used to establish a banana culture.

In banana tissue culture the block of tissue containing the growing point is extracted from the centre of the banana sucker. The block of tissue approximately 10 × 25 mm and containing a small amount of corm at the base with the central growing point enclosed by the pseudostem above is cut out using straight smooth cuts. This block of tissue is placed into a sterile, labelled bottle in preparation for the next step in the initiation process for shoot-tip or meristem culture using an aseptic technique in a laminar flow cabinet.

- Sodium hypochlorite (2.5% active ingredient diluted with distilled water and a few drops of surfactant (e.g. Tween 80) is poured onto the trimmed block of tissue and shaken periodically.
- After 15–20 min the sodium hypochlorite is removed and the tissue rinsed three times with sterile distilled water.
- The 1–2-mm-deep outer layer of bleached tissue is trimmed away to produce individual shoot-tip blocks (7 × 5 mm)
- For shoot-tip culture, a small sterilised block of tissue containing the growing point is used to establish the culture and a cross can be cut into the top of the block to allow the growing point to emerge more easily.
- For meristem culture, layers of tissue surrounding the growing point can be carefully removed until only the apical dome with up to three ensheathing leaf bases remains. A dissecting microscope is needed for meristem culture as the meristem will be smaller than 2 mm and should be inserted into the media pointing up.
- Cultures should be kept in darkness for 5–7 d after initiation to reduce phenolics.

6.6 Indexing banana for bacteria

Bacterial contamination can be a problem in banana tissue culture particularly since the growing point used for culture initiation grows below soil level. Bacterial contamination may be introduced through the explant tissue itself from surface bacteria, as endophytic bacteria residing within the plant tissue, or from the laboratory environment during tissue culture preparations. Contaminants may either grow immediately, making control relatively easy or in many cases remain latent for long periods of time, making the identification and management of contamination extremely difficult. Banana tissue culture can be produced free from bacteria if the source explant (particularly using meristem culture) is indexed at the time of initiation (Hamill, 2008; Hamill and Rames, in press). Small pieces of tissue taken from the explant or adjacent to the explant under sterile conditions should be inoculated onto nutrient broth agar (beef extract, 3 g L^{-1}; peptone, 5 g L^{-1}; sodium

Published by Burleigh Dodds Science Publishing Limited, 2018.

chloride, 8 g L^{-1}; yeast extract, 2 g L^{-1} and bacteriological agar, 15 g L^{-1}) by four methods: streaking the base of the explant onto nutrient agar, extracting two 1–2-mm^2 sections of tissue adjacent to the meristem, macerating with a sterile scalpel and inoculating to both nutrient agar and nutrient broth (without agar) and observing initiated cultures for contamination. Bacteria should be incubated at 26–28 °C, with nutrient broth shaken at 100 rpm and maintained for several weeks. Visual observation of cultures is also required. By using meristem culture with all four screening methods, any cultures initiated using explant tissue with culturable bacteria can be identified and removed to eliminate bacteria in subsequent production of plants.

6.7 After culture initiation

Banana plants in tissue culture commonly produce phenolics that result in blackening of the media due to oxidation. Phenolics can be especially problematic with certain cultivars and also when tissue is wounded such as at culture initiation. To reduce the level of phenolic production the newly initiated cultures should be covered to exclude light and after 5 d cultures should be placed under light for 16 h of day length. Blackening of the culture due to phenolics may continue to be a problem in some cultivars in the initiation phase, and if blackening occurs subculturing every one to two weeks with limited wounding may help reduce phenolic production. Antioxidants such as ascorbic acid (0.05–0.5 mM) and L-cysteine have been used to reduce this problem (Vulysteke and De Langhe, 1985) and 0.25–1.0% activated charcoal has been effective for this purpose (Kariyana and Nisyawati, 2014; Mazinga et al., 2014). Production of phenolics usually reduces after two months. If blackening is not a problem subculture onto fresh media should occur every three weeks during the initiation phase.

6.8 Multiplication

Most commercial varieties respond well to subculture at 4- to 6-week intervals on standard MS media. For both initiation and multiplication stages, 6-Benzylaminopurine (BAP) at 2.5 mg L^{-1} is the most commonly used cytokinin. Although not essential, indole acetic acid (IAA) is often included in MS media at 0.175–1.0 mg L^{-1}. Levels of BAP may be increased to 5 mg L^{-1} for difficult-to-multiply varieties, but it is normally used at the lowest level that will induce multiplication and to manage somaclonal variation. Rooting of plantlets can be achieved on MS media with no growth regulators. Some authors describe Benzyladenine (BA) instead of BAP, but they are the same (6-Benzylaminopurine can also be described as N^6-Benzyladenine $C_{12}H_{11}N_5$). TDZ is a less commonly used cytokinin in banana multiplication. It has been shown to increase multiplication compared to BAP but has also been linked to an increase in abnormal shoots (Shirani et al., 2009). Other cytokinins such as kinetin and N^6-(2-isopentyl) adenine (2ip) are used but are not as effective for multiplication as BAP for banana (Arinaitwe et al., 2000).

The plants propagate by producing shoots from the base of each of the plantlets. The shoots should be separated when they are at least 1–2 cm long. Clusters of shoots and buds are removed aseptically from the container and excess leaf and roots are trimmed away. Smaller shoots and buds should not be separated but kept together as clumps. Trimming the side of the explant where the corm meets the pseudostem should be avoided as this is where the shoots originate from. Phenolics may seal the underside of the corm and can be seen as a blackening and should be cut away to expose clean tissue to allow the explant to absorb nutrients. Blackened tissue can be carefully removed so as to not damage the new buds. The individual shoots are trimmed to approximately 2 cm high, leaving the base containing the growing point intact. If the variety begins to multiply excessively on the standard media so that there are mostly buds and lack of shoot elongation and differentiation, the cultures must be alternated between the standard multiplication media and MS media with no plant growth regulators. This alternation to media with and without plant growth regulators will allow very small shoots and buds to elongate into plantlets so that more controlled multiplication is achieved, and somaclonal variation managed.

However another common problem is the lack of multiplication. Some cultivars may start to multiply well but multiplication gradually declines and after 3–4 months multiplication almost stops. Considering the implications of somaclonal variation and the relationship between high cytokinins and adventitious budding, a minimum level of cytokinins is used in media to reduce off-types. If multiplication begins to slow, a number of steps can be taken to increase multiplication; however the impact of these steps on somaclonal variation is unknown. When cultures reach the stage that they have stopped producing shoots and are mainly thick single-stemmed plantlets the cultures should be re-initiated. If this is not practical then alternatively apical dominance can be removed by splitting the plants in half longitudinally and placing the stem sections containing the growing point directly onto the media surface. The cytokinin (BAP) level can be increased to 5 mgL^{-1} to induce buds only as a last option and then reduced as soon as possible. However, even with the adoption of these strategies the plants may not reach reasonable levels of multiplication. To ensure better quality of plants, cultures should be re-initiated annually.

7 Liquid culture systems

Research into systems of *in vitro* banana propagation that improve commercial production efficiency began in the late 1990s when different types of bioreactors, liquid systems with and without agitation and aeration and temporary immersion systems (TISs) were developed. TISs have been the main technology that has proven to be most effective in *in vitro* plant production including for banana. A useful review of TISs and methods has been compiled by Watt (2012). TISs can improve multiplication of plants *in vitro* in a labour-saving way but may be more prone to losses from contamination due to the increased number of plants produced per vessel. The use of TISs requires more advanced equipment than for conventional semi-solid micropropagation methods. Sometimes combinations of methods are used to produce banana plants; for example, banana may be multiplied using TISs but placed onto semi-solid media to elongate and root so that plants have a more uniform architecture and straight stems. Some recent research on performance of banana in TIS has been described (Roels et al., 2005; Wilkin et al., 2014).

8 Banana off-types and how to reduce them

Plant tissue culture is a process that causes stress to the plant which induces somaclonal variation (Larkin and Scowcroft, 1981), producing plants known as off-types. There have been a number of reviews of the factors contributing to the production of banana off-types

Published by Burleigh Dodds Science Publishing Limited, 2018.

(Smith, 1988; Cote et al., 1993; Israeli et al., 1995). The main studies have considered whether growth regulators of a certain concentration or type play a role. Higher levels of growth regulators were demonstrated to be linked to increased off-type production in banana (Bairu et al., 2006). Some research suggests that off-types more commonly occur in adventitious buds compared to axillary buds and that lowest levels of cytokinins should be used to reduce adventitious budding (Damasco et al., 1998).

While there has been some conflicting evidence of the role that plant growth regulators play in somaclonal variation, there have been several consonant studies demonstrating that off-type events increase the longer the time that plants are multiplied in tissue culture and the more times the plants are subcultured (Damasco et al., 1998; Borse et al., 2011). There is also increasing evidence that some cultivars are more predisposed to producing off-types. For example the main cultivars of global trade 'Williams' and 'Grande Naine', both AAA Cavendish types, are predisposed to producing dwarf off-types. The process of tissue culture gives rise to some unavoidable stress to the plants, and somaclonal variation is a result of the plant responding to stress, generally considered epigenetic. Epigenetic changes may cause temporary or inheritable variations that do not entail a change in the DNA which makes practical identification difficult. The somaclonal variation is an adaption to stress and the off-types usually multiply and grow better than the normal plantlets under the conditions where they developed.

There are several papers that have monitored occurrence of off-types during the micropropagation cycle that consistently report off-type events occurring at any time during the tissue culture process and that there were differences in multiplication rate and numbers of off-types produced in suckers from the same mother plant (Smith and Hamill, 1991; Damasco et al., 1998). The frequency of off-type events was found to increase from the eighth subculture cycle onwards and increased the longer the time the plants remain in culture (Borse et al., 2011). It has also been demonstrated that the off-type plants are more adapted to tissue culture conditions and multiply at a higher frequency than normal plants. This means if bananas are kept in culture for long periods, the frequency of off-types increases and this combined with off-type plants multiplying at a faster rate than normal plants results in off-types becoming the major population of plants produced. Off-types have a wide range of different traits, which include dwarf off-types, change in plant morphology, fruit quality changes and low vigour, and these have been described by various authors (Israeli et al., 1991; Daniells et al., 1999; Msogoya and Grout, 2012). Most off-types have negative traits, although somaclonal variation is used in banana improvement, including in a Taiwanese research programme to select Giant Cavendish with improved tolerance to FOC Tropical Race 4, one of the most serious banana diseases (Hwang and Ko, 2004; Lee et al., 2011), and use of somaclonal variation for crop improvement has been reviewed (Krishna et al., 2016).

The production of off-types in tissue culture cannot be eliminated but should be managed to reduce the numbers produced. The following steps assist in reducing off-types in banana micropropagation.

- Source suckers from plants that have not been produced in tissue culture (to reduce compounding effects due to epigenetic variation).
- Source a large number of suckers from many different mother plants.
- Use low levels of growth regulators and limit multiplication.
- Avoid undifferentiated growth (discard affected explants).
- Reduce wounding.

- Limit time in culture – that is limit the number of subculture cycles (maximum eight cycles, less than a year) and associated number of plants produced per growing point (500–1000).

As the incidence of off-types varies between suckers it is important that a large number of strong suckers are taken from many different good selections of mother plants. Large numbers of tissue culture plants should not be produced from only a few suckers. As it has been shown that the frequency of off-types increases by the eighth subculture cycle most laboratories restrict time in culture or limit production to 500 plants per sucker. Banana tissue culture laboratories initiate banana cultures annually to ensure they produce good-quality plants and this means that growers need to order plants 12–18 months ahead to have plants available for their target planting date.

8.1 Elongation and rooting

When the required number of plants have been multiplied they must move into elongation and root development phase. When the multiplication phase is completed, plantlets will be of varying sizes (single stems or many shoots). For the rooting phase it is important that plants are of uniform size and height. Smaller shoots and buds may need to be subcultured for an extra cycle on growth regulator-free MS media to allow them to elongate and stop multiplying. Shoots that are elongated and single-stemmed can be put directly into the hormone-free media for rooting. It is important that plants produced for the nursery are uniform and have 100% survival with a vigorous, well-developed root system. The shoots must be graded to ensure containers contain plants of similar size. Plantlets must be evenly spaced with largest shoots placed in the centre of the vessel and the smaller plants on the outside of the vessel to prevent the larger plants from growing quickly and shading the smaller ones. The containers should be placed on a single layer on shelves to ensure the plantlets have maximum exposure to light and that uniformity is not compromised as increased light is beneficial to plants at the rooting stage. Plantlets are separated into individual shoots and subcultured into large single-use plastic disposable containers – usually food grade, commonly available and able to withstand autoclaving. MS (1962) medium is used without plant growth regulators to induce roots. In the unlikely situation that a cultivar fails to produce roots 1 mg L^{-1} IAA may be added to induce roots. The plantlets will take 2–3 weeks to produce good root systems and will be ready for transportation to the nursery for deflasking and hardening-off.

8.2 Nursery production

An important step in banana tissue culture production is the weaning or acclimatisation phase of an *in vitro*–produced plant between rooting in the laboratory and establishment in a nursery. Tissue culture banana plants that have been rooted under aseptic conditions in the laboratory are extremely tender and sensitive and must be handled with care and attention at this stage. The processes in successful deflasking and establishment of tissue culture plants has been described by Robinson and Galán Saúco (2009a,b).

Facilities are needed that will provide protection from wind, shade and irrigation for watering and high humidity at first stages of acclimatisation. The facilities need to be removed from areas that harbour banana diseases and that might be transmitted to plants in the nursery. Nurseries should not be established in banana production areas that contain diseases such as Panama disease, BBTV or bacterial diseases that are rapidly transmitted by insect vectors. The nursery should be established in a position where dust is not blown onto the plants. The banana plants in the nursery must be on benches off the ground so that they will not be contaminated by soil or water splash.

All materials used in the nursery must be clean and free from disease. Water must be chlorinated or come from a deep bore or sterilised to ensure it does not carry banana pathogens such as spores of Panama disease. Soil-free potting mix is the best type of material for disease-free production of banana plants. Various types of inorganic potting media (e.g. perlite, vermiculite) or processed organic potting media (e.g. processed coconut husk fibre, composted pine bark, composted coffee husks, rice husks) can be used as long as they do not contain banana plant composts or soil because unless potting mix is pasteurised, poorly composted and soil-based media can present a very high risk of introducing disease. Potting mix should be stored away from soil and water run-off that might contaminate it.

On receipt of rooted laboratory plants in flasks, all equipment, including labels, should be available for immediate use. Plants should be deflasked in a shaded place protected from the wind. The banana plantlets should be gently separated and removed from the media and placed directly in clean water at room temperature. The plantlets can be teased apart from each other and gently washed to remove most of the adhering agar. At this stage it is important to rate the plants for size and grade them so that there is uniformity in batches of plants and so that plants will be of the same size and reduce shading and competition resulting in mixed batches in the nursery. It is important to label the plants at potting so that the cultivar and history can be traced back.

Potting mix should be prepared prior to deflasking the banana plants and should be moist. For the first few weeks of acclimatisation the plants will need to be grown under high humidity which can be achieved using high or low technology methods. Commercial nurseries commonly use fogging and misting irrigation systems, but simple hand watering and growing plants under plastic covers over benches can also be effective. The acclimatisation stage is always carried out under a shade structure and the plants will need constant monitoring and watering. The plants need clean and free draining potting mix and should not be grown in a pool of water. Simple processes for post-flask management of micropropagated bananas have been described in detail (Daniells and Smith 1991; Robinson and Galán Saúco, 2009a,b).

8.3 Temperature

The growing temperature should be maintained to ensure the banana plant has the optimum balance between photosynthesis and plant development. Temperature around 22°C is optimal for photosynthesis and dry matter accumulation, whereas temperature around 32°C favours plant development. If temperatures are continually below 25°C, the plant develops short internodes and small leaves but with a strong root system. Physiologically the plant is healthy and strong, but extra time is needed to achieve the required transplant size. Alternatively if temperatures are continually high, the stems become elongated, with large floppy leaves and light green colouration and the root system may be weak. With temperatures in the high 30s, respiration is dominant over photosynthesis, causing the plant to become physiologically weak and prone to dying. Optimum banana plant growth temperatures should be between 25 and 32°C to produce plants with good root systems and sufficiently developed rhizome that are more durable at planting.

Young banana plants being acclimatised should not be exposed to wind at any stage during the hardening-off period. Wind, or even a breeze, removes the boundary layer of high humidity from the leaf surface which causes leaf surface drying, increased temperature and leaf damage. Application of controlled release of fertiliser should provide for the nutritional needs of the young plants while in pots but foliar fertiliser can also be used at 7- to 10-d intervals.

The time taken from transplanting to readiness for field planting will be 12 to 16 weeks depending mainly on temperatures but also on pot size and management. Banana plants need to be sun hardened several weeks before they are due to plant. The holes can be dug the day prior to planting and irrigation is needed at the time of planting. Ideally plants should be brought to the field in the early morning or during overcast weather and planted deeply. Planting should be avoided in the heat of a sunny day especially in hot areas. Plants should not lay on their side in the sun since this can burn the roots through the plastic bags. At planting the bottom of the hole should be filled with loose topsoil and the plant positioned on this layer so that the top of the potting medium is well below the level of the surrounding soil (100–150 mm below the soil surface). Soil should be lightly firmed around the plant and watered as soon as possible. It is more effective to thoroughly water the planting hole well before planting and to give gentle irrigation after planting. Special attention should be given to the plants for the first three months after transplanting and after planting, and during very hot conditions the leaves and surface of the soil should be irrigated at least twice a day to cool the leaves and create a humid microclimate.

8.4 Screening and roguing for off-types

Off-types cannot be visibly identified in culture and there is no practical way to detect off-types *in vitro*. Molecular methods have been developed that can detect Cavendish Dwarf off-types. In the dwarf off-type there is a missing 1.5-kb band compared to the normal 'Williams' Cavendish (Damasco et al., 1996; Ramage et al., 2004); while this is useful in research it is not practical or cost effective as a screening tool as individual plants would need to be tested.

Obvious off-types such as distorted, mosaic or variegated plants can be removed in the nursery. There are ways that specific off-types can be rogued in the nursery phase of tissue culture production. Smith and Hamill (1993) found that the Cavendish Dwarf off-types, when acclimatised and growing in the nursery at the 5–8 leaf stage, had both a shorter petiole and a smaller length:breadth ratio as the leaf length was shorter compared to normal plants giving the off-type leaf a more round appearance with a short petiole. Based on the leaf shape and length at that 5–8 leaf stage the off-type plants can be rogued. However roguing plants cannot be accurately identified after that period or if the plants are growing in crowded conditions. A combination of off-type strategies used during the plant micropropagation cycle (i.e. initiating from many mother plants and suckers, using low levels of growth regulators, controlling multiplication, limiting time in culture) combined with roguing in the nursery reduces off-types reaching the grower.

9 Future trends and conclusion

There are driving economic forces that require large companies to improve efficiencies to remain viable and this has been one of the reasons that banana tissue culture has been so readily adopted around the world. Efficiencies in micropropagation methods will continue

Published by Burleigh Dodds Science Publishing Limited, 2018.

to improve with the aim to reduce labour costs. There are also growing threats to the banana industry due to diseases, so that use of clean planting material will become more important to both the small shareholder and multinational companies. Bits and suckers will continue to be used by the smaller grower but tissue culture plants of improved disease-resistant banana cultivars will be needed to face the challenges of disease pressure matched with more efficient production for growers to remain viable. Long-term commercial viability will also require development of environmentally sustainable banana production practices. The use of Integrated Pest and Disease Management and use of beneficial bacteria and fungi are exciting and growing areas that will become an integral part of banana production used both in tissue culture plant propagation and conventional banana farming. Increasingly, chemicals are being withdrawn from agricultural use and excessive use of chemicals is being restricted. Modern farming practice will further develop using holistic approaches, improving soil health and growing a diverse range of cultivars to reduce pest and disease problems associated with monoculture. Future research will need to develop farming systems that are not only productive but also using ecologically compatible practices considering long-term sustainability with the aim to produce a more nutritious fruit product with less impact on the environment.

10 Where to look for further information

This chapter describes the current best practices in banana propagation, and more detailed information on this topic can be found in the references included. Banana and plantain are an important global commodity, and there are several global networks established to assist in further research, collaboration and education. Useful sites for publications, research, meetings and contacts in banana propagation are as follows:

- The 'ProMusa' network shares news, knowledge and information on bananas and links scientists together in specific areas of interest in working groups. The ProMusa website also links to three databases: *Musalit,* the largest repository of references on banana; *Musarama*, the image bank on all things banana and *Musacontacts*, the place to find people working on banana.

  http://www.promusa.org/http://www.promusa.org/Musapedia
  http://www.promusa.org/Banana+sucker
  http://www.promusa.org/article10-Working-groups

  ProMusa Research networks have also been established in three regions: BAPNET is the Banana Asia-Pacific Network, MUSALAC is the banana research and development network for Latin America and the Caribbean and BARNESA is the Banana Research Network for Eastern and Southern Africa.

  http://www.promusa.org/BAPNET
  http://www.promusa.org/MUSALAC
  http://www.promusa.org/BARNESA

- Food and Agriculture Organization of the United Nations (FAO) 'World Banana Forum' brings all those in the banana supply chain together including researchers.

  http://www.fao.org/world-banana-forum/en/

- CGIAR (formerly the Consultative Group for International Agricultural Research) brings people together to develop solutions with emphasis on integrated agri-food systems–based approaches. CGIAR has a research programme on Roots, Tubers and Bananas 2017-2022 in parts of Africa, Latin America and the Caribbean and Asia.

  http://www.cgiar.org/about-us/our-programs/cgiar-research-program-on-roots-tubers-and-bananas-2017-2022/

11 References

Arinaitwe, G., Rubaihayo, P. R. and Magambo, M. J. S. (2000) Proliferation rate effects of cytokinins on banana (*Musa* spp.) cultivars. *Scientia Horticulturae* 86, 13–21.

Bairu, M. W., Fennell, C. W. and van Staden, J. (2006) The effect of plant growth regulators on somaclonal variation in Cavendish banana (*Musa* AAA cv. *'Zelig')*. *Scientia Horticulturae* 108, 347–51.

Bhende, S. S. and Kurien, S. (2015) Sucker production in banana. *Journal of Tropical Agriculture* 53(2), 97–106.

Blomme, G., Ploetz, R., Jones, D., Langhe, E. de, Price, N., Gold, C., Geering, A., Viljoen, A., Karamura, D., Pillay, M., Tinzaara, W., Teycheny, P. Y., Lepoint, P., Karamura, E. and Buddenhagen, I. (2013) A historical overview of the appearance and spread of *Musa* pests and pathogens on the African continent: Highlighting the importance of clean *Musa* planting materials and quarantine measures. *Annals of Applied Biology* 162(1), 4–26.

Borse, N., Chimote, V. P., and Jadhav, A. S. (2011) Stability of micropropagated *Musa acuminata* cv. Grande Naine over clonal generations; A molecular assessment. *Scientia Horticulturae* 129, 390–5.

Colbran, R. C. (1964) Effects of treatments of banana 'bits' for nematode control on emergence and yield. *Queensland Journal of Agricultural Science* 21, 237–8.

Côte, F. X., Sandoval, J. A., Marie, P. H. and Aubiron, E. (1993) Variations in micropropagated bananas and plantains: Literature survey. *Fruits* 48, 15–22.

Côte, F. X., Galzi, S., Folliot, M., Lamagnere, Y., Teycheney, P.-Y. and Iskra-Caruana, M.-L. (2010) Micropropagation by tissue culture triggers differential expression of infectious endogenous *Banana streak virus* sequences (eBSV) present in the B genome of natural and synthetic interspecific banana plantains. *Molecular Plant Pathology* 11(1), 137–44.

Coyne, D., Wasukira, A., Dusabe, J., Rotifa, I. and Dubois T. (2010) Boiling water treatment: A simple, rapid and effective technique for nematode and banana weevil management in banana and plantain (*Musa* spp.) planting material. *Crop Protection* 29, 1478–82.

Cronauer, S. S. and Krikorian, A. D. (1985) Reinitiation of vegetative growth from aseptically cultured terminal floral apex of banana. *American Journal of Botany* 72(10), 1598–601.

Damasco, O. P., Graham, G. C., Henry R. J., Adkins S. W., Smith M. K. and Godwin, I. D. (1996) Random amplified polymorphic DNA (RAPD) detection of dwarf off-types in micropropagated Cavendish (*Musa* spp. AAA) bananas. *Plant Cell Reports* 16, 118–23.

Daniells, J. W. and O'Farrell, P. J. (1987) Effect of cutting height of the parent psuedostem on yield and time of production on the following banana. *Scientia Horticulturae* 31(12), 89–94.

Daniells, J. W. and Smith M. K. (1991) *Post Flask Management of Tissue-cultured Bananas*. ACIAR technical Reports No. 18, Australian Centre for Agricultural Research, Canberra.

Daniells, J. W., Smith, M. K. and Hamill, S. D. (1999) *Banana Offtypes: An Illustrated Guide*. Information series QI99019, Department of Primary Industries, Queensland. ISSN: 0727–6273. Agdex 231/40.

De Langhe, E. (1961a) La phyllotaxie du bananier et ses consequences pour la comprehension du système rejectonnant. *Fruits* 16, 429–41.

De Langhe, E. (1961b) Multiplication vegetative acceleres in plantation du bananier plantain 'Bousa'. *Bulletin* Information *of* INEAG 10, 69–90.
De Langhe, E., Vrydaghs, L., De Maret, P., Perrier, X. and Denham, T. P. (2009) Why bananas matter; an introduction to the history of banana domestication. *Ethnobotany Research and Applications* 7, 165–77.
Denham, T. P., Haberle, S. G., Lentfer, C., Fullagar, R., Field, J., Therin, M., Porch, N. and Winsborough, B. (2003) Origins of agriculture at Kuk swamp in the highlands of New Guinea. *Science* 301(5630), 189–93.
Drew, R. A. and Smith, M. K (1990) Field evaluation of tissue cultured banana in south-eastern Queensland. *Australian Journal of Experimental Agriculture* 30, 569–74.
Eckstein, K. and Robinson J. C. (1995) Physiological responses of banana (Musa AAA; Cavendish subgroup) in the subtropics. *(IV) Comparison between tissue culture and conventional planting material during the first months of development. Journal of Horticultural Science* 70, 549–59.
Ghasemali, F., Farahani, F., Sheidai, M. and Satari, T. N. (2015) Induced somaclonal variation in regenerated banana plants (*Musa acuminata* l.) cv. Dwarf Cavendish. *Gene Conserve* 14(56), 32–55.
Hamill, S. D. (2008) *Plant Tissue Culture: Providing Strategic Support for the Banana Industry*. Final report BA04007, Horticulture Australia Limited, Sydney, Australia, pp. 52–60. ISBN: 0 7341 1705 1.
Hamill, S. D. (2016) Processes, costs and traits of plants produced in tissue culture must be considered to develop effective crop production systems. In: *XXIX IHC – Proc. Int. Symp. on Micropropagation and In Vitro Techniques*. M. Lambardi and S. Hamill (Eds). *Acta Horticulturae* 1113, 85–92. doi:10.17660/ActaHortic. 2016.1113.12.
Hamill, S. D. and Rames, E. (2018) An effective indexing method for banana tissue culture provides long-term freedom from bacterial contamination. In: M. Lambardi and R. Paiva (Eds.), *Proceedings of the Ist International Symposium on Vegetative Propagation and In Vitro Culture of Tropical Plants*. Acta Horticulturae, ISSN 0567-7572 (print), ISSN 2406-6168(electronic) (In press).
Hamill, S. D., Moisander, J. A. and Smith, M. K. (2009) Micropropagation of vegetatively propagated crops: Accelerating release of new cultivars and providing an important source of clean planting material. In: *Proceedings of the VIth International Symposium on In Vitro Culture and Horticultural Breeding*. R. J. Geijskes, P. Lakshmanan and A. Taji (Eds). *Acta Horticulturae* 829, 213–17.
Hamill, S. D., Sharrock, S. L. and Smith, M. K. (1993) Comparison of decontamination methods used in initiation of banana tissue cultures from field-collected suckers. *Plant cell, Tissue and Organ Culture* 33, 343–6.
Hwang, S. C. and Ko, W. H. (2004) Cavendish banana cultivars resistant to Fusarium wilt acquired through somaclonal variation in Taiwan. *Plant Disease* 88(6), 580–8.
Israeli, Y., Lahav, E. and Reuveni, O. (1995) *In vitro* culture of bananas. In: *Bananas and Plantains*. S. Gowen (Ed.). Chapman and Hall, London, pp. 147–78.
Israeli, Y., Reuveni, O. and Lahav, E. (1991) Qualitative aspects of somaclonal variations in banana propagated by *in vitro* techniques. *Scientia Horticulturae* 48, 71–88.
Kabunga, N. N., Dubois, T. and Qaim, M. (2014) Impact of tissue culture banana technology on farm household income and food security in Kenya. *Food Policy* 45, 25–35.
Kariyana, K., Nisyawati (2014) Effect of ascorbic acid, activated carbon and light duration on explant browning of banana cultivar Barangan (*Musa acuminata* L.) *in vitro* culture. *International Journal of Research and Reviews in Applied Sciences* 16(1), 118–23.
Kennedy, J. (2009) Bananas and people in the homeland of Genus *Musa*: Not just pretty fruit. In: *History of Banana Domestication. Special Issue of Ethnobotany Research and Application*. T. Denham, E. De Langhe and L. Vrydaghs (Eds). Ethnobotany Research & Applications, North America, 7, 179–198 www.ethnobotanyjournal.org/vol7/i1547-3465-07-179.pdf.
Krishna, H., Alizadeh, M., Singh, D., Singh, U., Chauhan, N., Eftekhari, M. and Sadh, R. K. (2016) Somaclonal variations and their applications in horticultural crops improvement. *3 Biotech* 6(1), 54. DOI 10.1007//s13205-016-0389-7, Published online https://www.ncbi.nlm.nih.gov/pmc/articles/PMC4752953/pdf/13205_2016_Article_389.pdf.

Kumar, P. L., Selvarajan, R., Iskra-Caruana, M. L., Chabannes, M. and Hanna, R. (2015) Biology, etiology, and control of virus diseases of banana and plantain. *Advances in Virus Research* 91, 229–69.

Larkin, P. J. and Scowcroft, W. R. (1981) Somaclonal variation- a novel source of variability from cell cultures for plant improvement. *Theoretical and Applied Genetics* 60, 197–214.

Lee, S. Y., Su, Y. U., Chou, C. S., Liu, C. C., Chen, C. C. and Chao, C. P. (2011) Selection of a new somaclone cultivar 'Tai-Chiao No. 5' (AAA, Cavendish) with resistance to Fusarium wilt of banana in Chinese Taipei. *Acta Horticulturae* 897, 391–7

Ma, S. S. and Shii, C. T. (1972) *In vitro* formation of adventitious buds in banana shoot apex following decapitation. *Journal of the Chinese Society of Horticultural Science* 18, 135–42.

Mazinga, K. M., Jara, M. G., Koninckxloo, M. van, Nyembo, K. L., Kasongo, L. M. E., Ntumba, K. B. and Baboy, L. L. (2014) Effects of activated charcoal in the rooting culture medium on the induction of FHIA-01 (*Musa* AAAB) *in vitro*. *Journal of Applied Biosciences* 76, 6352–60.

Meng, L., Gao, X., Chen, J. and Martin, K. (2012) Spatial and temporal effects on seed dispersal and seed predation of *Musa acuminata* in southern Yunnan, China. *Integrative Zoology* 7, 30–40.

Moradi, Z., Farahani, F., Sheidai, M. and Satari, T. N. (2017) Somaclonal variation in banana (*Musa acuminata colla* cv. Valery) regenerated plantlets from somatic embryogenesis: Histological and cytogenetic approaches. *Caryologia* 70 (1), 1–6

Msogoya, T. J. and Grout, B. W. W. (2012) Altered fruit eating quality in tissue culture derived off-type banana. *Journal of Applied Biosciences* 49, 3388–93. ISSN: 1997–5902.

Msogoya, T. J. and Mwakisitu, J. (2014) Effect of thidazuron on *in vivo* shoot proliferation of popular banana (Musa spp. L) cultivars in Tanzania. *Journal of Applied Biosciences* 81, 7214–20.

Nhut, D. T., Hong, L. T. A., Watanabe, H., Goi, M., and Tanaka, M. (2002) Growth of banana plantlets cultured *in vitro* under red and blue light-emitting diode (LED) irradiation source. *Acta Horticulturae* 575(1), 117–24. Proceedings of the International Symposium on tropical and subtropical Fruits. ISHS. R. Drew (Ed.). ISBN: 90 6605 885 4

Njukwe, E., Tenkouano, A., Amah, D., Sadik, K., Muchunguzi, P., Nyine, M. and Dubois, T. (2005) *Macro-propagation of Banana and Plantain*. Training manual, International Institute of Tropical Agriculture, Ibadan, Nigeria, 23p.

Ogelsby, R. P. and Griffiths, J. L. (1986) Commercial *in vitro* propagation in plantation crops. In: *Tissue Culture as a Plant Production System for Horticultural Crops*. R. H. Zimmerman, R. J. Griesbach, F. A. Hammerschlag and R. H. Lawson (Eds). Martinus Nijhoff Publishers, Dordrecht, pp. 253–7.

Perrier, X., Bakry, F., Carreel, F., Jenny, C., Horry, J-P., Lebot, V. and Hippolyte, I. (2009) Combining biological approaches to shed light on the evolution of edible bananas. Ethnobotany research and applications 7:199–216.

Perrier, X., De Langhe, E., Donohue, M., Lentfer, C., Vrydaghs, L., Bakry, F., Carreel, F., Hippolyte, I., Horry, J. P., Jenny, C., Lebot, V., Risterucci, A. M., Tomekpe, K., Doutrelepont, H., Ball, T., Manwaring, J., de Maret, P. and Denham, T. (2011) Multidisciplinary perspectives on banana (*Musa* spp.) domestication. *Proceedings of the National Academy of Sciences of the USA* 108(28), 11311–18

Pillay, M. and Tenkouano, A. (2011) *Banana Breeding Progress and Challenges*. CRC Press Inc, Taylor and Francis Group, Boca Raton, 383p. ISBN: 9781439800171.

Ploetz, R. (2015) Fusarium wilt of banana. *Phytopathology* 105, 1512–21.

Promusa website Musapedia. 'banana sucker', www.promusa.org/Banana+sucker. Contributors to page Anne Vézina, David W. Turner and Jane Gibbs. Page last modified 20 September 2017 by Anne Vézina.

Ramage, C. M, Borda, A. M., Hamill, S. D. and Smith, M. K. (2004) A simplified PCR test for early detection of dwarf off-types in micropropagated Cavendish bananas (Musa spp. AAA). *Scientia Horticulturae* 103(2004), 145–51.

Robinson, J. C. and Galán Saúco, V. (2009a) Weaning (acclimatisation) of *in vitro*-produced banana plants. *Fruits* 64, 325–32.

Robinson, J. C. and Galán Saúco, V. (2009b) Nursery hardening of *in vitro*-produced banana plants. *Fruits* 64, 383–92.

Published by Burleigh Dodds Science Publishing Limited, 2018.

Robinson, J. C. and Galán Sáuco, V. (2010) *Bananas and Plantains*, 2nd Edition. Crop production science in horticulture series: no.19, CABI International, Wallingford Oxfordshire, UK, pp. 125–47. ISBN-13978 1 84593 658 7

Rodriguez, C., Cayón, G. and Jairo Mira, J. (2006) Influence of the harvested mother plant psuedostem on follower sucker growth and yield in banana (*Musa* AAA Simonds). *Agronomia Colombiana*, 24(2), 274–9.

Roels, S., Escalona, M., Cejas, I., Noceda C., Rodriguez R., Canal M. J., Sandoval, J. and Debergh, P. (2005) Optimization of plantain (*Musa* AAB) micropropagation by temporary immersion system. *Plant Cell, Tissue and Organ Culture* 82, 57–66.

Sardos, J., Perrier, X., Dolezel, E. Hribová, Christelová, P., Van den houwe, I., Kilian, A. and Roux, N. (2016) DArT whole genome profiling provides insights on the evolution and taxonomy of edible bananas (*Musa* spp.). *Annals of Botany* 118, 1269–78.

Shirani, S., Mahdavi, F. and Maziah, M. (2009) Morphological abnormality among regenerated shoots of banana and plantain (*Musa* spp.) after *in vitro* multiplication with TDZ and BAP from excised shoot-tips. *African Journal of Biotechnology* 8(21), 5755–61. ISSN 1684–5315.

Smith, M. K. (1988) A review of factors influencing the genetic stability of micropropagated bananas. *Fruits* 43, 219–23.

Smith, M. K. and Hamill, S. D. (1991) Use of tissue culture for the propagation and improvement of bananas and plantains. In: *Banana Diseases in Asia and the Pacific*. R. V. Valmayor, B. E. Umali and C. P. Bejosano (Eds). INIBAP, Montpellier, France, pp. 158–73.

Smith, M. K. and Hamill, S. D. (1993) Early detection of dwarf off-types from micropropagated Cavendish bananas. *Australian Journal of Experimental Agriculture* 33, 639–44.

Smith, M. K., Hamill, S. D., Becker, D. K. and Dale, J. L. (2005) *Musa spp. Banana and Plantain In; Biotechnology of Fruit and Nut Crops*. Biotechnology in Agriculture No. 29. R. Litz (Ed.). CAB International, Wallingford, Oxfordshire 0X10 8DE UK, pp. 365–91. ISBN: 0 85199 662 0.

Smith, M. K., Whiley, A. W., Searle, C., Langdon, P. W. and Pegg, K. G. (1998) Micropropagated bananas are more susceptible to Fusarium wilt than plants grown from conventional material. *Australian Journal of Agricultural Research* 49, 1133–9.

Stirling, G. R. and Pattison, A. B. (2008) Beyond chemical dependency for managing plant-parasitic nematodes: Examples from the banana, pineapple and vegetable industries of tropical and subtropical Australia. *Australasian Plant Pathology:APP*; 37(3), 254–67.

Stover, R. H. and Simmonds, N. W. (1987) *Bananas*, 3rd Edition. Longman, London, 468p.

Strosse, H., Van Den Houwe, I. and Panis, B. (2004) Banana cell and tissue culture-review. In: *Banana improvement: Cellular, Molecular Biology, and Induced Mutations*. Mohan Jain, S. and Swennen, R. (Eds). Science Publishers Inc., New Hampshire, USA, pp. 1–12. ISBN 1-57808-340-0.

Thiemele, D. E. F., Issali, A. E., Traore, S., Kouassi, K. M., Aby, N., Gnonhouri, P. G., Kobenan, J. K., Yao, T. N., Adiko, A. and Zakra, A. N. (2015) Macropropagation of plantain cultivars Pita 3, FHIA 21, Orishele and Corne 1: Effect of benzyaminopurine (BA)) concentration. *Journal of Plant Development* 22, 31–9.

Thomas, J. E. (Ed.). (2015) *MusaNet Technical Guidelines for the Safe Movement of Musa Germplasm*, 3rd Edition. Bioversity International, Rome. https://drive.google.com/file/d/0B6WMCDtu_Ljpb2ZDUHhVb0hLWUk/view

Thomas, J. E., Smith, M. K., Kessling, A. F. and Hamill, S. D. (1995) Inconsistent transmission of banana bunchy top virus in micropropagated bananas and its implications for germplasm screening. *Australian Journal of Agricultural Research* 46, 663–71.

Vieira, L. do N., Fraga, H. P. de F., Anjos, K. G. dos, Puttkammer, C. C., Scherer, R. F., Silva, D. A. da, and Guerra, M. P. (2015) Light-emitting diodes (LED) increase the stomata formation and chlorophyll content in *Musa acuminata* (AAA) 'Nanicão Corupá' *in vitro* plantlets. *Theoretical and Experimental Plant Physiology* 27(2),91–8.

Vuylsteke, D. R. (1998) *Shoot-tip Culture for the Propagation, Conservation, and Distribution of Musa Germplasm*. International Institute of Tropical Agriculture, Ibadan, Nigeria, 82p.

Vulysteke, D. R. and De Langhe, D. (1985) Feasibility of *in vitro* propagation of bananas and plantains. *Tropical Agriculture (Trinidad)* 62, 323–8.

Watt, Paula M. (2012) Review: The status of temporary immersion systems (TIS) technology for plant micropropagation. *African Journal of Biotechnology* 11(76), 14025–35. Available online at http://www.academicjournals.org/AJB DOI:10.5897/AJB12.1693. ISSN: 1684–5315.

Wilkin, D., Jiménes Gonzalez, E., Gerth, A., Gómez-Kosky, R., Schumann, A. and Claus, D. (2014) Effect of immersion systems, lighting and TIS designs on biomass increase in micropropagating banana (*Musa* spp. Cv. 'Grande naine' AAA). *In Vitro Cellular and Developmental Biology-Plant* 50, 582–9. DOI: 10.1007/s11627-014-9605-5

Published by Burleigh Dodds Science Publishing Limited, 2018.

Chapter 7

Conserving banana germplasm through field genebanks

Mike Smith, Queensland Department of Agriculture and Fisheries, Australia

1 Introduction
2 Objectives of the field genebank
3 Choosing the site
4 Sourcing material
5 Planting
6 Crop management
7 Quality control
8 Dissemination of quality planting stock
9 Case study
10 Future trends and conclusion
11 Where to look for further information
12 Acknowledgements
13 References

1 Introduction

Bananas and plantains are members of a monocotyledonous family of giant perennial herbs that grow throughout the humid tropics and subtropics. Because most of the cultivars and landraces important for human uses are parthenocarpic and do not produce seeds, they are propagated vegetatively. Traditionally, bananas and plantains were maintained and conserved *in situ* on home, village or farm gardens (Saad and Rao, 2001; Daniells et al., 2016; Ocimati et al., 2016) where the livelihood of these farmers still depends on the availability of a wide range of cultivars that are adapted to local environmental conditions and for a variety of uses. Over the past decades, banana diversity has been in decline, and concerted action at a global level has been needed to establish and maintain genebanks to conserve and better utilise the genetic diversity inherent in the genus *Musa*. Recently it has been estimated that *Musa* germplasm is conserved in about 60 national collections hosting more than 6000 accessions (Channelière et al., 2011). These are represented by

http://dx.doi.org/10.19103/AS.2017.0020.13
Published by Burleigh Dodds Science Publishing Limited, 2018.

field, *in vitro* and cryopreserved genebanks and comprise a wide range of morphological and genetic diversity in *Musa*.

One of the largest field genebanks was launched early in 1959 as part of a major banana-breeding programme by the United Fruit Company in Honduras. Plant explorers were sent to the Western Pacific and Southeast Asia to gather and assemble in Honduras *Musa* germplasm for their breeders and geneticists and included wild species, landraces and cultivars (Rosales et al., 1999). Nearly 800 accessions were collected, including about 200 diploid accessions (Ortiz, 2011). The breeding programme, now managed by the Fundación Hondureña de Investigación Agrícola (FHIA), was supported by a field genebank of more than 450 accessions in 1987, and diploid sources of resistance to major banana pathogens such as Fusarium wilt (*Fusarium oxysporum* f.sp. *cubense*), black sigatoka (*Mycosphaerella fijiensis*) and burrowing nematode (*Radopholus similis*) were identified in the collection (Rowe, 1987). Realising the genetic potential of *Musa* germplasm in breeding for disease resistance, yield increase and quality improvement continue to be a priority and many genetic and crop improvement programmes around the world continue to rely on field and *in vitro* genebanks for a supply of plants, as do farmers for clean planting material.

This chapter discusses the key issues in the selection, establishment and management of a field *Musa* germplasm collection and how best to provide a reliable source of true-to-type, pathogen-tested planting stock. It is also recognised that field collections not only complement *in vitro* collections, but by working together they build synergy in safeguarding and disseminating valuable germplasm. Whether a scientist or a farmer, it is important that the planting material they receive is a true representative of the cultivar they seek, that it is robust and that it does not harbour pests and diseases.

2 Objectives of the field genebank

There are a number of advantages in establishing a collection of plants in the field (Table 1). First and foremost material can be evaluated and characterised while still being conserved. Certain banana and plantain cultivars are also prone to producing variants or off-types, particularly if they have been multiplied through tissue culture. For instance, dwarfism is common amongst Cavendish clones while some plantains undergo specific inflorescence

Table 1 Advantages and disadvantages of a field germplasm collection

Advantages	Disadvantages
Material can be characterised and evaluated while being conserved	Material is subject to the vagaries of weather, pests & diseases, theft & vandalism
Off-types (somaclonal variants) can be easily identified and removed	Large areas of land required, access to good quality soils and water
Easy access for research and extension	High maintenance costs, including experienced labour
Suckers can be collected from true-to-type plants and be pathogen tested and used for *in vitro* and conventional propagation	Slow multiplication of material for distribution
Requires limited technology	

Published by Burleigh Dodds Science Publishing Limited, 2018.

changes (Smith et al., 2005). Unwanted off-types can easily be identified in the field and removed from the collection. Nevertheless some variants have useful characteristics and it is worth careful observation and record-keeping, ensuring these are retained for further evaluation.

A field collection complements an *in vitro* collection in several important ways. By the curators working together an accession can be verified as true to type in the field before being initiated in culture, and micropropagated plants grown in an approved nursery are the best planting material to use to ensure they are free from pest and disease right from the start. Micropropagated plants, when cared for at planting, have 100% establishment rate and have greater uniformity, precocity and yield than those established from suckers (Robinson, 1996).

Unlike a tissue culture laboratory, a field collection does not require complex technical resources, but it does require large areas of land and experienced staff to manage the plants. A field collection is also subject to the vagaries of weather, pest and disease incursions, theft and vandalism. Another disadvantage is the multiplication of materials for distribution, which is much slower in the field compared to *in vitro* multiplication rates.

A field collection can be used for experimental purposes but it is important from the outset to establish clear objectives for collecting and management. Reed et al. (2004) set out the considerations needed when establishing a collection. There is clearly a need for long-term funding support and resources, including land and trained personnel, to ensure that the genetic potential of new germplasm for disease resistance, yield increase and quality improvement can be fully utilised. From a breeding perspective, there is a need for sufficient genetic diversity as important genes needed for traits of importance today may not be suitable tomorrow as climate change and new disease and pest incursions may require a new suite of genes. The choice of the site is also very important as the site's climate and soils must be conducive to maintenance of the collection.

The intended function of the collection will influence the germplasm to be established as well as the acquisition policy and degree of characterisation needed. For instance evaluation for yield potential or determining the level of resistance to a particular disease may require a greater number of plants or a special field layout suitable for replicated experiments compared to a field genebank engaged in characterisation and supply of germplasm.

3 Choosing the site

Banana field collections can be maintained indefinitely if curators have the resources to restore and rotate accessions to ensure pests and diseases do not accumulate and compromise the collection, as the goal is to have a source of a true-to-type, disease-free material for research, development and extension. Therefore the choice of site is very important. Considerations that need to be taken into account when choosing a site include:

- Meet the climatic requirements of the crop. Bananas and plantains are primarily crops of the humid tropical lowlands, which are areas characterised by less than 10° latitude, less than 100 m altitude, not less than 19°C mean minimum temperature and more than 100 mm rain every month (Robinson, 1996). Of course, the banana-growing areas of the world are geographically situated between the Equator and latitudes 30°N and 30°S; therefore the crop can tolerate wide temperature fluctuations and lower rainfall that is also poorly distributed. Where water is not limiting, and the crop has a high water demand (25 mm per week is regarded as a minimum for satisfactory growth), the rate of banana growth and development is determined by temperature. Robinson (1996) states that the mean temperature for an optimal balance between growth and development is 27°C.

- Choose soils with adequate drainage and where water-logging is not a problem. The best soils for banana are deep, well-drained fertile loams with high water-holding capacity and organic matter content (Tomekpe and Fondi, 2008).
- Avoid areas prone to known pests and diseases. Select sites that were not under banana during the previous two years and were covered by a rotation crop that is a non-host to important banana pathogens. Where Fusarium wilt is concerned, thought needs to be given to planting susceptible sentinel plants at least 12 months before the field genebank is to be established. If plants show signs of disease then alternative sites may need to be considered.
- Assure the long-term maintenance of the field genebank. This includes allowing for adequate space for rotation and expansion. Tomekpe and Fondi (2008) recommend ensuring that twice as much space as occupied for the collection for fallowing is essential for proper maintenance of accessions. Replanting at periodic intervals also allows the site to be realigned and plants maintained at an optimal density as followers emerge at variable distances from the parent stand.
- Ensure that adequate water resources are available to meet the demands of the crop.
- Good access to the site is necessary for frequent monitoring and maintenance operations, but this must be balanced with the need for site security. Security is not only important to avoid theft and destruction of valuable materials, but biosecurity is also needed to avoid unwanted contamination of the site by soil-borne pathogens introduced by footwear, vehicles and farm implements. Fencing is recommended but not always possible; however, introducing biosecurity measures such as quarantine signage (Fig. 1), footbaths, and wash-down and run-off areas at strategic sites are necessary.
- Plant duplicates elsewhere, either in a field genebank in a different location or in an *in vitro* collection.

Figure 1 An example of biosecurity signage erected at a banana field genebank.

Published by Burleigh Dodds Science Publishing Limited, 2018.

4 Sourcing material

The best source of material for establishing a field collection is a nursery-grown planting stock that has itself been sourced from an *in vitro* collection. Furthermore the micropropagated plants should have been indexed for diseases to ensure it is pathogen free (Thomas, 2015) and a field verification process implemented to ensure the material has not been mislabelled, misclassified or has undergone somaclonal variation during its time in culture (Chase et al., 2016).

The International Transit Centre (Bioversity International, Katholieke Universiteit Leuven, Belgium) currently holds the largest collection of *in vitro Musa* germplasm in the world, partly available to researchers, breeders and the larger *Musa* community via the online ordering system in the *Musa* Germplasm Information System (MGIS), Bioversity International. The MGIS is the most extensive source of data on *Musa* genetic resources and contains information on 540 accessions covering 63% of the collection available for distribution, including passport data, botanical classification, morpho-taxonomic descriptions (Table 2; IPGRI/INIBAP/CIRAD, 1996) and characteristics such as agronomic traits, disease resistance, stress tolerance, biochemical or molecular data, photographs and geographic information systems to produce maps illustrating the distribution of *Musa* diversity (Chase et al., 2016; Ortiz, 2011). In addition there are a number of *in vitro* collections in other parts of the world that can be used as a source of planting material and

Table 2 List of the minimum set of descriptors used to characterise *Musa* developed by the MusaNet Taxonomy Advisory Group (TAG, ON). For full set of 121 descriptors refer to IPGRI/INIBAP/CIRAD (1996). Characters should be recorded when the first ripe fruit develop from the ratoon crop and from at least three plants growing near each other

Pseudostem height (m)	Male bud shoulder
Main underlying colour of the pseudostem	Bract apex shape
Blotches at the petiole base	Bract imbrication
Blotches colour (petiole base)	Bract behaviour before falling
Petiole canal of the third leaf	Colour of the bract external face
Petiole margins winged	Colour of the bract internal face
Petiole margins clasping	Compound tepal main colour
Petiole margin colour	Lobe colour (tip of tepal) of compound tepal
Edge of petiole margin (rim)	Anther colour
Pigmentation of outer surface of cigar leaf	Number of hands on whole bunch
Bunch position	Fruit length (cm) at maturity
Bunch shape	Fruit shape (longitudinal curvature)
Male rachis position	Fruit apex
Male rachis appearance	Remains of flower relicts at fruit apex
Male bud shape	Fruit pedicel length (mm)
Male bud length (cm)	Fusion of pedicels

have international recognition owing to the richness of their collection or to the expertise, services or capacity building they provide.

There will be occasions when wild species, landraces or local selections of known cultivars collected from the field require establishment in a field genebank, but always be aware of the quarantine regulations that apply to movement of material. It is always recommended that they should first be established in an *in vitro* collection, isolated from other accessions in the tissue culture laboratory and indexed by a trained plant pathologist (see previous chapter). If this is not possible, and they are permitted to be planted, then selection of planting material becomes even more critical.

The best material is sword or maiden suckers (these do not develop broad leaves until they are more than 1 m high) collected from the accession to be planted (Fig. 2). Suckers removed from the mother plant must be pared to remove all the roots and trimmed to obtain a 'clean' rhizome free of darkened tunnels, dead or discoloured

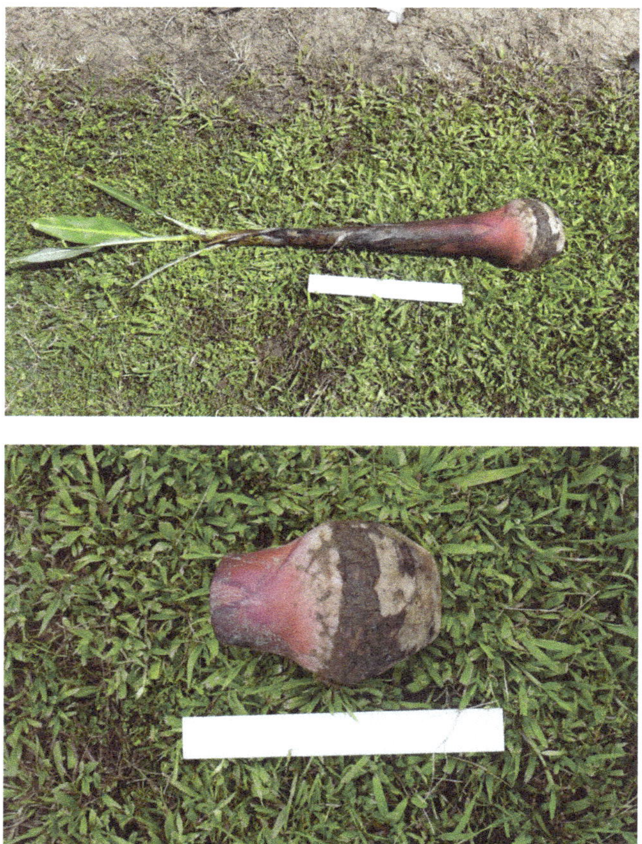

Figure 2 Sword suckers (top) are used as planting material (bottom) or for re-initiation *in vitro*. Scale = 30 cm.

areas. Cut the pseudostem in cross section 10–15 cm above the rhizome. Always use cutting tools that have been disinfested to avoid spreading disease. It is also advisable to allow the prepared suckers to suberise at a site away from field-grown bananas to limit the risk of insect pests laying eggs on the material to be planted. In some cases hot water treatment (30 seconds in boiling water or 20 minutes in water at 50°C) may be used as a further precaution to kill eggs and nematodes (Tomekpe and Fondi, 2008).

Materials entering the field collection should be inspected before bunches ripen by those experienced with *Musa* morphology, pests and diseases. Those preparing the suckers for planting should also know what signs to be looked out for. The introduction of major pests and diseases should be avoided at all costs. Insect pests such as *Cosmopolites sordidus*, nematodes such as *Radopholus similis* and *Pratylenchus* spp., bacteria such as *Ralstonia solanacearum* (phlotype II), fungi such as *Fusarium oxysporum* f.sp. *cubense* and viruses such as banana bunchy top virus and banana bract mosaic virus can easily be carried in planting material collected from the field.

It should go without saying, but each accession should be correctly and clearly labelled at each stage of its propagation cycle and each accession number should be unique and never be reassigned in case the accession is lost. Passport data and other information on the accession should be stored in a suitable database and backed up at regular intervals and preferably stored at a remote site to guard against loss from fire, computer failure and tampering.

5 Planting

The planting layout of a banana collection needs to take into account the purpose to which the collection will be put to use. If it is to be used only for conservation then fewer plants are required per accession because each accession is essentially a clone. Yunus (2001) suggests from 3–5 plants should suffice. If the collection is to be used as a source of planting material or to initiate into tissue culture for further dissemination, then larger numbers of each accession are needed as dictated by demand. In these types of field genebanks the genomic constitution or the use type of the accessions may be taken into account when grouping together for ease of comparison.

However if agronomic evaluation or pest and disease resistance screening is planned then 10–20 plants per accession are recommended and the layout usually follows a completely randomised design or a randomised complete block design (Orjeda, 1998; Carlier et al., 2003). New guidelines for conducting banana germplasm evaluation trials for the international community are soon to be released (Crichton and Van den Bergh, 2016). The reason for a randomised layout is that soil moisture, soil texture, fertility and other differences in the microenvironment are not homogenous in a field situation and the researcher attempts to minimize the heterogeneity of these factors as much as possible so that the accessions are fairly evaluated. In these layouts estimation of the experimental error is feasible and this is very important for the test of significance of difference of accessions in statistical analyses. For agronomic evaluation it is particularly important to conduct trials in several locations for two to three years to take into account climatic factors within a country or region.

With regard to planting density it is recommended to use a spacing of 3 m between rows and 2 m within each row in a rectangular-spaced arrangement which makes for more efficient use of the land and radiation available (Robinson, 1996; Tomekpe and Fondi, 2008). However climatic conditions and degree of mechanisation at each site will play a role in determining the planting density required. Planting should be done during the season most favourable to plant establishment and early growth.

Soil preparation for planting depends largely on topography. On flat land soil should normally be ripped, ploughed and disced before planting, while on steeper sites consideration needs to be given to preventing sheet erosion. Provision for adequate drainage is an important aspect as high rainfall events are often encountered. It is advisable to take a composite soil sample from each block and have it analysed to determine soil pH and nutrient levels. This should provide a reliable recommendation for pre-planting application of lime (dolomitic or calcitic), potassium and phosphate. Weeds can initially be controlled with a post-emergent, non-selective herbicide (e.g. glyphosate) two weeks before planting.

With adequate soil preparation it is only necessary to dig a planting hole slightly bigger than the planting material. Where slope or rocks or lack of access to equipment prevents mechanical cultivation, large planting holes must be prepared by hand. In this case it is recommended to plant into square holes with sides of 40 cm and a depth of 40 cm (Robinson, 1996; Tomekpe and Fondi, 2008). When using a pared sucker, 10–20 cm of soil should cover the cut pseudostem surface while for sun-hardened nursery stock derived from tissue culture, the soil level in the container (removed) should be at least 10 cm below the surface level of the field. Infilling and pressing of soil around the approximately 20 cm high plant provides further support. Post-plant management of these micropropagated plants is extremely important as any constraints to their growth during the first few months will cause the inherent advantages of uniformity and vigour to be lost. Early post-plant management of micropropagated and conventional plants is discussed in more detail by Robinson (1996).

Clearly label plants, make a field plan and keep both a hard copy and an electronic version of it.

6 Crop management

Having got off to a good start it is important to continue to manage the plants well up to bunch harvest, and beyond for subsequent ratoon crops. Nutrition, irrigation, mulching, weed control, sucker management, leaf removal, bunch management and control of pests and diseases are the subjects of other chapters in this volume and need not be discussed in this chapter. Also refer to Robinson and Galán Saúco (2010) for further reading on good crop management of bananas and plantains.

Data collected from a field collection that is well managed is more reliable than from poorly managed plants and there is more confidence in the information that is relayed to farmers and the end users about the qualities of any given variety. When plants are being evaluated for their agronomic performance a number of parameters are usually measured (Table 3) and comparisons are best made from vigorous, healthy plants. Fruit harvested from the trial and used for post-harvest and consumer trials should likewise be from plants growing at their best. Forward planning and good crop husbandry is fundamental to producing quality fruit and planting stock.

Table 3 List of characters important in evaluating the agronomic performance of banana and plantain cultivars (IPGRI/INIBAP/CIRAD, 1996; Crichton and Van den Bergh, 2016). The list is a guide only and observations should be made at harvest for both the plant crop and first ratoon

Time from planting to harvest – days
Plant height (from base of pseudostem to point of bunch emergence) – m
Plant girth (measured 1 m from base of pseudostem) – cm
Height of following sucker (measured from base of pseudostem to last leaf axil) – cm
Number of functional leaves at bunch emergence
Number of functional leaves at harvest
Leaf emission rate
Bunch weight (bunch stalk is cut above first hand at the level of the last scar and immediately below the last hand) – kg
Stalk weight – kg
Fruit weight (bunch weight minus stalk weight) – kg
Number of hands
Number of fingers (fruits)
Fruit length (recorded on the central external fruit of the middle hand) – cm
Fruit diameter (recorded on the central external fruit of the middle hand) – mm
Productivity – t/ha/yr

7 Quality control

Quality control starts and ends with attention to detail and good record-keeping.

Plants entering a collection must be correctly identified and all passport data, including any information about the accession when it was originally collected, should be retained and made available by the curator (IPGRI-INIBAP/CIRAD, 1996). It is important to have a documentation system in place before the first accession arrives on site. For a field genebank, environmental and site-specific parameters also need to be recorded as these are important when the accessions are characterised or evaluated. Verification that accessions are correctly identified and true to type should be done by crop experts or taxonomists. Access to knowledgeable crop experts is very important. An accession is declared true to type if its traits match those of the known reference and the *Musa* Germplasm Information System (MGIS) is useful for this purpose. There are over a hundred characters that are used to describe the diversity within banana (IPGRI-INIBAP/CIRAD, 1996). Molecular characterisation is increasingly being used to complement morphological data (Christelová et al., 2011).

A labelling and numbering system needs to be in place and each accession should be identified separately. Labels should be legible and robust in design for field use. Field maps should be developed before planting and kept up to date regularly. Old maps should be retained and dated for reference. Maps should be used during planting, harvest and evaluation. Always have spare copies and backup.

Each collection should have standard operating procedures for all work done. A manual containing work and safety procedures is useful as a training document, as a reference manual and as a documentation of procedures to provide continuity during staff changes. Protocols should be written by staff that actually perform the work to ensure the procedures are practical, but they must be detailed enough so that new workers can easily make use of them. Of course basic hands-on training should also be given to staff so they can carry out tasks accurately and with confidence.

8 Dissemination of quality planting stock

Guidelines for the dissemination of planting stock need to be established and communicated to end users. The guidelines need to establish who are eligible recipients, justification for the request, the number of accessions and suckers allowed, availability of material from a seasonal perspective and whether there are associated costs. By the same token a record of the date of request, accessions requested, requester's name and address and shipping dates and costs should be kept on file.

As stated in previous sections of this chapter, it is mandatory that the field collection is regularly inspected by crop specialists to ensure the plants are true to type and disease free. This is particularly important when planting stock is being collected. Depending on the quarantine requirements of the area the material is being moved, sections of the plant may have to be removed for disease indexing (see previous chapter for procedures recommended for material to be initiated *in vitro*). This is particularly true of important banana viruses where a range of sensitive diagnostic tests are available for their detection. Always be aware of and comply with any restrictions applying to the movement of plants from one country or district to another. Technical guidelines for the safe movement of *Musa* germplasm have recently been updated and mainly apply to micropropagated plants and seeds (Thomas, 2015).

9 Case study

This section discusses the key issues in the selection, establishment and management of a field *Musa* germplasm collection at the Queensland Department of Agriculture and Fisheries, Ayr Research Facility (19°36′S 147°22′E) which is situated in the dry tropics. It was thought that a second field genebank was needed as a backup to the one located in the wet tropics at South Johnstone Research Facility (17°38′S 145°59′E) centred in Australia's major banana production region. Both are supported by an extensive *in vitro* collection and quarantine laboratory at Maroochy Research Facility (26°38′S 152°56′E).

The rationale for the second field genebank included:

- The Ayr site is not located in a major banana production area so the biosecurity risks are lessened compared to the South Johnstone site. The establishment of this field collection has become even more important due to the outbreak of TR4 (*Fusarium oxysporum* f.sp. *cubense*, Tropical Race 4) in the Tully Valley approximately 50 km from the South Johnstone site. In contrast Ayr is approximately 300 km from Tully and is centred in a major sugarcane production area. Bananas had not been grown widely at the Ayr Research Facility because it was established primarily for grain and pulse crop R&D. Bananas are often grown around people's homes in the district but the only fruit crop of significance on site was mango.

- Security of and access to the Queensland government facility at Ayr is very good and farm staff were available on-site to manage the crop and monitor traffic entering the site.
- The Burdekin Basin is an important agricultural region with access to good quality water for irrigation. The Australian banana industry has been looking to relocate in other regions due to the frequency of cyclones in the Tully/Innisfail area (two Category 5 cyclones within five years of one another) and currently the TR4 threat. The frequency and intensity of cyclones in the Burdekin is significantly less than further north. It was also thought that valuable lessons could be learned about banana production in this region by observing how well the different banana accessions established and grew.
- The primary purpose of the field genebank, however, was to verify the trueness to type of selected accessions from the *in vitro* collection. Following inspection by a crop specialist familiar with the many different cultivars and *Musa* species, as well as by biosecurity staff to declare disease freedom, pathogen-free plants were to be established *in vitro* either by the curator at Maroochy Research Facility or by private laboratories. Therefore the field genebank was necessary to compliment the *in vitro* genebank by conserving *Musa* germplasm and to service the needs of R&D and the banana industry through the provision of quality planting stock (i.e. suckers) needed by the tissue culture laboratories for the micropropagation of cultivars in demand.

Having considered Ayr as the likely site for the field genebank, it was necessary to discuss arrangements for the genebank's establishment with facilities managers, industry leaders and the funding body. It was important to have a strong business case in these discussions.

A block on the site was selected that was partly protected from the strong prevailing winds for which the region is noted and had irrigation lines already installed. Furthermore it was sited in an area where further expansion of the collection and crop rotation was possible. Because one of the requirements was to have a site free of *Fusarium oxysporum* f.sp. *cubense*, it was decided to plant a scattered stand of Race 1 susceptible 'Ducasse' (ABB, Pisang Awak subgroup) throughout the block to act as sentinels (Fig. 3) in the event the plants succumbed to the pathogen. It should be noted that at the time of planting the sentinel block the TR4 outbreak in Tully had not been reported, otherwise 'Williams' (AAA, Cavendish subgroup) would have also been planted as it is resistant to Race 1 but susceptible to TR4.

No symptoms were detected in the 'Ducasse' after 12 months (none of the plants have yet developed symptoms after five years) and the entire block was planted to a range of accessions (416 plants from 53 banana and plantain accessions) to coincide with the wet season. Only 3–4 sun-hardened nursery grown micropropagated plants (Fig. 4) were planted for those accessions that had been in the Maroochy *in vitro* genebank for an extended period of time, while for those cultivars in demand up to 15 plants were planted. The spacing used was 1.5 m between plants within a row and 5 m between rows to facilitate movement of heavy farm machinery (Fig. 5).

We observed that plants along one of the boundaries were experiencing slow growth attributed to water stress, particularly those along the guard row which was established to protect the germplasm block from the prevailing winds. The water stress situation

Figure 3 A sentinel planting of 'Ducasse' (ABB, Pisang Awak subgroup) susceptible to *Fusarium oxysporum* f.sp. *cubense* Race 1. No plants developed symptoms of Fusarium wilt at Ayr Research Facility but at another site selected as a Race 1 screening area in the subtropics, all susceptible sentinel plants developed symptoms within 12 months. The subtropical site had not grown 'Lady Finger' (AAB, Pome subgroup) for 32 years after it was abandoned due to Fusarium wilt, thus demonstrating the longevity of the pathogen in the soil and the effectiveness of a sentinel planting. (Note: temporary windbreak to protect sentinel plants during establishment. For the field genebank a guard row was established to provide some protection from the winds).

Figure 4 Fully sun-hardened, pathogen tested micropropagated plants ready for planting in the field. These plants were provided by a private, accredited nursery and were typical of the plants used by farmers to establish their plantations.

Figure 5 A portion of the field genebank at Ayr Research Facility. Note that plants are not desuckered as per commercial practice as the purpose is to have suckers for re-initiation *in vitro*.

was remedied by changes to the irrigation schedule and by deep ripping along one boundary where a row of eucalyptus trees had roots that had invaded the block and were intercepting the water before the banana plants could fully utilise the irrigation provided. In a new phase of the project (typically funded for five years) support for a field experimentalist with banana experience is warranted. In any case, Ayr field staff have gained experience and confidence in the management of a banana field genebank. Plants within the collection will not only be important in conserving valuable germplasm; it will also be an important source of material for revitalising *in vitro* collections.

In conclusion the Ayr field genebank, working together with government and private tissue culture laboratories, represents a strategic investment by the banana industry in establishing a secure site for the provision of quality planting stock needed for on-going research, development and extension. The banana industry also gains further benefit by access to true-to-type, pathogen-tested stock to meet its planting material needs. Lobbying efforts for long-term support for this site continue.

10 Future trends and conclusion

The genetic potential of the diversity within the *Musa* genepool in breeding for disease resistance, yield increase and quality improvement is as important today as it was in 1959 when the United Fruit Company launched a major banana breeding programme. This diversity needs to be conserved and utilised for future generations. Genes that are needed today may not meet future threats caused by emerging pests and diseases, as well as climate change, and therefore genebanks are an important component of the conservation strategy. Whether they are field, *in vitro* or cryopreserved collections, they all complement one another.

Field genebanks will always be important. For a start they are the only genebank where a range of fully grown plants and their fruit can be seen, touched and tasted. Comparisons between cultivars are possible for important traits and significant differences can be measured. It is possible to select the best-performing cultivars for a specific region and use.

As discussed in this chapter, field genebanks complement *in vitro* collections by acting as an alternative site for germplasm conservation while at the same time verifying that the material in culture is true to type and has not been mislabelled or misclassified. The field genebank, when managed properly, also provides quality pest- and disease-free planting stock that can be used to rejuvenate *in vitro* stock. It is micropropagated plants produced through this process that are recommended to provide superior planting material for research purposes or commercial practice. However, even if only conventional planting material is to be used, the guidelines presented in this chapter should ensure quality planting stock is produced.

Moving forward, characterisation of field collections will continue to involve crop experts but it will be supported by advances in imaging technology and computerisation where images of an accession can be relayed to multiple experts around the globe. This is already happening through the *Musa* Germplasm Information System. Phenotyping germplasm for important traits such as drought tolerance and nutritional value will supplement the information already collected on disease resistance and agronomic performance. Metabolomics information will also feature more prominently when germplasm is characterised. Molecular fingerprinting will continue to assume more importance in characterisation and perhaps we will even reach a stage when off-types can routinely be identified in the laboratory using these approaches.

Likewise assessing collections for pest and disease freedom will improve with advances in diagnostics which in turn rely on advances in molecular biology. However it should never replace an experienced eye. The establishment of field collections in areas free of major disease threats will also be supported by improvements in surveillance and imaging technology.

Finally while *ex situ* conservation is very important every effort should be made to protect those areas where wild *Musa* species exist. The genetic diversity in cultivated plants is derived from wild ancestral species and yet commercial crops contain less than 10% of the total genetic diversity available to a genus (Reed et al., 2004). Knowledge of the wild diploid ancestors of the many cultivated bananas and plantains, combined with recent advances in *Musa* genomics, is allowing breeding approaches that mimic the sequences of crosses and selections that have occurred over several millennia (Bakry and Horry, 2016). Using this breeding approach allows gene introgression strategies to develop new banana and plantain varieties to meet the challenges of the future.

11 Where to look for further information

Bioversity International has a mission to deliver scientific evidence, management practices and policy options to use and safeguard agricultural biodiversity to attain global food and nutrition security. Bioversity and its previous iterations (i.e. International Plant Genetic Resources Institute, International Network for the Improvement of Bananas and Plantains, and International Board for Plant Genetic Resources) have published several important articles on management of genebanks, including field genebanks. The reader is encouraged

Published by Burleigh Dodds Science Publishing Limited, 2018.

to visit Bioversity's website (www.bioversityinternational.org), and a list of the more pertinent publications from these organisations can be found at the end of this section.

Another website is the *Musa* Germplasm Information System (www.crop-diversity.org/mgis) that provides information on a large number of *Musa* species, landraces and cultivars with links to organisations that host germplasm collections or conduct research on bananas and plantains.

For those wanting to keep up to date with banana and plantain research developments, note that the upcoming conferences and workshops, news and views, and publications are updated in the Pro*Musa* website (www.promusa.org) and register your interest in joining a network of people working on banana. The Pro*Musa* Secretariat, together with the International Society for Horticultural Science and host institutes, regularly holds conferences around the globe and publishes proceedings of papers and posters presented at these events in *Acta Horticulturae*.

Finally for those interested in the scientific principles underlying banana production practices it is hard to go past the book of Robinson and Galán Saúco (2010).

For further reading:

Carlier, J., De Waele, D. and Escalant, J. V. (2003), *Global evaluation of* Musa *germplasm for resistance to Fusarium wilt,* Mycosphaerella *leaf spot diseases and nematodes. Performance evaluation,* Vézina, A. and Picq, C. (Eds), INIBAP Technical Guidelines 7, The International Network for the Improvement of Banana and Plantain, Montpellier, France.

IPGRI-INIBAP/CIRAD. (1996), *Descriptors for banana* (*Musa* spp.), International Plant Genetic Resources Institute, Rome, Italy/International Network for the Improvement of Banana and Plantain, Montpellier, France/Centre de coopération internationale en recherche agronomique pour le développement, Montpellier, France.

Reed, B. M., Engelmann, F., Dulloo, M. E. and Engels, J. M. M. (2004), *Technical guidelines for the management of field and* in vitro *germplasm collections, IPGRI Handbooks for Genebanks No. 7*, International Plant Genetic Resources Institute, Rome, Italy.

Robinson, J. C. and Galán Saúco, V. (2010), *Bananas and plantains, 2nd Edition*, CAB International.

Saad, M. S. and Ramanatha Rao, V. (2001), *Establishment and management of field genebank, a training manual*, IPGRI-APO, Serdang, Malaysia.

Tomekpe, K. and Fondi, E. (2008), 'Regeneration guidelines: banana', *Crop specific regeneration guidelines [CD-ROM]*, Dulloo, M. E., Thormann, I., Jorge, M. A. and Hanson, J. (Eds), CGIAR System-wide Genetic Resource Programme, Rome, Italy.

12 Acknowledgements

Thanks to D. Peasley for his support, helpful discussions and for critically reading the manuscript. Thanks also to J. Daniells for provision of photos for Fig. 2. The financial support provided by the Australian banana industry and Horticulture Innovation Australia is gratefully acknowledged.

13 References

Bakry, F. and Horry, J.-P. (2016), 'Advances in genomics: Applications to banana breeding', *Acta Hort. (ISHS)*, 1114, 171–80. https://doi.org/10.17660/actahortic.2016.1114.24.

Carlier, J., De Waele, D. and Escalant, J. V. (2003), *Global evaluation of* Musa *germplasm for resistance to Fusarium wilt,* Mycosphaerella *leaf spot diseases and nematodes. Performance evaluation* (Vézina, A. and Picq, C. (Eds)), INIBAP Technical Guidelines 7. The International Network for the Improvement of Banana and Plantain, Montpellier, France.

Christelová, P., Valárik, M., Hřibová, E., Van den houwe, I., Channelière, S., Roux, N. and Doležel, J. (2011), 'A platform for efficient genotyping in *Musa* using microsatellite markers', *AoB*.

Crichton, R. and Van den Bergh, I. (2016), 'Publication of the results of the International *Musa* testing Programme, Phase III (IMTP III)', *Acta Hort. (ISHS)*, 1114, 191–5.

Daniells, J. W., Sachter-Smith, G. and Taylor, M. (2016), 'Bananas adrift in time – a case study in the Solomons', *Acta Hort. (ISHS)*, 1114, 27–33.

IPGRI-INIBAP/CIRAD (1996), *Descriptors for banana (*Musa *spp.)*, International Plant Genetic Resources Institute, Rome, Italy/International Network for the Improvement of Banana and Plantain, Montpellier, France/Centre de coopération internationale en recherche agronomique pour le développement, Montpellier, France.

Ocimati, W., Karamura, D., Sivirihauma, C., Ndungo, V., De Langhe, E., Adheka, J., Dheda, B., Ntamwira, J., Muhindo, H., Ragama, P. and Blomme, G. (2016), 'On-farm banana (*Musa*) cultivar diversity status across different altitudes in North and South Kivu provinces of eastern Democratic Republic of Congo', *Acta Hort. (ISHS)*, 1114, 35–44.

Orjeda, G. (1998), Evaluation of Musa germplasm for resistance to Sigatoka diseases and Fusarium wilt, *INIBAP Technical Guidelines 3.* International Plant Genetic Resources Institute, Rome, Italy.

Ortiz, R. (2011), 'Musa'. In *Wild Crop Relatives: Genomic and Breeding Resources Tropical and Subtropical Fruits* (Kole, C. (Ed.)). Springer-Verlag Berlin Heidelberg, pp. 97–128.

Reed, B. M., Engelmann, F., Dulloo, M. E. and Engels, J. M. M. (2004), Technical guidelines for the management of field and in vitro germplasm collections, *IPGRI Handbooks for Genebanks No. 7.* International Plant Genetic Resources Institute, Rome, Italy.

Robinson, J. C. (1996), *Bananas and Plantains.* CAB International. Wallingford, UK.

Rosales, F. E., Arnaud, E. and Coto, J. (1999), *A Tribute to the Work of Paul H. Allen: A Catalogue of Wild and Cultivated Bananas.* International Network for the Improvement of Banana and Plantain, Montpellier, France.

Rowe, P. (1987), 'Banana breeding in Honduras'. In *Banana and Plantain Breeding Strategies* (Persley, G. J. and De Lange, E. A. (Eds)). *ACIAR Proceedings No.* 21, pp. 74–7.

Saad, M. S. and Ramanatha Rao, V. (2001), *Establishment and Management of Field Genebank: A Training Manual.* IPGRI-APO, Serdang, Malaysia.

Smith, M. K., Hamill, S. D., Becker, D. K. and Dale, J. L. (2005), '*Musa* spp. Banana and Plantain'. In *Biotechnology of Fruit and Nut Crops* (Litz, R. E. (Ed.)). CABI Publishing. Wallingford, UK, pp. 366–91.

Taxonomic Advisory Group (TAG) (2016), *Minimum DescriptorLlist for Musa*, Bioversity International, Montpellier, France.

Thomas, J. E. (2015), Musa*Net Technical Guidelines for the Safe Movement of Musa Germplasm, 3rd Edition.* Bioversity International, Rome.

Tomekpe, K. and Fondi, E. (2008), 'Regeneration guidelines: banana'. In *Crop Specific Regeneration Guidelines [CD-ROM]* (Dulloo, M. E., Thormann, I., Jorge, M. A. and Hanson, J. (Eds)). CGIAR System-wide Genetic Resource Programme, Rome, Italy.

Yunus, A. G. (2001), 'Introduction to field genebank'. In *Establishment and Management of Field Genebank: A Training Manual* (Saad, M. S. and Ramanatha Rao, V. (Eds)). IPGRI-APO, Serdang, Malaysia,pp. 59–63.

Chapter 8

Good agricultural practices: an end point or a starting point for more sustainable banana production?

Charles Staver, Bioversity International, France

1 Introduction

2 Why GAP in banana production?

3 GAP-related certification schemes in banana production

4 Is sustainable banana production being operationalized through GAP certification?

5 Improving GAP for more sustainable banana production

6 Scientific frontiers in bananas and implications for GAP

7 Conclusion: mutual lessons between GAP certification and approaches to sustainable banana production

8 Where to look for further information

9 References

1 Introduction

Good Agricultural Practices (GAP) are on-farm practices to ensure food safety, low-impact use of natural resources and the safety of workers (COAG 2003). Schemes to certify GAP implementation reflect societal concerns about the safety and sustainability of how our food is produced. The food industry, advocacy groups for the environment, fair prices and consumer choice, together with the participation of governments, have put into place numerous certification frameworks for agricultural products. GAP have also been proposed to guide policies and actions for the economic, environmental and social sustainability of agriculture. In this chapter we examine the short-term nature and impact of value-chain-oriented GAP in banana production and their significance for how banana production can be made more sustainable.

Movements for food safety and more environmentally sound agriculture have a long history. The establishment of the Codex Alimentarius in 1962 by the Food and Agriculture Organization (FAO) and the World Health Organization marked the acceleration of the global standardization of food quality parameters (Trienekens and Zuurbier 2008). Rachel

http://dx.doi.org/10.19103/AS.2017.0020.14

Carson's book *Silent Spring* on the far-reaching impacts of pesticides in 1962, Earth Day in 1970 in the United States which contributed to the founding of the Environmental Protection Agency, as well as a global Earth Day in 1990, gave impetus to the Earth Summit in Rio de Janeiro in 1992 and a growing global consensus on the importance of sustainability. In 1995 the World Trade Organization (WTO) adopted Codex Alimentarius standards for use in resolving trade disputes. In response to these developments, as well as to broader consumer concerns about food safety and methods of food production, a leading GAP standard, GlobalGap, was founded in 1997 as EurepGAP by an alliance of supermarkets as an independent certification framework. In this same period, in response to broader consumer concerns, alternative certification frameworks were also implemented. Organic certification provided market recognition of production techniques not based on synthetic pesticides and chemical fertilizer, while FairTrade certification addressed concerns about the equity in commodity supply chains through standards for a fair price and social bonus to smallholder growers organized for group marketing and the right to organize with additional social benefits to workers on larger farms (Raynolds 2000). The International Trade Center, operating with a mandate from the United Nations (UN) and the WTO, maintains a web site which details over 200 different certification schemes (www.standardsmap.org) with a descriptive framework organized under five categories – environment, social, management, quality and ethics. Seventy-one of those listed apply to banana with the five categories being divided into 24 subheadings and 268 specific themes. These 268 specific themes are 40% environmental, 32% social, 15% quality, 11% management and 3% ethics related (Table 1).

In 2003 FAO set up an initiative to promote and expand the use of GAP at each stage in the production process to ensure food safety, market chain governance, better natural resource use and improved worker safety (http://www.fao.org/prods/gap/). Many countries around the world have set up national GAP standards based on the FAO initiative (FAO 2017). These national standards provide the framework for national public agencies to accredit third-party agencies which certify private sector and non-governmental organization GAP frameworks. In the FAO approach, GAP are non-prescriptive, voluntary and compatible with other regulatory codes such as the Codex and conventions on World Animal Health and International Plant Protection, summarized as practices that address

Table 1 Breadth of subjects covered in certification standards of agriculture and food products (www.standardsmap.org)

Theme	Subjects
Environment	Soil, forests, inputs, biodiversity, livestock, waste, water, energy, climate-carbon
Social	Human rights and local community, labour: conditions of work and social protection, labour: employment and employment relationships, labour: human development and social dialogue
Economic	Economic viability, sustainability management criteria, supply chain responsibilities
Quality	Product/service quality management: hazard and risk, process documentation, testing and response, traceability; food management systems: food processing and handling routines, traceability, documentation, corrective action documentation, facilities, input and waste management
Ethics	Corruption/anti-bribery, compliance with international and national conventions and laws – environment, religion, culture, gender

environmental, economic and social sustainability. The FAO–GAP initiative proposes a dual function for GAP: short-term consumer-oriented guarantees of practices for food safety, producer and worker safety and justice and low impact on natural resources, as well as long-term guidance and indicators of sustainability which are oriented towards public development policy, investment and innovation. The FAO definition of sustainability suggests four pillars of GAP:

- Food safety and quality.
- Economic viability.
- Environmental sustainability.
- Social acceptability.

A National Research Council task force on sustainable agriculture for the twenty-first century in the United States (NRC 2010) proposed goals of agricultural sustainability with only three mentions of GAP in their lengthy document, suggesting that sustainable agriculture is much more than GAP. However, their proposed goals of sustainable agriculture, which overlap closely with the FAO pillars, are as follows:

- Satisfying human food, feed and fibre needs, and contributing to biofuel needs.
- Enhancing environmental quality and the resource base.
- Sustaining the economic viability of agriculture.
- Enhancing the quality of life for farmers, farm workers and society as a whole.

This chapter on GAP in banana discusses the justification for GAP certification frameworks in banana production and the range of GAP-related certification schemes used in the sector. The effectiveness of certification systems is then assessed in terms of the broader aims of sustainable agriculture. Are the current GAP certification frameworks an effective checklist of the broader agenda for sustainable banana production and, if not, how can they be redeployed and strengthened for greater effectiveness? This assessment is then applied to smallholder organic export banana production, particularly addressing opportunities for ecological intensification (EI) with economic benefits. The last section surveys scientific frontiers in banana research for their potential contribution to sustainable banana production and compatibility with GAP certification. The conclusions synthesize the lessons on how GAP can promote more sustainable banana production and what the agenda for sustainable banana production can learn from societal experiences with GAP.

2 Why GAP in banana production?

Bananas as a crop consist of numerous groups of cultivars belonging to the genus *Musa*. They are the eighth most important crop in terms of cultivation area and sixth in terms of total production after the major grain crops, potatoes and cassava (calculations from FAOStat). Bananas are commonly cited as the fourth most important crop in terms of value. Bananas are grown in 146 countries from subtropical to tropical zones, from sea level to mid-altitudes above 1500 m and in climates ranging from very rainy to desert. In each particular zone, growers deploy locally adapted techniques to address abiotic challenges and manage pests and diseases to produce bananas which satisfy specific market demands. The large diversity of cultivars and production systems is a challenge for the applicability of generalized GAP in diverse local contexts.

One large subgroup, Cavendish bananas, makes up around 40% of all banana production (Lescot 2012). Over 35% of Cavendish production is exported, representing around 14% of the total volume of bananas grown globally. Bananas were the first perishable fruit or vegetable traded from the tropics to temperate, higher-income countries with trade well established and expanding by the 1880s. As a result, in the early 1900s bananas were cheaper than apples in US cities (Cumo 2015). Bananas are still the fruit with the highest percentage of production exported globally (Liu 2008).

Although bananas have a long-established tradition of production and trade, they have not been immune from consumer concerns about food safety and the environmental and human health impacts of their production methods. The expansion of banana production in the humid coastal tropical lowlands of the Caribbean basin, then the coastal region of Ecuador and, more recently, in the coastal lowlands of Philippines and West and Central Africa, resulted initially from the clearing of tropical forest for the establishment of large-scale perennial monocultures. These monocultures have a high production potential of up to 60 tons fruit/ha/year. However, to ensure a blemish-free fruit which can withstand a trip to market of up to 3 weeks, growers use several hundred kilograms of nitrogen (N) and potassium (K) fertilizer per hectare as well as frequent aerial fungicide applications to maintain an abundance of healthy, disease-free leaves and, in some regions, multiple applications of nematicides annually to protect the root system (Henriques et al. 1997). The high levels of pesticide and fertilizer use can contaminate surrounding areas as well as pose a potential safety risk for workers applying them.

The high yields of blemish-free fruits also have a high labour requirement both in production and post-harvest operations. Banana production is labour intensive with many practices done manually. Each banana mat (flowering stalk, vegetative stalks and smaller suckers arising from a single rhizome) needs to be tended twice weekly year after year for leaf pruning, sucker thinning, weeding and bunch management to ensure big, blemish-free bunches. The bunch is bagged very soon after flower emergence with at least two successive visits for the placement of protectors and colour-coded flagging to identify projected harvest date and for deflowering. The combination of high labour use and pesticide applications generates potentially high exposure of workers to toxic substances (Henriques et al. 1997). From the early 1990s onwards banana workers in Costa Rica, Philippines and Nicaragua sued Dow Chemical and Dole for damages resulting from the use of the pesticide DBCP in bananas for a decade after its use was prohibited in the United States (Bingham and Monforton 2013). Similarly banana workers in Guadalupe and Martinique sued for damages from problems of sterility resulting from the use of the nematicide chlordecone in banana production. Significant ground water contamination and ongoing health risks continue even to the current day (Multigner et al. 2007).

As well as these problems of pesticide use and worker safety, banana production faces challenges from the spread of disease. The threat of Fusarium wilt Tropical Race 4 (TR4) has raised concerns among both producers and consumers about security of supply, production methods and the sustainability of banana production (Ploetz 2015). *Fusarium* spores remain in the soil for decades, effectively prohibiting the replanting of banana. This concern is justified by previous disease outbreaks which have affected large-scale banana culture. The initial export dessert banana Gros Michel was abandoned by export growers and replaced by the resistant cultivar Cavendish in the 1950s due to the spread of Fusarium wilt Race 1. More recently, black leaf streak, a virulent leaf disease which also accelerates fruit maturation rates, has spread across banana-growing regions and has led to increased use of fungicides to ensure production of a transportable fruit (Marin et al. 2003).

Ironically, the global banana market is also chronically perched on the brink of oversupply with a complex system of spot prices to channel any surplus into consuming countries where consumers are very sensitive to banana prices (Liu 2008). In the late 1990s and through 2005, for example, prices dropped and remained low due to oversupply and fiscal crises in Asia and Eastern Europe which affected demand. Liu (2008) further suggests that this oversupply potential may lead to pressure on production costs and the use of risky cost-cutting measures and labour practices which increase consumer suspicion of banana integrity and lower demand, generating further potential for oversupply. The GlobalGap certification framework originated with supermarkets as a strategy to gain market advantage through improved consumer confidence. Other certification approaches, although originating from civil society, also operate to provide market advantage and price stability. This marketing strategy continues to the present with even low-cost supermarkets selling FairTrade bananas in the United Kingdom (The Guardian 2016). In the next section we characterize the alternative GAP certification frameworks.

3 GAP-related certification schemes in banana production

Standards which facilitate trade and commerce are fundamental to current human society. For banana trade, the long distance between producers and consumers, the need for predictability in product response to storage, handling and ripening and the concerns of both marketers and consumers about how the product was produced which may not be apparent in the product itself have made standards central. In this section we will focus on five certification frameworks – GlobalGap, Rainforest, EU Organic, FairTrade and SA8000, which orient banana growers on production processes and ensure buyer confidence.

All of these frameworks draw elements from three quality assurance systems in the food sector (Trienekens and Zuurbier 2008). They are based on GAP which offer orientation on practices at each stage of the process. They also include elements from hazard analysis of critical control points (HACCP) which is applied not only in agricultural production, but also in food processing and in many other industrial processes. In HACCP systems, production processes are broken down into steps for risk analysis with the identification of critical points which determine product safety and quality. Once critical control points are identified, they can be monitored to measure compliance. Production processes are broken down into steps for risk analysis with the identification of critical points. Finally, the frameworks integrate management procedures which are independent of the product itself and serve to guide meeting customer satisfaction, to orient monitoring and self-improvement and to ensure that non-compliance is addressed. Organizations of small growers which are certified as a group through their organization represent an important challenge in management procedures to ensure internal control procedures which keep all growers in compliance.

These schemes are all third-party certification frameworks which depend on three elements to be operational (Loconto and Busch 2010). A set of standards are proposed for certification representing the specific interests of proposing group. Meeting these standards brings access or benefits which means certifiers are also needed which are independent from those proposing the standards, from those producing the product and from other direct actors in the supply chain. Finally, accreditation agencies are needed to govern the certifiers. These have taken both public and private forms, but more importantly international agreements have served to ensure that accreditation is recognized across national borders, an essential element in global supply chains.

An overview of the emphasis of the five different certifications illustrates the evolving reach of safety and traceability, environmental impact and social justice in food production. Below, each of the five certification frameworks is described in terms of its origin and major categories of interest following the method used by the International Trade Centre (ITC) which is mandated by the UN and the WTO to provide technical advice to promote global trade. The ITC has developed a Standards Map which compares over 200 standards used in global trade (Fig. 1).

GlobalGap was founded in 1997 as an independent certification framework to address the issue of brand protection for large international supermarkets in light of consumer concerns about food safety and methods of production. The certification is based on 157 criteria of the 268 in the ITC comparator, representing all the categories except ethics (Fig. 1). Certification is required of all producers who supply the large multi national supermarkets. Inspections are annual and include an unannounced visit and most criteria must be met for initial certification. GlobalGap controls use of inputs such as pesticides, which are governed by national and international regulations, and promotes risk management to ensure food safety and reduce human and environmental exposure to pesticide use. Certification procedures on farm and in the post-harvest stage are based on a review of practices, while the destination countries also sample for Maximum Residue Limits. GlobalGap certification is not allowed to be used in food labelling for consumers.

Rainforest/Sustainable Agriculture Network certification began in 1990 as Eco-OK labelling applied only in banana with a focus on reducing the environmental impact of agricultural production. The Rainforest Alliance label was introduced in 2001 and currently certifies a broader range of products, including pineapple, coffee and cocoa. Rainforest certification has been used in branding, having been adopted by the major banana producer,

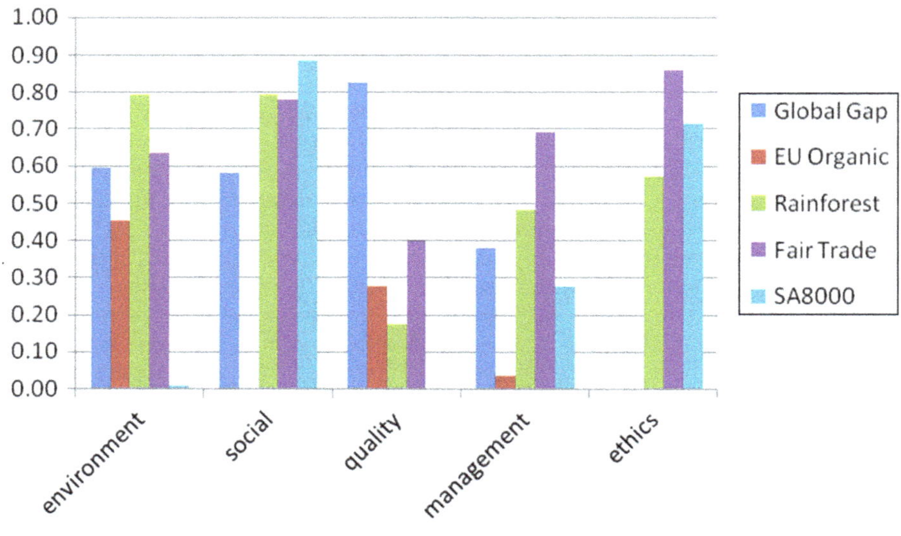

Figure 1 Percentage of ITC criteria specified for each GAP certification approach.

Chiquita, for all its banana production, as well as by some major supermarkets which only sell Rainforest-certified bananas in response to consumer pressure on conservation issues. The Rainforest label covers all five themes identified by the ITC, addressing 177 of the 268 themes detailed by the ITC (Fig. 1). Certification involves a minimum set of immediate criteria to be met and regular inspection for demonstration of improvement in the case of a large number of other criteria. The Rainforest organization maintains its own list of prohibited inputs based on diverse international conventions.

The European Union (EU) Organic Agriculture label was established in 2007 by the European Council of Ministers to promote organic agriculture and to reassure consumers interested in consuming food produced organically. The EU organic label focuses primarily on environmental criteria through the use of inputs of natural origin and quality factors covering 60 of the 268 criteria employed by the ITC (Fig. 1). A conversion period of 2–3 years is required, after which certification is annual with planned and unannounced inspections.

The FairTrade label was set up in 2002, bringing together numerous national organizations focused on more equitable supply chains offering a fair price for small growers and fair wages for workers. The FairTrade label covers the relationship between growers and the supply chain and guarantees that growers have received a fair price and that funds from each contract are invested in community improvement. The FairTrade label currently covers all five themes (Fig. 1), addressing 176 of the ITC 268 criteria.

Similar to the FairTrade label, SA8000 is a label addressing the fair treatment of workers, and is managed by Social Accountability International which was set up in 1997. The label is based on the Universal Declaration of Human Rights and International Labour Organization conventions. SA8000 covers four themes in the ITC analysis (Fig. 1), with major emphasis on social and ethical issues, covering 90 of the 268 ITC criteria. The certification is granted to the work place or company, but cannot be displayed on individual products.

The five certification schemes can be compared by looking at the percentage of the criteria in each theme which each scheme addresses (Fig. 1). Whilst GlobalGap covers environmental and social criteria, the certification scheme is more detailed in its requirements for quality. The label promotes good practices of conventional agriculture based on inputs covered by national authorities and food safety through risk analysis. The Rainforest label has a few key criteria for initial certification based on their own list of prohibited substances and suggested practices for forest and natural ecosystem conservation. Although their standards cover many areas, the primary basis for ongoing certification is demonstration of improvement from initial conditions. The EU Organic scheme specifies the use of inputs for organic production such as natural or naturally derived substances but does not address many other areas. The FairTrade label's focus on an agreed-on price and a fund for community improvement from each contract provided the initial core of this label and FairTrade still addresses a greater percentage of the social, management and ethical criteria than the other labels. They have also included more environmental criteria in recent versions. SA8000 differs from the other labels in being a workplace certification process with an almost exclusive emphasis on social and ethical issues.

The different weights given to the five categories in the labels represent contrasting approaches to ensuring safer food and lower environmental impacts in agriculture – a transitional or gradual change and transformational change based on modifications of fundamental relationships. GlobalGap strategizes a transitional approach to sustainable banana production with incremental improvements as national standards for inputs and labour are improved. Rainforest GAP are more ambitious in forest and ecosystem preservation, and prohibited substances and human and community rights. However, the proposed change is transitional based on relatively loose initial criteria with requirements for ongoing improvement in mutually agreed criteria. The EU Organic label is based on a transformation of production processes based on organic inputs, but has no proposed criteria for labour standards or socio-economic benefits to the community. FairTrade's mechanisms of minimum price, contract and premium for community improvement seek a transformation towards more equitable supply chains.

The area of banana production for export and volume exported continues to expand, although from year to year production fluctuates (FAO 2017) with current estimated volume at slightly more than 17 million tons, increasing from 11 million tons in 2000. Over the same time period, the price per ton almost doubled, although since 2009 prices have been fluctuating, but are relatively stable. Per capita consumption in Europe, the United States and Japan is considered steady, while much potential exists for increased per capita consumption in the Middle East, Eastern Europe and Central Asia which is expected to drive expansion. In those markets with stable per capita consumption, however, demand for organic and FairTrade bananas continues to increase. Globally, the trends for the different certification frameworks described in this section are harder to track, since country export and import statistics are maintained only by commodity. Lernoud et al. (2016) surveyed certification levels in multiple commodities and estimated that for banana around 300000 ha are certified, which is about 50% of export banana area and under 10% of total global banana area. Many producers may have multiple certifications complicating the estimation of total area certified. Nearly 250000 ha had GlobalGap certification followed by Rainforest, Organic and FairTrade with 90000, 48000 and 39000 ha, respectively. Over the period of Lernoud et al.'s (2016) survey, the area certified increased for the Rainforest, Organic and FairTrade labels, while that certified by GlobalGap fluctuated. With the growing importance of certification, it is important to examine more closely the contribution of GAP certification to banana sustainability.

4 Is sustainable banana production being operationalized through GAP certification?

The GAP certification frameworks under analysis here are a set of mechanisms which can be analysed using a theory of change about how agriculture can be made more sustainable (Oya et al. 2017). Adapting the four goals from the National Research Council (2010) presented earlier to banana, GAP should:

1. satisfy demand for safe bananas of desired qualities;
2. enhance environmental quality and resource base of banana lands and surrounding landscapes;
3. contribute to the economic viability of banana growing and its supply chains; and
4. enhance the quality of life of growers, their households, their communities and the larger society.

A thorough criteria-by-criteria review of the five themes and 268 specific criteria employed by ITC's standardsmap.org is beyond the scope of this chapter. We propose to focus on two aspects of sustainability which can be operationalized in the time span of certification procedures. First, to what extent do the certification frameworks contribute to the economic viability of banana farms? Certification is primarily oriented to satisfy the supply chain and consumers; but to address sustainability, it should also have value to growers and their organizations in order to improve their profitability, competitiveness and management capacity. Second, do labels build on ecological processes which underpin agricultural production systems (Phelan 2009)? While sustainability is addressed in the time frame of future generations, progress on the economic viability and synchronization of production with ecological processes can be implemented and monitored on the annual and multi-year certification cycle (Staver et al. 2017). In this section, we contrast four labels (not including SA8000) in terms of their contribution to grower management capacity as a key element in economic viability and their basis in ecosystem processes, and then in the following section propose opportunities for operationalizing sustainability through GAP certification for organic banana-growing smallholders.

Two reviews provide evidence about the role of certification in the economic viability of banana production. In their review of studies of the impact of certification frameworks on grower and community income and wellbeing, Oya et al. (2017) formulated an overall theory of change which identified four types of interventions – capacity building to growers and their organizations, market access and price, premiums to the community and labour standards. The four labels of interest here for banana are relatively similar in their approach to capacity building (Fig. 1), although the specific content of their GAP may differ. A review by ITC (2016) of user-friendliness of certification frameworks expanded further on grower capacity building. They analysed the provision of documents, technical assistance to meet requirements, technical assistance beyond certification, financial assistance and social learning. In all certification frameworks, growers and organizations increase their capacity to follow certification procedures and to implement a process of monitoring and improvement. Only 25% of labels reviewed (ITC 2016) offered technical assistance addressing productivity, profitability and market access. In the other three areas of intervention proposed by Oya et al (2017) (Table 2), FairTrade is the only label which addresses equity in the supply chain with a specified minimum price. The organic label has a potential, although not guaranteed, of higher price, while GlobalGap and Rainforest simply ensure access to certain markets without a price benefit.

A closer examination of the content of certification criteria, formats and procedures shows that GlobalGap has 220 critical control points to document for compliance, EU Organic 52, Rainforest 126 and FairTrade 35 along with other forms and checklists to be completed routinely. This paperwork requirement, often viewed as a burden (Steering Committee of the State-of-Knowledge Assessment of Standards and Certification 2012), is usually assigned in grower marketing organizations to a certification department which operates separately from the technical assistance department. Commonly, growers and the organization's field staff document production practices and inputs used weekly with less frequent review of checklists of risk analysis in the management of labour, water and waste. This documentation is oriented towards ensuring food safety and production methods for supply chain buyers and consumers.

However, the documentation only tangentially serves the business enterprise management concerns of growers and grower organizations. GAP certification frameworks contain only minimal criteria measuring profitability and production efficiency. In unpublished recent surveys in Peru and Ecuador, growers were not familiar with the costs of many inputs or services negotiated by their organizations such as foliar spraying for

Table 2 Types of interventions present in four certification schemes to promote socio-economic outcomes in banana

	Types of interventions			
Label	Capacity building	Price/market access	Premium to community	Labour standards
GlobalGap	Understanding of Conventional GAP and certification, monitoring and correction process for grower and organization	Obligatory for certain markets with potential to qualify for numerous markets	NA	Safe, legal employment following national laws
EU Organic	Understanding of Organic GAP and process of certification, monitoring and correction for grower and organization	Potential price increase for organic label	NA	NA
Rainforest	Understanding of Rainforest GAP, process of certification and expectation for upgrading for grower and organization	Obligatory for certain markets with potential to qualify for more markets	Criteria on human rights and local communities monitored for improvement	Safe, legal employment with link to international treaties
FairTrade	Understanding of supply chain governance, FairTrade GAP and process of certification and improved compliance	Guaranteed minimum price, contract with supply chain	Premium paid with each contract and used following plan set up by organization and community	Safe, legal employment with link to international treaties

disease control or the provision of fertilizers. Grower organizations cited production costs in a general category, not grower by grower. In spite of extensive paperwork, certification procedures do not contain calculations of cost of production/box of bananas, a breakdown of cost/box into labour and production inputs or routine monitoring of labour or input productivity – bunches bagged/day or kg N applied/box. This is a particular challenge for small growers who are highly dependent on larger organizations with a separate certification department responsible for recordkeeping. These departments are primarily oriented towards meeting certification standards rather than measuring economic efficiency. Whilst these systems may ensure certification, they do not provide a basis for individualized strategies to increase farm productivity, efficiency and profitability. A major challenge to sustainable banana production is the leveraging of the data collection in GAP for enterprise management as proposed by Bloom et al. (2012).

The socio-economic impact of certification, addressed in a mega-review by Oya et al. (2017) of 43 studies of the socio-economic impact of certification, of which around half dealt with coffee and fruit, found a small positive effect on income from certified product, price received and schooling for children in certified households compared with non-certified households. However, no significant differences were found for yields, overall household income or assets and illness in the household. Wages paid to labour were found to be slightly lower on certified farms. In spite of the large efforts in documentation and recordkeeping linked to certification, these results suggest that improved economic viability is coming primarily from price premiums and improved market access with only minimal leveraging of improved enterprise management.

In the framework of sustainable banana production, the economic managerial effectiveness of a banana farm which appears to be underemphasized in certification criteria must be paralleled by management effectiveness of the banana farm as an agroecosystem with both planned and unplanned biological components which interact with abiotic factors (Phelan 2009). The export banana agroecosystem has the inputs of natural ecosystems, sunlight and water, but also numerous other inputs (Fig. 2). The inputs (solid line arrows in Fig. 2) are targeted to specific components in the agroecosystem, most often the population of banana mats, but also pests, pathogens and weeds which reduce banana yield. The basic internal interactions among the specific components of a banana field, shown by dashed line arrows, represent the food web from plants to herbivores to consumers and the decomposition cycle with dead biomass processed through the decomposition food web to make nutrients available again to banana plants and other organisms in the system. These ecosystem processes operate in the banana field regardless of whether management takes them into account. In addition to bunches of bananas, the banana field outputs reflect both the type and level of inputs, but also the internal efficiencies in the agroecosystem. Management to increase and to build interaction among these ecological processes as the underlying basis of production should contribute to the optimization of the use of external inputs, labour, nutrients, water and sunlight; the reduction of losses to abiotic and biotic stress and the reduction in negative off-field impacts (Staver et al. 2017). We will refer to this as EI or ecological intensification throughout the chapter.

The ITC Standards Map overview of the environmental component of labels covers nine categories relevant to lower-impact practices and food safety – soil, forests, chemical or organic inputs, biodiversity, animals/livestock, waste, water, energy and climate/carbon – involving 106 criteria. We contrasted the ecological basis of certain criteria by first mapping

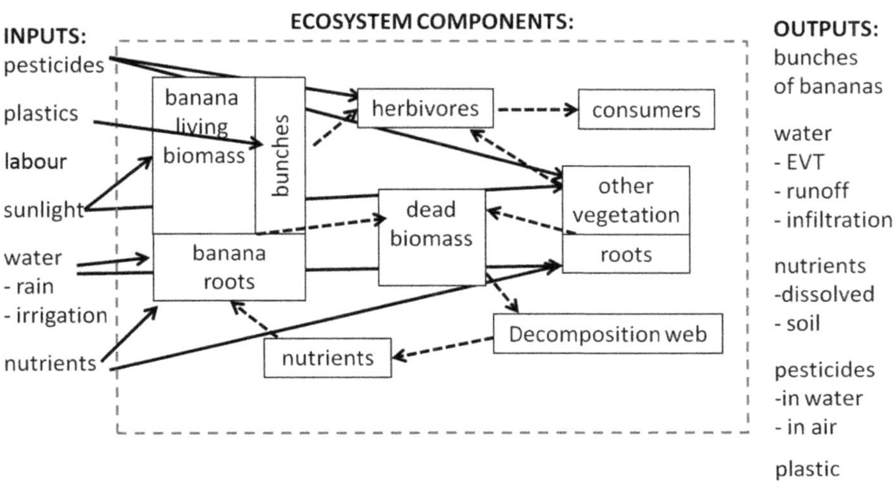

Figure 2 Inputs, outputs and main ecosystem components and processes.

selected GAP criteria onto the agroecosystem diagram (Fig. 3) – pesticide impacts, banana waste management, plastics, nutrients and soil cover – and then comparing the four certification frameworks (Table 3).

In the interest of food safety and the safe use of pest control products, the different certification frameworks address the type of inputs used, their handling and storage and the disposal of containers. In Fig. 3 the effects of pest control products on workers, aquatic organisms and soil biology are identified. The type of input permitted varies by label. GlobalGap follows national standards in the country where the crop is produced, while Rainforest and FairTrade have independent lists drawn from important international treaties with an orientation towards reduced use of more toxic products with time. The labels appear not to take into account possible effects on aquatic life which has been an important issue in banana-growing regions (Castillo et al. 2000) or soils (Bünemann et al. 2006). Mendez et al. (2018) have suggested that pesticide use, which focuses on minimizing residues in marketed bananas, still poses significant environmental risks.

The management of banana wastes and plastics receives attention from the four labels (Table 3). Requirements for management of banana wastes, primarily bunch stems which have a high K content, are addressed with specific requirements in the four labels. Plastic bags used to protect bananas for 11–13 weeks from flowering to harvest are a highly visible waste product. In conventional production, bags are impregnated with insecticides, a practice which reduces overall insecticide use, but also creates a hazardous waste. Three of the four labels target their recycling.

Lastly, in the areas of nutrient and soil management, all codes require documentation of both the amounts of fertilizers and soil amendments applied and the amount of banana produced. In addition, growers are required to conduct soil analyses. For small growers, economic factors and availability may be more important in the decisions on fertilizer application than soil analysis. The nutrient input–output calculation has both ecological

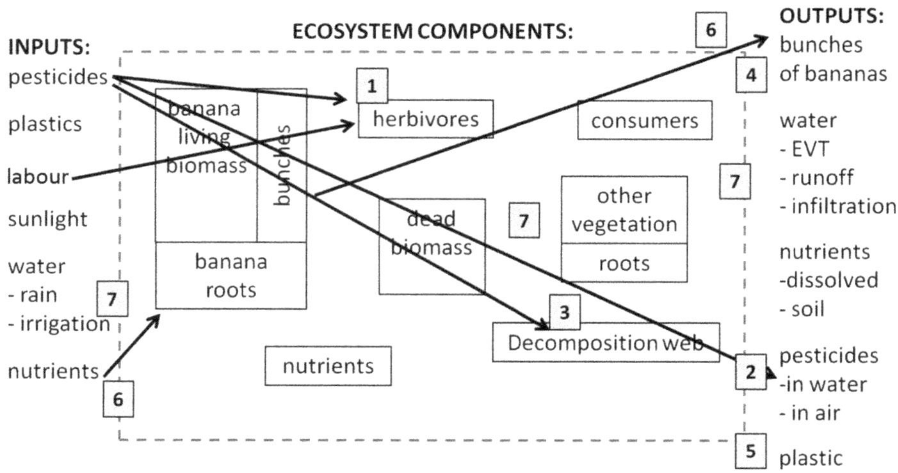

Figure 3 The relation between certification criteria and ecological processes in a banana agroecosystem. 1 – pesticides–labourers; 2 – pesticides–aquatic life; 3 – pesticides–soil biology; 4 – banana waste; 5 – plastics waste; 6 – nutrient balance; 7 – soil cover in waterways and banana field.

Table 3 Comparison of four certification schemes with reference to banana production and ecological processes

	GlobalGap	EU Organic	Rainforest	FairTrade
Pesticides – human	Determined by national codes; Safe handing, application and disposal critical	Natural products required with list of exceptions, Little emphasis on application, storage and disposal	Prohibited list as well as national codes; Safe handling, application and disposal critical; Training required	Prohibited list as well as national codes; Safe handling, application and disposal critical; Training required
Pesticides – aquatic life	Not a specific criterion in type of product or amount	Not a specific criterion in type of product or amount	Not a specific criterion in type of product or amount	Not a specific criterion in type of product or amount
Pesticides – soil biology	Not a specific criterion in type of product or amount	Not a specific criterion in type of product or amount	Not a specific criterion in type of product or amount	Not a specific criterion in type of product or amount
Banana waste	Composting specified	Composting to serve soil biology	General principle of recycling	Compost and mulch as recommended practice
Plastic bags	Plan for minimization and limited risk disposal	Not mentioned specifically –	Waste disposal plan addressing hazards for types of wastes	Waste disposal plan addressing hazards for types of wastes
Nutrient input–output	Not calculated	Not calculated	Not calculated	Not calculated
Soil cover – banana field, waterways	General principle	General principle, herbicides prohibited	General principle, herbicides prohibited	General principle, herbicide prohibited

and economic implications and the numbers needed to do the calculation are collected in the certification framework. However, the calculation is not required, a point we will take up in the next section. Seufert et al. (2017) identify the lack of criteria on nutrient use efficiency in their analysis of numerous organic codes. All codes also highlight soil erosion and protection, although often structured more towards annual crops rather than a perennial crop like banana.

This limited check of the ecological dimension of certification suggests that those related to direct hazard, safety and health due to use of external inputs are addressed in detail, but the question can still be asked about the contribution of GAP to sustainable production. Melo and Wolf (2005) examined the effectiveness of ecocertification among Ecuadorian banana growers. They constructed four indices of environmental management in banana, based on certification data – land, agrochemical, waste and water quality – and then compared smallholder FairTrade-certified production with non-certified production, and large-grower Rainforest-certified with non-certified production. They found that in all scores, the certified farms had superior indices to the non-certified farms based on data from 2003. They observed that non-certified farms were often in violation of national environmental laws which were poorly enforced. Lockie et al. (2015) made a similar observation in Philippines where the dependence of GlobalGap on documentary compliance with national regulations was considered ineffective compared to Rainforest certification.

In their scoring, Melo and Wolf (2007) used categorical (high risk reducing, intermediate risk reduction or risky practices) or binary scoring (presence or absence of risk reducing practices) to evaluate risks. For example, banana waste management was classified as either return or no return of bunch stalks to the field, while weed management was categorized as cultural control (avoiding use of herbicides), legal herbicide or banned herbicide use. This highlights a point made by Oya (2017) that certification frameworks are weakest in their capacity building on how to improve productivity. Numerous techniques are available for the recycling of banana bunch stems. The simple coding by Melo and Wolf (2007), which overlooks opportunities for more effective management beyond binary or simple categorical characterizations, illustrates a central challenge to GAP certification and integrated crop management (ICM) for sustainable production – can we develop approaches based on simple observational criteria of the potential for EI? We also observed that the certification schemes offer little direct orientation on economic effectiveness to growers. These two dimensions – indicators of EI which contribute to improved economic viability of banana production – present a double challenge to GAP data formats, collection and analysis. In the next section, we will use smallholder organic export banana as an example to explore the potential of this approach.

5 Improving GAP for more sustainable banana production

Growers and their organizations spend increasing time and effort in registering information required for certification. This information has often been viewed as an extra bureaucratic burden, even though it is becoming increasingly essential for maintaining access to markets in general and to specific markets which may have price benefits. The paperwork linked to compliance with government regulations on agricultural production methods is also becoming more common, although seen by growers as a distraction from the real focus of their farms. While agricultural enterprises are different from other production enterprises, management informed by data collection and analysis to allow benchmarking and improvement is a common feature of any business (Lans et al. 2013). In this section, we examine the case of smallholder export banana growers and their organizations which often manage two or more of the labels covered in this chapter. A dual strategy is described – first the improvement of management capacity and, in particular, the use of data and, in parallel, the identification of indicators more relevant to the ecological and economic performance of the banana farm. We propose that this combination of economic and ecological mechanisms through GAP certification has potential to lead to improved strategies for sustainable banana production.

Key areas of management capacity for business performance have been identified by Bloom et al. (2012):

- long-term goal setting with short-term performance benchmarking,
- use of incentives based on process benchmarking to motivate team effort,
- collection and analysis of data to identify opportunities for improvement.

Noteworthy in Bloom et al.'s framework is the role of data in each of the three areas. Two contrasting approaches for data use are either 'business process re-engineering' based on radical changes or 'continuous improvement' which focuses on small changes integrated into production processes based on observations of all members of the production team (Ohno et al. 2009). For small growers of organic export banana, who cannot easily sacrifice regular cash flow to completely uproot and redesign their banana production practices, a process of making continual small changes (known by the Japanese term 'kaizen') is more appropriate, provided it is adapted to local conditions (Hosono 2009). At the end of this section, we describe a more radical redesign strategy for banana production based on crop rotation for greater sustainability and lower pesticide use under development in Guadeloupe and Martinique (Risède et al. 2010a,b).

The adaptation to smallholder banana production of 'kaizen' business improvement tools should lead to a 'lean' organization based on the reduction of waste. The five analytical tools known as '5S' address the following:

1 value identification;
2 value stream mapping to eliminate waste;
3 process efficiency;
4 improved final product orientation; and
5 continuous improvement.

This approach has been applied, for example, by Hartman (2015) to improve the efficiency of a diversified vegetable family farm. He describes tools such as flowcharts with clear descriptions of procedures at each step and spaghetti diagrams designed to trace labour movement and tool use in production processes which are useful to eliminate waste in existing routines. Gains included organization of workers for more efficient production and harvest processes, tool storage and input supplies. Our observations in Section 3 on the minimal contribution of certification frameworks to farm management suggest that the ability of management to use observation, data collection and analysis for enterprise improvement is an essential element in the path to more sustainable banana production. However, Section 3 also concluded that ecological processes linked to improved productivity were underemphasized in the certification frameworks. In the following paragraphs we illustrate how opportunities for EI can be identified and then tested for contributing to improved productivity, labour efficiency and profitability, based on appropriate data or indicators and 'kaizen' type small improvements.

To map and analyse the potential for EI, which contribute to more short-term banana productivity, the simple ecological processes (Fig. 3) were expanded (Fig. 4) to highlight the interactions which need to be taken into account by growers, field technicians, scientists and potentially certifiers. These interactions describe processes in all banana fields whether they are organic or conventional and provide the basis for a sequential analysis of the potential for banana EI (Staver et al. 2017) as follows:

- the presence of system-threatening biotic pests or pathogens;
- cropping systems yield gap as a function of available light, water and temperature regimes;
- nutrient input/output and recycling, soil conservation and biological health;
- pest and pathogen management interactions;
- other yield factors.

This framework utilizes yield gap analysis employed by agronomists (van Ittersum et al. 2013), although potential yield and the analysis of yield gap factors are largely absent from the GAP certification frameworks. The EI framework also provides clear and practical steps to build an ICM approach in banana which addresses the negative and positive feedbacks among and between biotic and abiotic management practices in a given climate and soil (Agra, C. E. A. S. Consulting 2002).

A project financed by FONTAGRO, which promotes agricultural development in Latin America and the Caribbean, illustrates the application of an EI framework to improve environmental and economic sustainability (https://www.fontagro.org/proyecto/fortaleciendo-pequenos-productores-de-banano-organico-integracion-de-actores-manejo-sostenible-de-plagas-y-estrategias-de-salud-de-suelos/). The project originally focused on management of red rust thrips and soil health in export banana production in Dominican Republic, Ecuador and Peru, but has subsequently expanded its scope. Here we describe five performance indicators used in the project, the basis for their potential contribution to crop performance and how each practice interacts with other components. We then look at whether sufficient data are captured in current GAP certification processes to use these indicators at the level of individual farms. The five indicators are:

1 Bunch weight
2 Stand density
3 Banana crop nutrient and residue management
4 Insect pests of the bunch
5 Fusarium Tropical Race 4.

5.1 Bunch weight

Crop yield among the export banana growers is measured in boxes/ha/week, since this is the basis for payment. However, in small surveys carried out by the project, few growers were able to provide their average bunch size. Follow-up studies showed that bunch size was highly variable (Fig. 4). Among four farms in Ecuador bunch weight ranged from 7 to 45 kg. Replacement of the 20% lowest-yielding stems with an average weight bunch

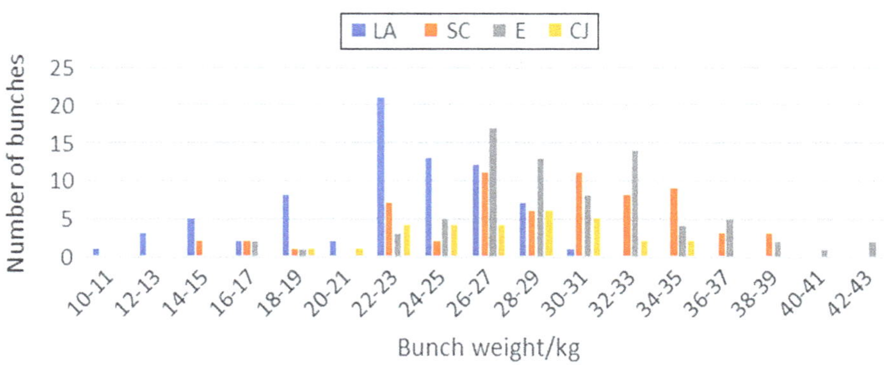

Figure 4 Bunch weights for individual bunches harvested from plots of 100 mats during a period of 8 months in four different farms in Ecuador.

would lead to an additional 2–3 tons of production per year. Unfortunately, the growth and productivity of single mats replanted into a semi-perennial stand are usually affected by low light caused by surrounding mats. Data on bunch weights by mat over several cycles along with spatial distribution of bunch weights would provide criteria to develop a strategy to address this yield gap through replanting in patches.

5.2 Stand density

Growers and their labourers actively manage the distribution of stems through the selection of suckers, elimination of defective mats and replanting. However, they did not readily have an estimate of mat density. Higher mat density can generate a shadier environment more suppressive of weeds and increase the number of bunches/ha. However, excessive stem density can result in smaller bunches and longer intervals between bunches on the same mat. High mat density correlated with high light interception also creates a less favourable environment for biomass accumulation and N fixation of leguminous cover crops (Tixier et al. 2011). A few small growers in Ecuador manage *Pueraria phaseoloides* or tropical kudzu as a cover crop. They suggested that the leguminous covers improved plantation growth and reduced costs for weeding (Corozo et al. 2016).

5.3 Banana crop nutrient and residue management

For every 10 tons of banana bunches harvested, 20 kg of N, 3 kg of phosphorus (P) and 67 kg of K leave the field. However, over an annual crop production cycle, a hectare of monocropped banana mats turns over more than 40 000 leaves through leaf pruning. More than 2000 pseudostems and 14 000 leaves are cut and left on to the soil surface when bunches are harvested. This biomass contains between 200 and 400 kg of K and smaller amounts of N and P. This biomass provides numerous opportunities to build positive interactions into banana crop management. Three practices are proposed to leverage nutrient and residue management for more sustainable production.

First, a simple calculation of nutrients applied and nutrients extracted, cross-checked for possible low levels of specific nutrients in the soil, can guide small growers to make simple adjustments of types and amounts of inputs (Rengifo et al. 2016). The large nutrient pools within the field based on biomass turnover can help to buffer short-term fluctuations in an organic system.

Second, the biomass from frequent leaf pruning provides abundant material to cover mineral soil. Instead of letting the leaf fall randomly, labourers can guide the leaf to the nearest patch of bare soil with a simple arm movement. A possible drawback of this practice is the spread of leaf diseases. However, in drier climatic zones where organic production is concentrated, in practice this risk is low.

Third, when bunches are harvested, all green leaves and at least a third of the pseudostem remain in the field largely intact on the soil surface where they fall. This biomass represents a resource to create a hot spot of nutrient availability, root activity and increased soil biological activity. The soil decomposition web has numerous groups of decomposers as well as fauna which feed on decomposers. The common grower practice of random residue management has been compared with strategic stacking of residue, not only for bunch size and return rate (Rengifo et al. 2017a), but also for soil health, using indices measuring the nematode community (Sanchez et al. 2017). By placing this plant biomass in front of the stem which forms the next bunch from the mat, the recycling time for nutrients is accelerated. Root proliferation and soil biological activity occur in response to the decomposing biomass, creating a site of healthy, active roots where organic fertilizers can be applied. Application of the slow-release organic fertilizers under the decomposing plant material covering the soil where roots are active creates conditions for more effective and efficient nutrient use.

5.4 Insect pests of the bunch

Banana bunches are infested by a number of insect pests which do not cause physical yield loss. Losses occur because there is zero tolerance of their presence on the bunch for export. Bananas with scale insects (*Dysmicoccus bispinosus*), mealy bugs (*Diaspis boisduvalii*, *Aspidiotus destructor*) and damage from red rust thrip (*Chaetanaphothrips signipennis* and *Ch. orchidii*) cannot be exported, although these reject bananas can be sold to the local market. Red rust thrips inflicted large losses to smallholder organic growers in Ecuador and Peru in 2011–12 which refocused interest in red rust thrips management. Losses due to red rust have been reduced with early bagging just as the flower emerges from the leaf whorl before the first bract opens. Organic pest control substances like Spinosad are also applied to the flower, leaf whorl and bunch stem 2–3 times, and some growers also apply this organic insecticide to the whole plantation or to stems which are close to flowering.

Unfortunately, although Spinosad is allowed for organic production, it is highly toxic for predatory mites and parasitoids which provide natural control of red rust thrips and is also highly toxic for aquatic invertebrates. Large predators have been identified in Ecuador with slow response to the build-up of thrips. Predatory mites, parasitoids and insect-killing fungi have also been identified which react more rapidly to increasing thrip numbers (Arias de López et al. 2017a). EI-based management of thrips and other bunch insect pests can be based on early bagging (Arias de López et al. 2017b), use of pest control products or repellents compatible with beneficial insects (known as beneficials) and the development of corridors, ground covers and field borders which offer flowers and insectories targeted for key beneficials (Fig. 5).

5.5 Fusarium Tropical Race 4

GlobalGap recently added a section to the certification procedures to address the risk of the spread of Fusarium Wilt TR4 (GlobalGap 2017). These standard procedures are proposed for application to all banana farms, independent of farm size, nearness of TR4 or applicability of other risk factors. In addition to training of all farm staff on the symptoms and the risk of TR4, the certification registration identifies various means for soil and plant material to enter the farm, and outlines strategies to avoid soil movement within the farm and to minimize water movement of TR4. This last point is problematic in many organic banana farms located in dry zones where flood irrigation is a common practice. While some banana growers in Taiwan, Australia and Philippines are familiar with phytosanitary procedures to limit the spread of infected material, the implementation of these procedures will present a challenge to countries in West Africa, Latin America and the Caribbean. The certification procedures on clean seed, soil conservation and efficient water management overlap with other ecological processes and with risk reduction for other pathogens like *Ralstonia* bacterial wilt and banana bunchy top disease which are also spreading into new zones. Building a TR4 readiness package which aligns with other

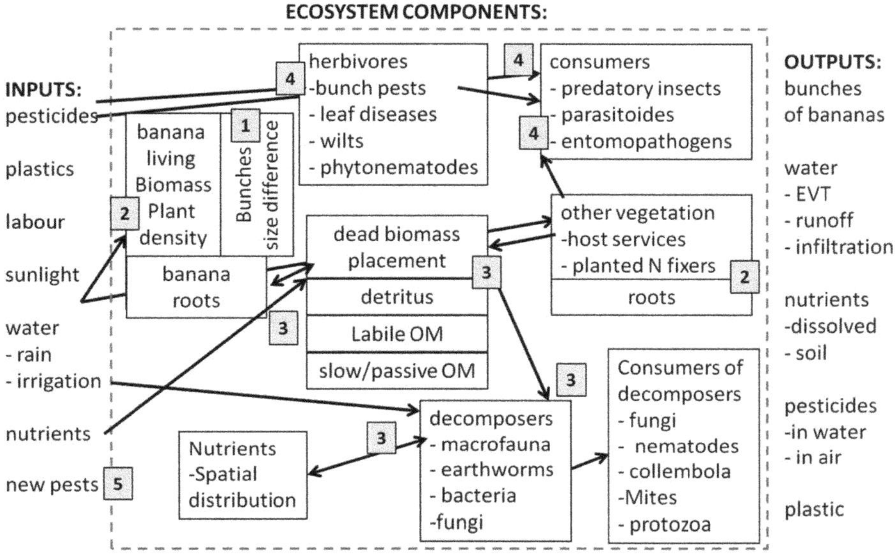

Figure 5 An expanded view of the banana ecosystem components and their relationship to selected EI-based practices. 1 – increased banana bunch size; 2 – banana stand density design; 3 – nutrient management and residue placement; 4 – banana bunch pest management; 5 – Foc TR4 dispersal risk.

EI-based practices may reinforce overall improved management instead of becoming just another GlobalGap checklist which many might regard as an administrative burden.

For the five proposed EI-based practices described above, data are needed in the form of indicators to be used to improve management decisions, whether through formal GAP formats or as part of individual farm recordkeeping. There is only one practice where information can be taken from current data registration formats – the toxicity of pest control substances for beneficial insects (Table 3). In other cases, data are collected, but calculations still need to be done to produce useful management information. In many other cases as shown in Table 4, improved decision-making depends on the collection of more data characterizing the status of the banana field.

Technical assistance and recommendations to improve productivity are commonly based on single improved practices which contribute to a single improved output. This approach contributes useful information for annual crop management and, occasionally, partial budgeting is available to show that the proposed practice will contribute to greater profit. However, management improvements are less straightforward in a semi-perennial crop like banana in which a practice may have effects on future yield potential, other components of the system and labour and input use. Addressing EI-based practices for more profitable and sustainable production clearly represents a challenge to management routines. Practices which strengthen ecological interactions for greater efficiency and productivity cannot easily be isolated into a simple partial budget analysis.

While the focus of GAP is primarily on metrics of food, labour and environmental safety, indicators to guide agroecosystem business management improvement must address productivity, profitability and EI. Metrics from manufacturing to guide continuous

Table 4 Certification data status of selected EI practices in smallholder organic export banana

	Current status of data needed to monitor use of practice			
Possible practices	Data available in current records	Data available, calculations needed	More data collection	Research to guide practice, data to collect
More uniform, higher bunch weight			X	
Optimize plant density for yield/ha, weed suppression, ground cover growth and harvest interval			X	
N-fixation from cover crop				X
Nutrient balance		X		
Leaf residue management to protect soil surface		X	X	
Placement of plant residue at harvest for quick nutrient recycling and stimulation of root activity and soil biology		X	X	
Fertilizer placement in zone of root activity			X	
Use of pest control products safe for beneficials	X			X
Habitat management to promote predators and parasitoids				X
Measures to prevent Fusarium TR4 entry		X	X	

improvement include factor productivity, floor space, defect rate, part travel time, throughput, lead-time, inventories and on-time delivery (Kosandal and Farris 2004; Singh and Singh 2012). Banana growers commonly cite boxes/ha/year, boxes/bunch and bunching interval to measure productivity. The number of healthy leaves at flowering and at harvest are also key indicators used to establish fruit marketability. Fewer than five leaves at harvest represent increased risks of premature ripening during transport.

Other indicators are needed both to determine current production efficiency and profitability and to allow continuous improvement strategies. Percentage of bananas rejected and the causes differentiates between physiological factors such as abiotic stress due to water, nutrients and temperature, physical damage due to deficient practices during bunch filling and harvesting, and the presence of damage due to pests and pathogens. Cost/box could be broken down by inputs and labour for nutrients, water, pest and disease management and mat management, while labour costs/ha/year could also be analysed by type of activity.

Opportunities for benchmarking ecological efficiency have already been mentioned earlier in this article (Table 3). Nutrient balance, N and K applied/box and insect pest control products applied/ha/year could be calculated based on current data taken in GAP certification. Measures such as pseudostem diameter, height of follower sucker, percentage of bare soil and percentage of stacked plant residues by next bunching pseudostem would require additional data collection. These and other variables were tested as indicators of soil health in the FONTAGRO project (Rengifo et al. 2017b) and in Australia (Pattison et al. 2008). Number of leaves at harvest along with pseudostem diameter would provide an indicator of biomass and nutrient recycled. Staver et al. (2017) identified advanced indicators of the status of EI in banana cultivation. Soil biological health indicators such as labile carbon status and the maturation and enrichment status of the nematode community (Ferris et al. 2012), efficiency of light and water conversion to biomass and habitat for beneficial organisms would require additional data collection.

The five indicators identified earlier (Fig. 5) arose initially out of research in the FONTAGRO project focused on bunch pests and soil health, with the aim to improve the production efficiency of banana production with a focus on EI. However, it is important not to neglect the need to improve economic efficiency and productivity. The challenges to quantify economic costs and benefits are illustrated in Table 5. Many costs and benefits related to use of labour and inputs and production lost or gained are relatively easy to estimate by growers. Other costs such as water use by cover crops, as well as benefits related to soil quality, nutrient cycling and optimizing habitat for beneficials are less easy to quantify.

Table 6 synthesizes the possible interaction between the EI-based practices and common banana mat and bunch labour routines. Improved practices which are compatible with existing efficient routines, equipment and inputs may be easier to implement than practices which are not. In some cases, EI practices have the potential to improve the efficiency of existing working practices. For example, reduced bunch weight variability means labour used for both bagging and harvesting generates more value. Mat density interacts with several labour routines. Lower mat density with bigger bunches, shorter bunching intervals and greater light penetration to stimulate N-fixing ground covers would appear to offer certain cost and labour advantages as well as improvements in soil biological status. However, additional costs associated with labour movement, propping bunches and increased water use need to be taken into account. Reorienting nutrient and banana plant residue management depends on changed labour routines which may increase labour costs, as well as involve additional implementation costs in training workers to adopt changed practices.

There are potential management solutions to reconcile EI practices with greater productivity. One solution is to use a continuous improvement approach to engage with workers to identify practices which are easier, more effective and rewarding. Bottom-up kaizen methods of management are oriented to using the knowledge of individual workers and groups, although these methods are more successful where labour stability is high (Singh and Singh 2012). Tools for a diagnostic analysis of labour management routines among smallholder banana growers are needed to determine the potential for greater worker participation in continual improvement.

In closing this section, a contrast can be made between continual improvement of permanent banana stands by smallholders with the re-engineering approach being developed in Guadeloupe and Martinique by French banana scientists. Motivated by a need to eliminate highly toxic nematicides which also control banana weevils, the system of permanent banana stands has been replaced with multi-year rotations of banana with sugarcane and pineapple or cleansing rotations with multi-functional fallows (Risède et al. 2010a,b; Tixier et al. 2010). Significant research has gone into the development of an integrated production approach with reduced pesticide use and risk of pest damage.

Table 5 Information needs for cost/benefit calculations for EI-based practices

Practices	Costs	Benefits
Improved bunch weight	Bunch weight georeferenced by mat Uprooting old mats and planting new mats Bunches not produced during replanting Support by propping for larger bunches	Increased bunch weight Increased yield/ha or same yield with fewer mats Efficiency of bunch management
Lower plant density with improved distribution	Increased weeding Support by propping to bunches Uprooting and replanting	Shorter interval between bunches on same mat Lower cost/ha of leaf pruning, bunch management, fertilizer application Increased light for cover crop growth with increased N-fixation
N-fixing cover crop	Seed, planting, establishment Increased water use Control practices to manage growth Slow-down of labour movement in routine practices Soil microrelief not visible Slower water movement during flood irrigation Increased water use	Reduced weed control Increased contribution of mulch and N Improved soil structure Improved soil protection Habitat for beneficial fauna Reduced evaporation from bare soil
Nutrient balance	Data on bunch weight, interval between bunches, mat density Data on amount of rejected bananas	Reduction in fertilizer applied
Leaf residue to protect soil surface	Longer time for leaf pruning to orient falling leaves to closest bare soil	Improved infiltration Lower soil surface temperature Reduced soil movement
Plant residue placement at harvest	Labour to chop and stack leaves and pseudostem	Faster nutrient re-absorption Healthier roots due to greater biological activity Increased root activity for application of fertilizers and soil amendments
Fertilizer placement	Move stacked plant residue to achieve better soil and root contact	More rapid nutrient uptake Increased fertilizer efficiency
Pest control safe for beneficial	Lost efficiency of alternative products	Reduced damage from insect pests in bunch due to higher populations of beneficial
Habitat for beneficials	Seed and costs of planting and establishment of appropriate plants Reduced mat density to improve light entry Prune habitat plants	Reduced pest control product use Reduced damage from bunch insect pests due to increased beneficial Possible benefits as cover crop
Measures to prevent Foc TR4 entry	Training for all staff and workers Foot baths, duplicate footwear Water management to avoid TR4 spread	Long-term risk reduction Improved soil management Improved water management

The effects of fallow grasses and legumes and crops on nematodes have been reviewed (Rhino et al. 2010). Research has also studied the movement of weevils out of recently uprooted fields as well as the use of effective trapping strategies (Vinatier et al. 2010). The

Table 6 Interaction of EI-based practices with common labour routines in banana

	Common banana management routines					
EI-based practices	Leaf pruning	Desucker stem cleaning	Weeding	Apply nutrients	Bunch bagging	Harvest
Improved bunch weight					Fewer, bigger bunches	Fewer, bigger bunches
Optimize plant density	Fewer stems to deleaf	Fewer mats to treat	Increased light	Fewer mats, increased dose	Fewer, bigger bunches	Fewer bigger bunches
N-fixation from cover crop	Slower, more careful walking	Manage cover crop biomass	Manage cover crop biomass	Reduce N applied		Slower, more careful walking
Nutrient balance				Altered dose		
Leaf residue to protect soil surface	Fallen leaf placement		Reduced bare soil for weed growth			
Plant residue placement at harvest		Residue chopping and placement				
Fertilizer placement				Placement in active root zone		
Pest control safe for beneficial					Altered products and rates	
Habitat for beneficials	Slower, more careful walking		Selective vegetation management		Increased bagging interval	Slower, more careful walking
Measures to prevent Foc TR4 entry	Footwear sanitation, tools left on farm, planned and controlled access to fields all represent possible new procedures to be implemented. Soil conservation and building soil biological activity will not prevent the disease from reaching the farm, but serve to reduce internal spread and yield collapse					

multiple functions of cover crops both in fallow and in association with recently established banana crops have been measured (Damour et al. 2014). Studies are underway to test the applicability of cover crops for smallholder banana growers in the Dominican Republic, but current practices with flood irrigation and very low rates of stand renovation have limited the establishment of cover crops within established banana stands.

6 Scientific frontiers in bananas and implications for GAP

GAP are a moving target, depending on whether the certification framework is organic, FairTrade, GlobalGap or Rainforest, and depending on changing consumer concerns about food safety, environmental issues, worker health and economic justice for growers and workers. The rules for organic amendments, for example, have become increasingly specific, based on stricter interpretation of what is organic. GlobalGap has recently developed an add-on for Fusarium TR4 which will place new demands on growers to control access to their banana stands. On the horizon are numerous new dimensions in GAP: measuring carbon and water footprints as well as greenhouse gas emissions, together with measuring on-farm biodiversity and other emerging pests and diseases.

At the same time global banana production remains particularly vulnerable to oversupply with new investment in areas such as Côte d'Ivoire, Ghana, Sudan, Mozambique and Angola. Mexico has recently re-entered the organic market, and countries such as India with large Cavendish production for their national market are seeking to expand through export markets. For growers and marketers, the challenges to remain competitive are often contradictory. On the one hand they need to lower costs and improve productivity; on the other they face the pressure to meet ever stricter, safety, environmental and worker welfare standards which must often be absorbed without major price benefits. New knowledge generated by scientific research may improve banana yield potential and generate targeted practices to reduce risks to the environment and human safety. In this section we review five areas of scientific advancements, which may contribute to increasing the yield potential for bananas and to achieving greater efficiencies through ICM (Fig. 6).

As is frequently the case, the five scientific frontiers may not result in viable production technology used on-farm. At the same time the new technologies may challenge the certification frameworks which simplify and codify often complex biological and ecological interactions. However, these areas represent the intersection of scientific advances and societal priorities for sustainability in agriculture. Figure 6 locates the five themes in the ecosystem system process framework proposed at the beginning of the chapter. The first area, new tools from gene sequencing and manipulation, focuses on the improvement of banana yield potential and resistance to pests and diseases. The second area, soil microbiology and ecology, addresses the challenges of building carbon sequestration capacity in agriculture and reducing greenhouse gas emissions. The third area is the emerging understanding of the plant and rhizosphere microbiome which may provide new opportunities to build plant defences to abiotic and biotic stresses. The fourth area is the integration of biodiversity into production systems with a potential to synergize positive interactions in nutrient cycling, water storage and availability as well as pest control. Finally, Fig. 6 highlights the area of system analysis, bringing together the previous four

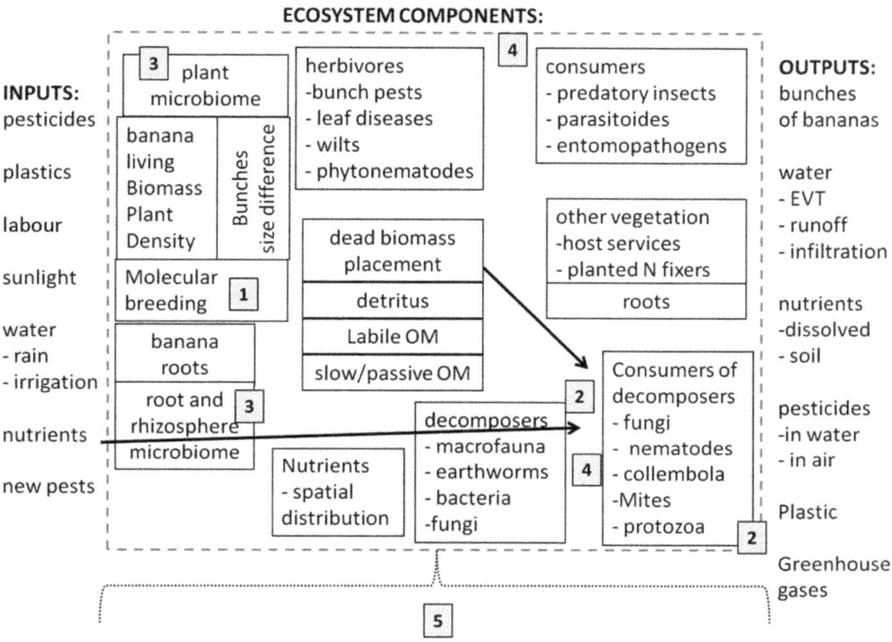

Figure 6 Banana ecosystem components and their relationship to areas on scientific frontier. 1 – banana cultivar improvement; 2 – input and soil management to reduce greenhouse gas emission; 3 – harnessing of plant and soil microbiome to reduce biotic and abiotic stress; 4 – services from more biodiverse plant communities; 5 – new tools and paradigms for integrated banana health and productivity.

points as well as other advances in research, which may create new paradigms for banana productivity and health and thus contribute more effectively to sustainability.

6.1 Potential for new banana cultivars to address biotic and abiotic stresses

Perhaps, unique among commercial horticultural crops, breeding has not made a major contribution to either yield potential or to pest and pathogen resistance of cultivars used for export or national production (DeBuck and Swennen 2016). The potential to address pest and disease losses through the use of banana diversity has been the subject of a recent review (Staver and Capra 2017) based on seven different approaches – cultivar substitution, cultivar mixtures, conventional breeding, trans and *cis* genetic modification, mutation breeding, clonal selection and cultivar-specific microbiome selection. Conventional breeders continue to search for approaches to manage ploidy and avoid banana streak virus which result in varieties with improved fruit pulp quality, agronomic and post-harvest traits and disease resistance to black leaf streak and *Fusarium* (Ortiz 2013).

The breeding programme at Embrapa in Brazil has released numerous cultivars with resistance to *Fusarium* which have found increasing acceptability in the national market

(Amorim et al. 2013). At the same time, advances in sequencing the banana genome, marker-assisted breeding and gene editing offer promise for improved cultivars, primarily focused on improved disease resistance with a lower priority given to abiotic stress resistance and yield potential (Ortiz and Swennen 2014). Dale et al. (2017a) suggest that gene editing has the potential to generate cultivars acceptable for organic certification and recently reported positive results for a GMO Cavendish with resistance to TR4 (Dale et al. 2017b). GMO crops are currently not acceptable in all certification schemes except GlobalGap. New banana cultivars based on *cis* transformation and gene editing, although not yet available, will challenge certification frameworks, but this may also depend on the changes in consumer attitudes to GMOs.

6.2 Managing the soil microbial community to reduce gas emissions and increase carbon sequestration

Conventional banana growers use high levels of fertilizer N, generating a high potential for denitrification. Field studies in Costa Rica have shown that between 6% and 9% of N applied to fields is converted to nitrogen oxide (Veldkamp and Keller 1997). Greenhouse gas emissions are not yet addressed in GlobalGap certification schemes. Organic banana growers who generally use lower rates of N and often in less soluble forms face the challenge of meeting the nutrient needs for commercial yields when they rely on often scarce and expensive organic inputs. These two contrasting situations come together in a single challenge to banana growers – meeting the short-term nutrient needs of the plant, while also building carbon in the soil and minimizing the release of greenhouse gases.

Less frequent tillage and high levels of crop residue returned to the soil should favour soil carbon accumulation in banana production (Singh et al. 2010). However, other elements favour a bacterial-dominated microbial community less favourable to carbon accumulation. Pattison et al. (2008) contrasted conventional and organic banana with forest and pasture showing loss of carbon in banana production. Conventional banana also had less carbon and higher soil bulk density than organic bananas, but no differences were found in the size of the nematode community. Recent unpublished results of nematode communities in organic banana cultivation in Peru, Ecuador and the Dominican Republic indicate that bacterivores dominate with a scarcity of fungivores and predator nematodes. In this largely unexplored field of study, numerous tools could provide insight into the management of organic and fertilizer inputs in banana production to ensure short-term N availability with low emissions of greenhouse gases, while also promoting fungi-dominated decomposition cycles (Bowles et al. 2014, 2015).

6.3 The banana plant and rhizosphere microbiome

In recent years, advances in nucleic acid sequencing and mass spectrometry have enabled scientists to learn more about the omnipresence of microorganisms on earth (White et al. 2016) and to begin to unravel their functions in different environments (Coleman-Derr and Tringe 2014). The banana pseudostem, corm, roots and rhizosphere harbour abundant microbes. Rossmann et al. (2012) found abundant bacterial communities in banana pseudostems, markedly different from rhizosphere microbes which were dominated by fungal endophytes. Dita et al. (2014) isolated and characterized 360 bacterial and 143 fungal endophytes from the roots of 20 accessions in the Musa collection at Corbana, Costa Rica. Certain endophytes were cultivar specific such as a *Trichoderma asperellum* isolate from cv. Yangambi Km5 which inhibited growth of pathenogenic *Fusarium* in Petri dish tests. Other isolates tripled root growth in Cavendish plants in greenhouse screening.

Field characterizations of the microbiome in the cultivar Gros Michel grown associated with coffee and leguminous shade trees highlighted both differences in cropping system and in location (Köberl et al. 2015). Xue et al. (2015) isolated a *Bacillus* bacterium from disease-suppressive soil and showed that it suppressed *Fusarium* infections in greenhouse trials. Dita et al. (2014) used clean suckers and endophyte inoculation to extend plantation life in *Fusarium*-infected soils compared to untreated suckers. Although as yet unexplored in bananas, cultivar capacity to recruit beneficial microbes has been tested in other crops. Trognitz et al. (2015) documented the capacity of different wheat cultivars to associate with specific microbes. Hoagland (2015) has detailed the conditions needed to develop a breeding programme which promotes cultivar capacity to recruit beneficial microbes either for resistance to disease or abiotic stress. The different pathways for harnessing the microbiome – single microbes or microbial cocktails effective across many banana cultivars or improved cultivars with the capacity to recruit beneficial microbes – represent different challenges to research addressing ICM and yield gaps. At the same time the technologies for harnessing the microbiome do not appear to conflict with the different certification frameworks discussed earlier.

6.4 Increased biodiversity for increased ecosystem services

Commercial banana production is widely characterized as a large-scale monoculture which is made even more vulnerable by the genetic uniformity of the Cavendish varieties on which it is based. In ecosystem terms, the vulnerability derives from the simple food webs both above- and below-ground, which are supported only by the biomass of the banana mat. An overview of the role of diversity on ecosystem functioning concluded that increased functional diversity is central to increased ecosystem services (Hooper et al. 2005). Greater plant diversity can increase resource use such as nutrient and water retention and provide energy and habitat for a more diversified food web which provides improved pest control. These ecosystem services not only contribute to productivity, but also make ecosystem processes, including production, more resilient. Species functional diversity ensures that, under fluctuating abiotic conditions, specific ecosystem processes like pest control are maintained.

Gaba et al. (2014) have proposed that managing biological interactions are key to EI. An example is the use of cover crops with Cavendish production which shows potential to contribute to both above- and below-ground interactions. Fruit quality standards show zero tolerance for damage or physical presence of several insect pests of the bunch, including red rust thrips (*Chaetanaphothrips signipennis* and *Ch. orchidii*), scale insects (*Dysmicoccus bispinosus*) and mealy bugs (*Diaspis boisduvalii*, *Aspidiotus destructor*). Bagging the bunch provides some protection, but these pests also have numerous natural predators. Recent unpublished studies in Ecuador identified a predatory thrip, a *Blaptostethus* pirate bug and *Megaphragma* parasitoid feeding naturally on the red rust thrips in organic banana plantations. The challenge to research is to minimize use of compounds detrimental to beneficials and find compatible plant species which promote their presence. At the same time cover crops also alter the below-ground biological community. Djigal et al. (2012) found that leguminous and grass cover crops in banana generally increased the abundance of nematodes of all trophic groups, except plant-feeding nematodes, as compared to bare soil. Paspalum grass cover crops had significantly higher abundance of omnivorous and fungivorous nematodes and lower plant-feeding nematodes, while the different leguminous cover crops stimulated different responses from the nematode community. The authors suggested that increased omnivores were responsible for decreased plant-feeding nematodes, a top-down effect by the grass cover crop with a more general bottom up-effect on the other trophic groups.

Screening the multiple effects of different plant communities in association with banana monocultures represents a scientific challenge for research which can build on an emerging understanding of the functional role of different species in ecological interactions (Gaba et al. 2014). Wardle et al. (2004) and Phelan (2009) highlighted potential interactions between above- and below-ground biodiversity which has clear implications for the EI of banana monocultures. At the same time, more diverse production systems are highly compatible with all the GAP labels, although they may present practical challenges to growers and the labour routines of mat management and harvesting.

6.5 Interactions among these elements: integrated banana health and productivity

The previous four themes, which have highlighted new areas of research of particular significance for banana sustainability, point to a fifth frontier – integrated banana health and productivity. This builds on the increasing understanding of biological and ecosystem processes and the interactions of crop management and inputs with temporal and spatial dimensions of banana mats and other plant species. The management practices to harness ecological interactions involving the above- and below-ground food web and the plant and rhizosphere microbiome, and their contribution on crop productivity, efficiency of conversion of production resources and off-farm environmental impacts call for innovative research approaches. We have already seen examples in the earlier sections. The comparison of soil processes in organic tomato farms (Bowles et al. 2015) has related productivity and input type to greenhouse gas emissions and carbon dynamics. Ferris et al. (2012) compared soil nematode communities from different commercial banana fields to test a hypothesis about the indirect relationship between amplifiable and target prey and predator nematodes. Field screening of cover crops shows a range of interactions with soil nematode communities (Djigal et al. 2012). While there are other studies addressing specific interactions, no long-term systems comparisons have been undertaken in commercial banana comparing alternative management paradigms as has been the case, for example, in conservation tillage (Xiao-Bin et al. 2006) or coffee agroforestry (Haggar et al. 2015).

Banana scientists have started to develop new approaches to address integrated banana health and productivity. Staver et al. (2017) highlight lessons learnt, bringing together scientists, farm households and value chains to build production scale prototypes incorporating EI. Field and farm diagnostic studies, experiments measuring component interactions, farmer test plots and simple models contribute to the development of improved integrated decision-making tools. The banana research team at CIRAD, taking a different approach, has developed a series of banana models, beginning with SIMBA (Tixier et al. 2008) to test sustainable banana cropping systems. Sub routines have been developed for banana weevils (Vinatier et al. 2009), the selection of cover crops (Tixier et al. 2011) and the effects of crop management on N and light partitioning (Ripoche et al. 2012). Model frameworks of food web-based models are under development (Tixier et al. 2013) to address complex interactions between plants, above-and below-ground food webs, management practices and abiotic factors. Contrasting plant communities have been compared using a modelling approach to address the trade-offs between plant competition and benefits from improved nutrition and pest regulation (Poeydebat et al. 2016). This succession of increasingly complex models provides an important basis to design long-term system comparisons and grower decision-making tools.

7 Conclusion: mutual lessons between GAP certification and approaches to sustainable banana production

This chapter has suggested that GAP-based certification frameworks to promote sustainable banana production are an imperfect tool. We have seen that a hodgepodge of GAP schemes reflect the interests of sub-groups of consumers and supermarkets promoting the reliability of their brand. These are supported by national public sector regulatory agencies which accredit certification agencies and regulate inputs, labour contracts and land use. These certification standards are both very demanding, depending on the specific scheme, while also being simplistic, potentially reducing a range of local enterprises, with widely differing challenges, to a globally oriented series of checklists. Although published evidence is sparse, the different certification frameworks have been effective in improving production techniques, labour practices, price negotiation and environmental conservation compared to non-certified production. To close this chapter, we propose to highlight the lessons from the broader sustainable banana agenda which suggest ways to modify GAP certification schemes in the pursuit of sustainable banana production.

Using the goals of sustainable agriculture (National Research Council 2010) adapted to banana production, three opportunities for GAP certification frameworks to contribute more clearly to sustainable banana production have been identified. First, although GAP certification procedures are considered a burden by growers and their organizations, their formats and checklists can easily be modified and re-oriented to become more useful for enterprise management decisions both for individual farms and the group marketing organizations which serve them. Many data management tools for use in improved decision-making are already in use and under development on larger farms, although small business management routines and data management and processing adapted for small growers are still major challenges.

Second, GAP certification processes should be improved to avoid perfunctory compliance with specific standards, but which do not contribute to more rigorous sustainability criteria. Seufert et al. (2017) analysed organic certification standards and concluded that a far greater effort was focused on the type of inputs than 'environmental' processes (which we have referred to as EI). Some versions of sustainability standards risk being the certification of business as usual. Among certain stakeholders in organic agriculture, creeping conventionalization is also lamented in which external inputs rather than ecological processes become the major focus of productivity (Brzezina et al. 2017). A focus on the integration of ecological processes into farm management and certification provides indicators which go beyond natural resource conservation for both conventional and organic banana production.

Third, while FairTrade, SA8000 and other labels address the social sustainability of banana production, current GAP labels are still limited in this dimension. Ellis and Keane

(2008) propose a 'Good for Development' certification framework which emphasizes the potential for positive impacts of export banana production for rural communities. Their aim is to achieve wider coverage of certification, while also encouraging marketers to share the costs of certification and develop mechanisms to contribute to local development goals in banana-producing communities.

With nearly half of export bananas, over 300 000 ha, under at least one GAP label, the experience with GAP implementation also provides three perspectives on pathways to more sustainable banana production. First, the different labels have taken both transitional and transformational approaches to changing banana production. On the transformational side, the market for organic labelled bananas has stimulated new production zones and production systems, in spite of the tendency for conventionalization. The FairTrade label has transformed value chains first to benefit smallholders and more recently to include workers on large farms, with an increasing presence in numerous national markets. On the transitional side, all labels have continued to expand their initial criteria, often under the influence of the presence of other labels on the market. FairTrade has incorporated more environmental issues, while organic labels have been pressured to become more social and ethical. While multiple labels may confuse the consumer, the exploration of multiple pathways towards more sustainable bananas appears beneficial. In an analysis of the future of organic agriculture in Europe, Brzezina et al. (2017) point out that both organic and conventional growers have incorporated more environmentally friendly practices over the same period. The scientific frontiers described earlier, for the most part, will contribute to either transformation or transition.

Second, a shared characteristic of all GAP frameworks is their routine of paperwork, checklists and inspections implemented through agencies for certification and accreditation. This raises a fundamental question about the mechanisms for guiding sustainability for banana production beyond just the export sector (Cocklin et al. 2007). The future of sustainable banana production would appear to be more paperwork, inspections and certifications. Although perhaps not considered a frontier in science, a research agenda on the minimum set of indicators for sustainable banana production, along with the appropriate data collection and inspection routines required to support these indicators, is needed for adaptation to the wide diversity of production systems for national markets.

Finally, among consumer concerns about their food is a growing emphasis on food miles, short food chains and locally sourced production. Brzezina et al. (2017) highlight the threat of imported organic products to the economic viability and expansion of European organic producers. They suggest that the incorporation of distance between supplier and consumer be integrated into certification standards. Is banana production, sustainable or not, certified or not, threatened by recent focus on carbon and water footprints and food miles? Adams and Salois (2010) conclude that a preference for local produce is a response to large-scale labelled and regulated agriculture. In the meantime, banana is still the world's most exported and consumed fruit under a both diverse and dynamic set of certification schemes. From the perspective of banana growers, marketers and scientists, this chapter suggests a greater engagement with two key elements – science-based inputs to more effective certification schemes, and tracking and monitoring to achieve more effective certification frameworks which address our unfolding understanding of the basis of sustainable banana production.

8 Where to look for further information

The World Banana Forum website on GAP in banana provides an overview as well as many additional links to related themes: http://www.fao.org/world-banana-forum/projects/good-practices/good-agricultural-practices/en/

The website BananaLink advocates for more attention to complex social and labour issues in the export banana sector: http://www.bananalink.org.uk/.

More details about different certifications and their focus can be found at: www.standardsmap.org; http://www.ecolabelindex.com/; and the websites for different individual labels.

9 References

Adams, D. C. and Salois, M. J. 2010. Local versus organic: A turn in consumer preferences and willingness-to-pay. *Renewable Agriculture and Food Systems*, 25(4), 331–41.

Agra, C. E. A. S. Consulting. 2002. Integrated crop management system in the EU. Amended final report for European Commission, DG Environment, pp. 1–157. Available from: http://ec.europa.eu/environment/agriculture/pdf/icm_finalreport.pdf

Amorim, E. P., dos Santos-Serejo, J. A., Amorim, V. B. O., Ferreira, C. F. and Silva, S. O. 2013. Banana breeding at Embrapa cassava and fruits. *Acta Horticulturae*, 986, 171–6.

Arias de López, M., Corozo-Ayovi, R. E., Vivas, L., Medrano, S., Sánchez, L., Alburqueque, D., Vera, T., Jines, A., Delgado, R., Rengifo, D., Rojas, J. and Staver, C. 2017a. Búsqueda de Enemigos Naturales a Chaetanaphothrips signipennis y C. orchidii en Banano Orgánico en Ecuador, Perú y República Dominicana. *IV Congreso Latinoamericano y el Caribe de Plátanos y Bananos*, 27–30 November, Tecoman, Colima, Mexico.

Arias de López, A., Rosa Corozo-Ayovi, R. E., Delgado, R., Osorio, B., Moyón, D., Rengifo, D., Suárez, P., Paulino, A. and Staver, C. 2017b. Reducción del daño Causado por Chaetanaphothrips signipennis y C. orchidii en Plátano con el uso de Fundas Simples y Aplicaciones de Productos Orgánicos en Ecuador y República Dominicana. *IV Congreso Latinoamericano y el Caribe de Plátanos y Bananos*, 27–30 November, Tecoman, Colima, Mexico.

Bingham, E. and Monforton, C. 2013. The pesticide DBCP and male infertility. In: Hansen, S. F. (Ed.), *Late Lessons from Early Warnings: Science, Precaution, Innovation*. European Environmental Agency Report. https://www.eea.europa.eu/publications/late-lessons-2

Bloom, N., Sadun, R. and Van Reenen, J. 2012. Does management really work? *Harvard Business Review*, 90(11), 76–82.

Bowles, T. M., Acosta-Martínez, V., Calderón, F. and Jackson, L. E. 2014. Soil enzyme activities, microbial communities, and carbon and nitrogen availability in organic agroecosystems across an intensively-managed agricultural landscape. *Soil Biology and Biochemistry*, 68, 252–62.

Bowles, T. M., Hollander, A. D., Steenwerth, K. and Jackson, L. E. 2015. Tightly-coupled plant-soil nitrogen cycling: Comparison of organic farms across an agricultural landscape. *PloS one*, 10(6), e0131888.

Brzezina, N., Biely, K., Helfgott, A., Kopainsky, B., Vervoort, J. and Mathijs, E. 2017. Development of organic farming in Europe at the crossroads: Looking for the way forward through system archetypes lenses. *Sustainability*, 9(5), 821.

Bünemann, E. K., Schwenke, G. D. and Van Zwieten, L. 2006. Impact of agricultural inputs on soil organisms – a review. *Soil Research*, 44(4), 379–406.

Castillo, L. E., Ruepert, C. and Solis, E. 2000. Pesticide residues in the aquatic environment of banana plantation areas in the north Atlantic zone of Costa Rica. *Environmental Toxicology and Chemistry*, 19(8), 1942–50.

COAG. 2003. Development of a framework for good agricultural practices. *FAO Committee on Agriculture – Seventeenth Session*, 31st March to 4th April 2003. FAO, Rome, Italy. http://www.fao.org/docrep/meeting/006/y8704e.htm

Cocklin, C., Mautner, N. and Dibden, J. 2007. Public policy, private landholders: Perspectives on policy mechanisms for sustainable land management. *Journal of Environmental Management*, 85(4), 986–98.

Coleman-Derr, D. and Tringe, S. G. 2014. Building the crops of tomorrow: Advantages of symbiont-based approaches to improving abiotic stress tolerance. *Frontiers in Microbiology*, 5, 283.

Corozo, R. E., Arias, M., Durango, W., Staver, C. and Turmel, M. 2016. Cover crops in organic-export banana systems: Farmer experiences and soil health in Ecuador. *Promusa Symposium 2016 – Agroecological Approaches to Promote Innovative Banana Production Systems*, 10–16 October, Montpellier, France. http://www.promusa.org/article142-2016-France-symposium#poster

Cumo, C. M. 2015. *Foods that Changed History: How Foods Shaped Civilization from the Ancient World to the Present.* ABC-CLIO.

Dale, J., Paul, J. Y., Dugdale, B. and Harding, R. 2017a. Modifying bananas: From transgenics to organics? *Sustainability*, 9(3), 333.

Dale, J., James, A., Paul, J., Khanna, H., Smith, M., Peraza-Echeverria, S., Garcia-Bastidas, F., Kema, G., Waterhouse, P., Mengersen., K. and Harding, R. 2017b. Transgenic Cavendish bananas with resistance to Fusarium wilt tropical race 4. *Nature Communications*, 8(1), 1496.

Damour, G., Dorel, M., Quoc, H. T., Meynard, C. and Risède, J. M. 2014. A trait-based characterization of cover plants to assess their potential to provide a set of ecological services in banana cropping systems. *European Journal of Agronomy*, 52, 218–28.

De Buck, S. and Swennen, R. 2016. *Bananas, the Green Gold of the South.* VIB fact series. IPBO, Ghent, Belgium, 55p.

Dita, M., Schilly, A., Vargas, J., Chaves, N., Guzman, M., Sandoval, J. and Staver, C. 2014. Endophyte microbiome of banana roots reveals high diversity and potential for agricultural uses. *International Conference on Research on Food Security, Natural Resource Management and Rural Development*, 17–19 September 2014, Tropentag, Prague, Czech Republic, p. 129. http://www.tropentag.de/2014/proceedings/proceedings.pdf

Djigal, D., Chabrier, C., Duyck, P. F., Achard, R., Quénéhervé, P. and Tixier, P. 2012. Cover crops alter the soil nematode food web in banana agroecosystems. *Soil Biology and Biochemistry*, 48, 142–50.

Ellis, K. and Keane, J. 2008. A review of ethical standards and labels: Is there a gap in the market for a new 'Good for Development' label? Working Paper 297. Overseas Development Institute, London.

FAO. 2017. Banana market review 2015–2016. Trade and Market Division, FAO, Rome. http://www.fao.org/economic/est/est-commodities/bananas/en/

Ferris, H., Pocasangre, L. E., Serrano, E., Muñoz, J., Garcia, S., Perichi, G. and Martinez, G. 2012. Diversity and complexity complement apparent competition: nematode assemblages in banana plantations. *Acta Oecologica*, 40, 11–18.

Gaba, S., Bretagnolle, F., Rigaud, T. and Philippot, L. 2014. Managing biotic interactions for ecological intensification of agroecosystems. *Frontiers in Ecology and Evolution*, 2, 29.

GlobalGap. 2017. Integrated Farm Assurance. TR4 Biosecurity Add-On for Bananas. Control Points and Compliance Criteria. http://www.globalgap.org/export/sites/default/.content/.galleries/documents/170427_TR4_BioSecurity_Add-On_Bananas_CPCC_V1-0_en.pdf

Haggar, J., Asigbaase, M., Bonilla, G., Pico, J. and Quilo, A. 2015. Tree diversity on sustainably certified and conventional coffee farms in Central America. *Bioversity and Conservation*, 24(5), 1175–94.

Hartman, B. 2015. *The Lean Farm: How to Minimize Waste, Increase Efficiency, and Maximize Value and Profits with Less Work.* Chelsea Green Publishing, White River Junction, VT.

Henriques, W., Jeffers, R. D., Lacher, T. E. and Kendall, R. J. 1997. Agrochemical use on banana plantations in Latin America: Perspectives on ecological risk. *Environmental Toxicology and Chemistry*, 16(1), 91–9.

Hoagland, L. 2015. Breeding for beneficial root-microbial relationships: How do we get there? In Akademie, A. (Ed.), *First EUCARPIA Workshop on Implementing Plant–Microbe Interaction in Plant Breeding.* http://www.cobra-div.eu/wp-content/uploads/2015/09/booklet-Breeding-for-plant-microbe-interaction-June-25th-26th-2015-v6_MM-final.compressed.pdf

Hooper, D., Chapin, F., Ewel, J., Hector, A., Inchausti, P., Lavorel, S., Lawton, J., Lodge, D. Loreau, M., Naeem, S., Schmid, B., Setälä, H., Symstad, A., Vandermeer, J. and Wardle, D. 2005. Effects of biodiversity on ecosystem functioning: A consensus of current knowledge. *Ecological Monographs*, 75(1), 3–35.

Hosono. A. 2009. Kaizen – quality, productivity and beyond. In: *Introducing Kaizen in Africa. GRIPS Development Forum.* National Graduate Institute for Policy Studies, Japan, pp. 23–38. http://www.grips.ac.jp/forum/pdf09/Introducing_KAIZEN_in_Africa.pdf

International Trade Centre and European University Institute. 2016. *Social and Environmental Standards: Contributing to More Sustainable Value Chains.* ITC, Geneva.

Köberl, M., Dita, M., Martinuz, A., Staver, C. and Berg, G. 2015. Agroforestry leads to shifts within the gammaproteobacterial microbiome of banana plants cultivated in Central America. *Frontiers in Microbiology*, 6, 91.

Kosandal, P. and Farris, J. 2004. The strategic role of the kaizen event in driving and sustaining organizational change. *Proceedings of the 2004 American Society for Engineering Management Conference*, 20–23 October 2004, Alexandria, VA, pp. 517–26.

Lans, T., Seuneke, P. and Klerkx, L. 2013. Agricultural Entrepreneurship. In: Carayannis, E.G. (Ed.), *Encyclopedia of Creativity, Invention, Innovation and Entrepreneurship.* Springer, New York, Heidelberg, Dordrecht and London, pp. 44–9.

Lernoud, J., Potts, J., Sampson, G., Voora, V., Willer, H. and Wozniak, J. 2016. *The State of Sustainable Markets-Statistics and Emerging Trends 2015.* ITC, Geneva.

Lescot, T. 2012. The genetic diversity of bananas. *FruiTrop*, 200: 75–8.

Liu, P. 2008. Certification in the value chain for fresh fruits: The example of banana industry. Trade and Markets Division. FAO, Rome.

Lockie, S., Travero, J. and Tennent, R. 2015. Private food standards, regulatory gaps and plantation agriculture: Social and environmental (ir) responsibility in the Philippine export banana industry. *Journal of Cleaner Production*, 107, 122–9.

Loconto, A. and Busch, L. 2010. Standards, techno-economic networks, and playing fields: Performing the global market economy. *Review of International Political Economy*, 17(3), 507–36.

Marin, D. H., Romero, R. A., Guzmán, M. and Sutton, T. B. 2003. Black Sigatoka: An increasing threat to banana cultivation. *Plant disease*, 87(3), 208–22.

Melo, C. J. and Wolf, S. A. 2005. Empirical assessment of eco-certification: The case of Ecuadorian bananas. *Organization & Environment*, 18(3), 287–317.

Melo, C. J. and Wolf, S. A. 2007. Ecocertification of Ecuadorian bananas: Prospects for progressive North–South linkages. *Studies in Comparative International Development*, 42(3–4), 256–78.

Mendez, A., Castillo, L. E., Ruepert, C., Hungerbuehler, K. and Ng, C. A. 2018. Tracking pesticide fate in conventional banana cultivation in Costa Rica: A disconnect between protecting ecosystems and consumer health. *Science of the Total Environment*, 613, 1250–62.

Multigner, L., Cordier, S., Kadhel, P., Huc-Terki, F., Blanchet, P., Bataille, H. and Janky, E. 2007. Pollution par le chlordécone aux Antilles Quel impact sur la santé de la population? *Environnement, Risques & Santé*, 6(6), 405–7.

National Research Council. 2010. *Toward Sustainable Agricultural Systems in the 21st Century.* National Academies Press, Washington DC.

Ohno, I., Ohno, K. and Uesu, S. 2009. *Introduction In Introducing Kaizen in Africa. GRIPS Development Forum.* National Graduate Institute for Policy Studies, Japan, pp. 1–12. http://www.grips.ac.jp/forum/pdf09/Introducing_KAIZEN_in_Africa.pdf

Ortiz, R. 2013. Conventional banana and plantain breeding. *Acta Horticulturae*, 986, 177–94.

Ortiz, R. and Swennen, R. 2014. From crossbreeding to biotechnology-facilitated improvement of banana and plantain. *Biotechnology Advances*, 32(1), 158–69.

Oya, C., Schaefer, F., Skalidou, D., McCosker, C. and Langer, L. 2017. Effects of certification schemes for agricultural production on socio-economic outcomes in low- and middle-income countries: A systematic review. 3ie systematic review 34. International Initiative for Impact Evaluation (3ie), London.

Pattison, A. B., Moody, P. W., Badcock, K. A., Smith, L. J., Armour, J. A., Rasiah, V., Cobon, J., Gulino, L. and Mayer, R. 2008. Development of key soil health indicators for the Australian banana industry. *Applied Soil Ecology*, 40(1), 155–64.

Phelan, P. L. 2009. Ecology-based agriculture and the next green revolution: Is modern agriculture exempt from the laws of ecology. *Sustainable Agroecosystem Management*, 97–135.

Ploetz, R. C. 2015. Fusarium wilt of banana. *Phytopathology*, 105(12), 1512–21.

Poeydebat, C., Carval, D., de Lapeyre de Bellaire, L. and Tixier, P. 2016. Balancing competition for resources with multiple pest regulation in diversified agroecosystems: A process-based approach to reconcile diversification and productivity. *Ecology and Evolution*, 6, 8607–16. doi:10.1002/ece3.2453

Raynolds, L. T. 2000. Re-embedding global agriculture: The international organic and fair trade movements. *Agriculture and Human Values*, 17(3), 297–309.

Rengifo, D., Perez, A., Rojas, J. C., Vega, U., Corozo, R. E., Durango, W., Staver, C. and Turmel. M. 2016. Challenges and opportunities for keeping nutrients in balance in organic-export banana systems. *Promusa Symposium 2016 – Agroecological Approaches to Promote Innovative Banana Production Systems*, 10–16 October, Montpellier, France. http://www.promusa.org/article142-2016-France-symposium#poster

Rengifo, D., Perez, A., Durango, W., Delgado, R., Corozo, R. E., Rojas, J. C., Vega, U. and Staver, C. 2017a. Distribución Estratégica de Residuos y Ajustes en la Zona y Cantidad de Fertilizantes para Mejorar la Nutrición del Banano Orgánico. *IV Congreso Latinoamericano y el Caribe de Plátanos y Bananos*, 27–30 November, Tecoman, Colima, Mexico.

Rengifo, D., Perez, A., Durango, W., Delgado, R., Corozo, R. E., Osorio, B., Rojas, J. C., Vega, U., Turmel, M., Staver, C. and Sanchez, S. 2017b. Indicadores de Salud de Suelo como Herramienta para Mejorar la Productividad del Banano Orgánico. *IV Congreso Latinoamericano y el Caribe de Plátanos y Bananos*, 27–30 November, Tecoman, Colima, Mexico.

Rhino, B., Dorel, M., Tixier, P. and Risède, J. M. 2010. Effect of fallows on population dynamics of Cosmopolites sordidus: Toward integrated management of banana fields with pheromone mass trapping. *Agricultural and Forest Entomology*, 12(2), 195–202.

Ripoche, A., Achard, R., Laurems, A. and Tixier, P. 2012. Modeling spatial portioning of light and nitrogen resources in banana cover-cropping systems. *European Journal of Agronomy*, 41, 81–91.

Risède, J. M., Lescot, T., Cabrera Cabrera, J., Guillon, M., Tomekpe, K., Kema, G. H. J. and Côte, F. 2010a. Challenging short and mid-term strategies to reduce the use of pesticides in banana production. Banana Case Study—Guide Number, 1. ENDURE.

Risède, J. M., Chabrier, C., Dorel, M., Dambas, T., Achard, R. and Quénéhervé, P. 2010b. Integrated management of banana nematodes: Lessons from a case study in the French West Indies. Banana Case Study—Guide Number, 4. ENDURE.

Rossmann, B., Müller, H., Smalla, K., Mpiira, S., Tumuhairwe, J. B., Staver, C. and Berg, G. 2012. Banana-associated microbial communities in Uganda are highly diverse but dominated by Enterobacteriaceae. *Applied and Environmental Microbiology*, 78(14), 4933–41.

Sanchez, S., Rengifo, D., Garcia, S., Maldonado, E., Rojas, J. C., Navias, D., Delgado, R. and Staver, C. 2017. Los Nematodos Como Indicadores Para Orientar el Manejo de Suelos en Sistemas de Producción de Banano Orgánico de Exportación. *IV Congreso Latinoamericano y el Caribe de Plátanos y Bananos*, 27–30 November, Tecoman, Colima, Mexico.

Seufert, V., Ramankutty, N. and Mayerhofer, T. 2017. What is this thing called organic? How organic farming is codified in regulations. *Food Policy*, 68, 10–20.

Singh, J. and Singh, H. 2012. Continuous improvement approach: State-of-art review and future implications. *International Journal of Lean Six Sigma*, 3(2), 88–111.

Singh, B. K., Bardgett, R. D., Smith, P. and Reay, D. S. 2010. Microorganisms and climate change: Terrestrial feedbacks and mitigation options. *Nature Reviews. Microbiology*, 8(11), 779–90.

Staver, C. and Capra, I. 2017. Banana diversity and the food and income threats of pest and pathogen losses: Priority research areas to deploy diversity to reduce pest and disease losses. Working document [n.p.]. Bioversity International, Montpellier, France, 55p. http://hdl.handle.net/10568/91529

Staver, C., Turmel, M., Siles, P., Mpiira, S., Bustamante, O., Calberto, G., Dita, M., Omondi, A., Alvarez, E. and Zheng, S. 2016. Step by step tools to identify ecological intensification alternatives for banana cropping systems. *Promusa Symposium 2016 – Agroecological Approaches to Promote Innovative Banana Production Systems*, 10–16 October, Montpellier, France. http://www.promusa.org/article142-2016-France-symposium#poster

Staver, C., Siles, P., Bustamante, O., Castellon, N., Cerdan, C., Garming, H. and Burkhardt, J. 2017. Can science contribute to smallholder management of complex systems? *Lessons from Agroecological Intensification of Multi-Strata Coffee with Banana*, 17–19 September, Tropentag, Prague, Czech Republic, p. 129. http://www.tropentag.de/2014/abstracts/links/Staver_qnOj1u4r.php

Steering Committee of the State-of-Knowledge Assessment of Standards and Certification. 2012. *Toward Sustainability: The Roles and Limitations of Certification*. RESOLVE, Inc., Washington DC.

The Guardian. 2016. Lidl is the latest to switch to sustainable bananas. Will it make a difference? https://www.theguardian.com/sustainable-business/2016/feb/06/lidl-sustainable-bananas-commitment-workers-dominican-republic-supermarket-prices

Tixier, P., Malézieux, E., Dorel, M. and Wery, J. 2008. SIMBA, a model for designing sustainable banana-based cropping systems. *Agricultural Systems*, 97(3), 139–50.

Tixier, P., Vinatier, F., Cabrera Cabrera, J., Padilla Cubas, A., Okolle, J., Chabrier, C. and Guillon, M. 2010. Integrated Pest Management of black weevil in banana cropping systems. Banana Field Study–Guide Number, 3. ENDURE.

Tixier, P., Lavigne, C., Alvarez, S., Gauquier, A., Blanchard, M., Ripoche, A. and Achard, R. 2011. Model evaluation of cover crops, application to eleven species for banana cropping systems. *European Journal of Agronomy*, 34(2), 53–61.

Tixier, P., Duyck, P. F., Côte, F. X., Caron-Lormier, G. and Malézieux, E. 2013. Food web-based simulation for agroecology. *Agronomy for Sustainable Development*, 33(4), 663–70.

Trienekens, J. and Zuurbier, P. 2008. Quality and safety standards in the food industry, developments and challenges. *International Journal of Production Economics*, 113(1), 107–22.

Trognitz, F., Mayerhof, H., Mitter, B. and Sessitsch, A. 2015. Genotype-dependent interaction with beneficial plant-associated bacteria. In: Akademie, A. (Ed.), *First EUCARPIA Workshop on Implementing Plant – Microbe Interaction in Plant Breeding*. http://www.cobra-div.eu/wp-content/uploads/2015/09/booklet-Breeding-for-plant-microbe-interaction-June-25th-26th-2015-v6_MM-final.compressed.pdf

van Ittersum, M. K., Cassman, K. G., Grassini, P., Wolf, J., Tittonell, P. and Hochman, Z. 2013. Yield gap analysis with local to global relevance – a review. *Field Crops Research*, 143, 4–17.

Veldkamp, E. and Keller, M. 1997. Nitrogen oxide emissions from a banana plantation in the humid tropics. *Journal of Geophysical Research: Atmospheres*, 102(D13), 15889–98.

Vinatier, F., Tixier, P., Le Page, C., Duyck, P. F. and Lescourret, F. 2009. COSMOS, a spatially explicit model to simulate the epidemiology of Cosmopolites sordidus in banana fields. *Ecological Modelling*, 220, 2244–54.

Vinatier, F., Chailleux, A., Duyck, P. F., Salmon, F., Lescourret, F. and Tixier, P. 2010. Radiotelemetry unravels movements of a walking insect species in heterogeneous environments. *Animal Behaviour*, 80(2), 221–9.

Wardle, D. A., Bardgett, R. D., Klironomos, J. N., Setälä, H., Van Der Putten, W. H. and Wall, D. H. 2004. Ecological linkages between aboveground and belowground biota. *Science*, 304(5677), 1629–33.

White III, R. A., Callister, S. J., Moore, R. J., Baker, E. S. and Jansson, J. K. 2016. The past, present and future of microbiome analyses. *Nature Protocols*, 11(11), 2049–53.
Xiao-Bin, W., Dian-Xiong, C. A. I., Hoogmoed, W. B., Oenema, O. and Perdok, U. D. 2006. Potential effect of conservation tillage on sustainable land use: A review of global long-term studies. *Pedosphere*, 16(5), 587–95.
Xue, C., Ryan Penton, C., Shen, Z., Zhang, R., Huang, Q., Li, R., Ruan, Y. and Shen, Q. 2015. Manipulating the banana rhizosphere microbiome for biological control of Panama disease. *Scientific Reports*, 5, 11124. doi:10.1038/srep11124.

Chapter 9

Challenges in cultivation of bananas in the subtropics

Víctor Galán Saúco, Instituto Canario de Investigaciones Agrarias, Spain

1 Introduction
2 The influence of climate in banana production
3 Main problems for banana cultivation in the subtropics
4 Solving cultivation problems
5 Other cultivation practices relevant to the subtropics
6 Greenhouse banana cultivation
7 Banana cultivars for the subtropics
8 Plant protection
9 Access to market
10 Conclusion
11 Where to look for further information
12 References

1 Introduction

The production of bananas, as with many other tropical crops, is highly influenced by the climate. It is commonly thought that banana production is more suited to a tropical climate than to a subtropical one. However, this only applies when temperatures are in a moderate range, which allows uniform plant development throughout the year and prevents physiological disorders. There are many other severe constraints associated with the tropics (Galán Saúco et al., 2012), which are nearly all due to the heavy rains and cloudy conditions that prevail in these regions, but these will not be discussed here. In the case of bananas, the relatively cool climate of the subtropical areas causes undesirable physiological phenomena and long cropping cycles. However, yields obtained in the subtropics are excellent, and those in countries like Israel or Spain (the Canary Islands), which are well inside the subtropical belt, can be considered among the highest in the world (Galán Saúco et al., 1992, 1998).

http://dx.doi.org/10.19103/AS.2017.0020.15

2 The influence of climate in banana production

From the phenological point of view, banana, *Musa acuminata* Colla, like the rest of the species of the genus *Musa*, is characterized by the existence of a vegetative phase before the onset of the reproductive phase, and although there is no correlation between both phases, the final weight of the bunch is directly associated with the vegetative rhythm of growth (Galán Saúco and Robinson 2013).

Although most of the banana-growing areas of the world are geographically situated between the equator and latitudes 20°N and 20°S, there are commercially profitable plantings at latitudes close to or even slightly higher than 30°N and 30°S. As indicated in the 'Introduction', it is precisely in areas with mild subtropical climates, such as the Canary Islands (27–29° North latitude), where the best yields for bananas are obtained, especially under greenhouse cultivation.

As for many other crops in the tropics, where enough water is available, temperature is the main limiting factor for banana growth. Studies conducted in South Africa for Dwarf Cavendish (Green and Kuhne, 1970; Robinson and Anderson, 1991), and probably also valid for close cultivars such as Grand Nain or Williams, indicate that the growth of young banana plants stops completely when temperatures fall below 14.5°C. Below 20°C, the influence of temperature is absolute, making it impossible to improve the growth of the plant by using any local cultivation practice unless the temperature is increased. Furthermore, leaf emergence stops in these cultivars, and probably in most banana and plantains, when the mean daily temperature threshold ([maximum + minimum] divided by 2] falls below 16°C, and dry matter assimilation also ceases at temperatures below 14°C (Robinson and Galán Saúco, 2010). In particular, Turner and Lahav (1981) indicated that in the cultivar Williams, the optimum temperatures (day/night) for leaf emission are in the range of 33/26°C to 25/18°C. Robinson (1990) reported 21–25°C as the optimum temperature for carbohydrate production, while 22°C is considered the optimum temperature for both growth and flower initiation, and approximately 31°C is the optimum temperature for leaf emergence. Leaf emission also ceases at temperatures higher than 38°C (Aubert, 1971; Ganry, 1973), with 47°C being considered as the thermal death point for banana leaves. Even in well-watered soil, the banana plant may experience heat stress and a 'temporary wilt' when the ambient temperature increases above 38°C because the ability of the plant to extract water from the soil is insufficient to meet the evaporative demand and transpiration losses. In general, around 27°C is considered to be the best overall mean temperature for optimum balance between growth (assimilation) and development (leaf emission rate, LER), when plants are grown under field conditions in the subtropics (Robinson and Galán Saúco, 2010).

Temperatures below 13°C may cause underpeel discolouration of fruit (see Fig. 1), and winter chilling damage in leaves – that is, progressive yellowing due to chlorophyll destruction – can also be seen when night temperatures fall below 6°C. Temperatures below 0°C cause leaves and the whole plant to die, but if the frost is of short duration and low intensity, the plant can recover from to the rhizomes (Champion, 1967). Low temperatures occurring in winter in the subtropics also cause damage, such as the uneven softening of fruits during ripening, a reduced shelf life and, when below 9°C, the coagulation of latex and the subsequent loss of the fruit's ripening capacity (Tai, 1977).

The effect of low temperatures is clearly shown in the LER. According to Oppenheimer (1960) and Aubert (1971), under ideal climatic conditions, a banana plant should emit one leaf

Figure 1 Underpeel discolouration in ripe banana fruits. Courtesy J. Robinson.

per week. Conventionally, the emission of more than three leaves per month is considered to be a sign of vigorous growth under subtropical conditions (Galán Saúco, 1992).

The slowdown of growth through low temperatures (and also dry conditions) can also give rise to chlorosis and foliar and floral malformations due to the reduced growth of the leaf sheaths and the reduction of internode distance. Even if no foliar obstruction occurs, flower differentiation inside the pseudostem during low winter temperatures results in small and malformed bunches. In extreme cases, the bunch has difficulty in emerging. It is then necessary to make a cut in the pseudostem to allow the bunch to emerge, but it is usually distorted and has no commercial value (Robinson and Galán Saúco, 2010). In the Canary Islands this practice is known as '*hacer la césarea*' (perform a caesarean) because of its similarity to the caesarean section operation for women who have birth-related problems (Galán Saúco, 1992).

The banana plant is a species that demands a great deal of water, both because of its rapid and continuous growth and because of its great foliar area. Banana demands 25 mm of water per week, which is considered to be the minimum for satisfactory growth (Robinson and Galán Saúco, 2010). This explains why adequate water supply, after temperature, is the most important factor for banana cultivation. It is probably the main factor responsible for growth when temperatures exceeding 20°C are constant (Green and Kuhne, 1970; Lassoudière, 1974). In the humid tropics, the water requirements of bananas are met totally by rainfall, but in the dry tropics, the subtropics and in Mediterranean areas, supplementary irrigation (see Fig. 2) is absolutely necessary to obtain good commercial yields.

Figure 2 Young banana planting under drip irrigation in the Canary Islands.

Other climatic problems which can affect banana production are the incidence of hail and wind. The latter can cause a variety of damage, depending on the wind intensity. Hurricane or cyclonic winds of more than 15 m/s (54 km/h) can partly or totally destroy the whole plantation. Regular strong winds (5–10 m/s) can cause leaf tearing, which may reduce yield, but even short-speed winds (2.5–5 m/s) can reduce fruit quality due to a phenomenon known as leaf and dust abrasion. Hot and dry winds induce water stress and may cause temporary wilt (Robinson and Galán Saúco, 2010). Banana cultivars, especially Dwarf Cavendish and Grand Nain, tolerate winds of up to 40 km/h well (Soto, 1991; Galán Saúco, 1992).

Light exposure is not usually important under normal cultivation practices (Oppenheimer, 1960) except when plantings are excessively dense, in which case the cycles are delayed. It should be noted that the photosynthetic rate (P_s) and the LER are reduced when the seasons are long and dry, the temperatures are high and periods of light are short (Sastry, 1988; Turner, 1998a,b). However, exposure to high light intensity, especially when accompanied by high temperatures, causes leaf burning. In Costa Rica, Soto (1991) considered 1500 hours of light per year (4 hours per day) as the minimum duration for obtaining a profitable banana crop.

According to Champion (1967), day length does not play a role in banana growth except under extreme conditions, although Turner et al. (2009) and Turner and Fortescue (2012) indicate that, pending experimental verification, the banana may be considered as a quantitative long-day plant. Unpublished data (Table 1) from a recent trial run by the author of this chapter in the Canary Islands under greenhouse conditions realized that prolonging the day length to 14 hours during the winter months (see Fig. 3) seem to confirm this theory. The results of this experiment showed a reduction in the length of the first cycle, for both the Grand

Table 1 Influence of prolonging day length on the first cycle length and bunch weight in Grand Nain (GN) and Dwarf Cavendish (DC) cultivars under greenhouse conditions (Galán Saúco et al., unpublished)

Treatment	Planting – harvesting (days)	Planting – bunch emergency (days)	Bunch emergency – harvesting (days)	Bunch weight (kg)
GN + prolonging light*	865.4[a]	684.5[a]	180.9[a]	40.57[a]
GN normal light	917.6[a]	699.8[b]	217.8[a]	35.12[b]
DC + prolonging light*	888.6[a]	691.8[a]	196.8[a]	43.20[a]
DC normal light	907.7[a]	706.3[a]	201.4[a]	34.40[b]

*Prolonging day length with an individual lamp (100 watt) per plant since 01 October 2011 until harvesting.
Means followed by different letters are significant at $p < 0.05\%$.

Nain and Dwarf Cavendish varieties, and an increase in bunch weight, which was always in favour of the plants exposed to a greater day length. However, the results regarding the flower to harvest interval were only significant for bunch weight in both cultivars, and only for Grand Nain regarding the planting to bunch emergency interval.

Finally, it should be mentioned that although the generally predicted climatic changes occurring across the world as a result of global warming (that is, towards increasing temperatures and the likelihood of drought) may be 'well suited', *grosso modo*, to banana production under irrigation in the subtropics, because it diminishes some of the main problems of banana cultivation, these changes may also bring about an increase of the most serious banana pests and diseases, which are not prevalent now (see below) but which will complicate banana cultivation in those areas of the world.

Figure 3 Trial on artificially prolonging day length in the Canary Islands.

3 Main problems for banana cultivation in the subtropics

As indicated by Robinson and Galán Saúco (2010), the banana can be considered primarily as a crop of the humid tropical lowlands. In the humid tropics, mean temperatures are generally within the optimum growth range of 22–31°C throughout the year, and there is no evaporative stress on the plant; no chilling occurs and irrigation is not necessary to improve production. In the subtropics, however, there are usually periods when either cold winter temperatures or excessive heat stress limit production. Furthermore, rainfall is usually low, erratic and seasonal.

The effect of climatic hazards in the subtropics, semi-arid tropics or in the Mediterranean region can range from barely noticeable to complete destruction, depending on the plantation site and the severity of the climatic conditions. Quite clearly, the choice of plantation site is vitally important for reducing the impact of these climatic hazards and the installation of an irrigation system is necessary to obtain profitable crops. The relatively cool climate of the subtropics causes undesirable physiological phenomena and long cropping cycles, despite which yields obtained in the subtropics are excellent and compare favourably with many other locations where banana is grown.

The specific and most common problems occurring in the subtropics include (Robinson and Galán Saúco, 2010):

a Frost damage.
b Winter chilling of leaves.
c Growth cessation – LER drops sharply below 20°C, reaching less than 1 leaf per month during winter in the subtropics compared with 4–5 leaves per month in the tropics or during summer in the subtropics. No visible damage appears on the plant, but cycle length is considerably delayed.
d Choke throat problems (see Fig. 4).
e Winter flower initiation, colloquially known as 'May flower' in the Northern Hemisphere or 'November dump' in the Southern Hemisphere, which can cause small and malformed bunches (see Figs. 5 and 6).
f Underpeel discolouration (see Fig. 1).
g Winter leaf sunburn as a combined effect of high (>30°C) daytime temperatures, low (<25%) relative humidity, high light intensity and a depleted feeder root system due to growth cessation.
h Other climatic problems, such as wind damage, hail and lightning, are also characteristic of subtropical climates, causing specific constraints to growth, yield and fruit quality.

4 Solving cultivation problems

Some of these problems, particularly (d), (e) and (f), may be avoided through the adoption of appropriate local cultivation practices orientated towards obtaining flower initiation and flower emergence during the summer and harvesting prior to the onset of low temperatures. Besides appropriate fertilization and irrigation and wind protection,

Figure 4 Choke throat problem. Courtesy J. Robinson.

adequate site selection may diminish some of these problems, as will the choice of planting date, planting density and appropriate sucker selection. However, the cultivation of banana under greenhouse conditions, which is now increasingly practised in different parts of the subtropics, is more significant for reducing these climatic constraints.

4.1 Site selection

Although, in the tropics, it is mainly soil type, and not climatic factors, that determines site selection, in subtropical and Mediterranean banana-growing areas, site selection for banana cultivation is only partly determined by soil type, and suitable areas for open-air cultivation have to be carefully selected according to temperature registers, avoiding places with minimum winter temperatures below 10°C and maximum summer temperatures regularly above 38°C. However, in places such as Western Australia and Saudi Arabia, with peak summer temperatures ranging between 40°C and 45°C, and in Israel, with winter temperatures sometimes in the range of 1–8°C, bananas can also be cultivated by modifying the management of the crop, in particular through protected cultivation.

Figure 5 Deformed bunch (May flower) due to winter flower initiation.

Within the suitable range of temperature in the subtropics, site selection should be based on other factors, with the most important, according to Robinson and Galán Saúco (2010) being the following:

- Sufficient elevation for frost protection.
- Shelter from prevailing winds.
- Avoiding known hail belts or storm areas.
- Avoiding areas prone to waterlogging or salinity.
- Ensuring proximity to perennial water supplies for supplementary irrigation.
- A suitable fertile and well-drained loamy soil.
- An appropriate aspect and slope according to hemisphere and latitude.

4.2 Planting date

The establishment of the best planting date for bananas is determined by two major factors, namely, timing the crop harvest to coincide with high market prices and timing the planting date to benefit from, or to avoid, certain climatic conditions. Very often, these two factors interact to severely limit the planting date range. Usually, this timing is only effective for the plant crop, and possibly for the first ratoon cycle, unless a strict desuckering policy is

Figure 6 November dump. Courtesy J. Robinson.

imposed. Paradoxically, it is more difficult to keep production in the tropics occurring at the time of better market prices, usually in winter in the western markets, because constant good weather in the tropics favours natural harvest spread and nullifies this benefit. Regarding temperature in the mild subtropics, the existence of well-defined seasons makes it easier to control timing through good cultivation practices and especially through appropriate desuckering.

Traditional plantings in Israel are initiated in the spring. Since the banana plant crop is small, the objective is to time planting for an optimum first ratoon sucker, undertaking sucker selection in June (summer), three months after planting, to allow flower initiation and flower emergence to occur in the warm weather of the spring/summer in the following year, and to harvest in November/December (autumn/early winter), which is ideal for good quality fruit and high prices. In the mild climate of the Canary Islands, planting is usually done at the beginning of summer, from 15 June (at higher altitude) to the middle of August (at coastal sites), in order to get the first harvest in the second winter after planting. Similarly, in the Southern Hemisphere subtropics of Kiepersol, South Africa, and in New South Wales, Australia, summer planting (December/January) has been identified as the optimum time to give high yields, good quality fruit and high local market prices from an autumn/winter harvest.

4.3 Planting density and spatial arrangement

Climate is the main determinant of planting density. While planting in the dry and hot tropical countries can hold ratoon densities of 2500–3000 plants per hectare, a lower density (1500–2000 plants per hectare) should be used in mild subtropical climates to: (a) allow light penetration, (b) enhance growing temperatures and (c) accelerate the cycle time. These latter densities do not differ much from those (around 1750 plants per hectare) generally used under the hot conditions and cloudy environment of tropical banana-exporting countries, where cultivars such as Grand Nain, Williams and Dwarf Cavendish are exported (Robinson and Galán Saúco, 2010), and are even bigger than those of around 1366–1600 plants per hectare used for Prata Ana in Brazil (Vilela Rodrigues et al., 2008). It is not only density but also spacing relationships which are important for banana productivity. Experiments conducted in South Africa over five years (Robinson et al., 1989) at a density of 1666 plants per hectare clearly demonstrate that overall productivity was higher with rectangular (3 × 2 m) spacing (see Fig. 7) than with hedgerow (4 × 1.5 m) or tramline (6 × 2 × 1.5 m) plantings, indicating that there is a loss of physiological efficiency when the inter-row became wider. Since differences in yield are not relatively small (usually not much more than a 7% yield reduction in the long-term period) for a given density, other circumstances, such as available labour force or mechanization, may determine the choice of the more cost-efficient tramline planting design. The following are some of the advantages of this last system: (a) easier and cheaper plant support, (b) less mechanical damage to bunches, (c) easier irrigation control and (d) more efficient

Figure 7 Dwarf Cavendish open-air planting (2 x 3 m parent crop) in the Canary Islands.

control of bunch development and complete accessibility to each plant by tractor and trailer (Robinson and Galán Saúco, 2010).

4.4 Sucker management

In any banana planting, this practice is critical not only to maintain initial plant density and spatial distribution during the entire plantation life, but also to obtain a good yield, both in terms of quantity and quality, at the desired moment of harvesting. Since sucker production is closely dependent on a healthy and vigorous root system, any stress affecting plant growth and development should be avoided. This is especially important in the subtropics where cold stress is always present.

Although undesired suckers must be removed frequently, thus preventing them from becoming excessively large and unmanageable, under subtropical conditions, suckers are not removed during winter because of the slow growth due to the low winter temperatures. Rather, this operation begins in the spring. Suckers are removed every 6–8 weeks during the spring and autumn and, as in the tropics, every 3–4 weeks during the summer, except in places in Australia where desuckering is only done twice a year due to the high labour cost (Robinson and Galán Saúco, 2010). To avoid competition with the selected follower, unwanted suckers should not be taller than 300 mm and leaves should still be in the thin, bract-like stage at the time of removal (Robinson and Nel, 1990).

Selection of the correct follower is one of the most critical operations in the banana plantation, especially when changing from the plant crop (P) to the first ratoon (RI). As indicated by Robinson and Galán Saúco (2010), two important general aspects of banana development – the stage of parent plant development and the direction of sucker selection – should be considered. These are summarized here with regard to the correct sucker selection practice under subtropical conditions.

4.4.1 Stage of parent plant development

Sucker growth is strongly influenced by the parent plant, with a vigorous parent producing a vigorous sucker and vice versa, while the mother plant inhibits sucker development by apical dominance effects (Eckstein and Robinson, 1999). Thus, a good balance should be established between plantation vigour and density to facilitate optimal sucker development. If the independence of the sucker from the mother plant occurs too early (as with a water sucker or by selecting too early), it will become weak and produce a small bunch. Conversely, if suckers that are too young are selected at the time of the parent harvest (late selection), the ratoon yield could also be inhibited.

4.4.2 Direction of sucker selection

Several factors, such as topography, irrigation, a planting system and, especially in the subtropics, climate conditions, influence the direction of sucker selection. In this sense, it is better to select the suckers orientated towards the East for sun exposure in the morning, when photosynthesis is highest, and for shade of the parent in the afternoon, when heat stress is likely. It is essential that all suckers in R1 and successive cycles are selected in more or less the same direction (see Fig. 8).

The axial sucker is initially selected as the follower in tropical countries, but in successive cycles it is the most vigorous sucker rather than the axial sucker that is selected, with the subsequent loss of the plantation symmetry in the successive ratoons. Under subtropical

Figure 8 Uniform sucker selection in the same direction. South Africa. Courtesy J. Robinson.

Figure 9 Selection of sucker number 9 (second flush) in the Canary Islands.

conditions, selection strategies are different. Under open-air cultivation in the Canary Islands, the axial sucker as well as suckers from the first flush (the first pentagon of De Langhe, 1961) produced on the parent plant during the winter are destroyed when they reach 300 mm high, and the final selection is made at the end of the spring based on sucker leaf number. Sucker number 9 from the 'second flush' (see Fig. 9) is the sucker preferred as the follower. In open-air plantings in Israel, similar to the Canary Islands, the specific month of sucker selection and size of sucker are critical every year, in order to avoid flower initiation or flower emergence in the winter. They apply the 5/6/6 system (5/8/8 in the first cycle), meaning that a sucker with five leaves is selected on the 6th day of the 6th month (June). With *in vitro* plants in the first cycle, the formula changes to 5/8/9 (Galán Saúco, 1992). In subtropical South Africa, RI sucker selection is normally made between five and ten months after planting depending on the interval from planting date to the onset of winter (Robinson et al., 1985).

In the case of *in vitro* plants, suckers from the first pentagon emerging after planting should be removed because these are set too low on the parent rhizome and may push the parent rhizome above ground, causing 'high mat'. One of the suckers 6–10 from the second pentagon in the correct direction of marching should be selected.

5 Other cultivation practices relevant to the subtropics

The practice of removing healthy leaves at flowering has proven to be useful for accelerating ratoon sucker growth in the subtropics, but a minimum of 12 healthy leaves should preferably be present at harvest in open-air plantings to obtain maximum bunch filling and green life extension (Turner, 1970; Robinson et al., 1992; Daniells et al., 1994).

The benefits of the use of bunch covers in subtropical banana-growing countries, including better pest protection, increased finger length, higher yield, shorter flower to harvest interval and reduced chafing from dust and leaves, have been reported elsewhere (Robinson and Nel, 1984; Daniells et al., 1987; Galán Saúco, 1992; Galán Saúco et al., 1996), and a summary is reported in Robinson and Galán Saúco (2010). According to the summary, blue covers, particularly translucent blue (see Fig. 10), which allow heat transmission but reduce sunburn damage, with a thickness of 35–50 µm are the most popular type of bunch cover used in the subtropics to protect against hail and wind, and may be used for two or three bunch cycles. Bunch covers are usually impregnated with ultraviolet protection to prolong their useful life.

6 Greenhouse banana cultivation

The protected cultivation of banana began in the 1980s in Morocco and the Canary Islands and has since been extended to other countries in the Mediterranean region, such as Israel and Turkey. At the end of the 2010s around 11 000 hectares of banana were under protected cultivation (Galán Saúco and Damatto Junior, 2012).

Experimental work in the Canary Islands by Galán Saúco et al. (1992) demonstrated clearly that banana plants were larger, bunches were heavier, cycle intervals were shorter and annual yield per hectare was significantly increased under plastic greenhouses compared with open-air cultivation (see Fig. 11). These effects were particularly pronounced on cool,

Figure 10 Bunch cover in bananas in Israel. Courtesy Y. Israeli.

north-facing slopes where climatic constraints are more severe. With such large increases in yield (up to 77%), the use of a plastic greenhouse was shown to be cost-effective, and the investment can be recovered within five years.

The relevant advantages of greenhouse cultivation for subtropical conditions were fully described by Robinson and Galán Saúco, (2010) and are summarized as follows:

1 An increased number of days with a mean temperature above 16°C.
2 Protection against wind and other weather conditions (sunburn, hail), especially from sea winds (see Fig. 12).
3 A reduction in water consumption, since evapotranspiration is reduced by up to 25%.

Figure 11 Bananas under greenhouse cultivation. Cv. Grand Nain. Canary Islands.

4 An increased leaf surface area, leading to a higher photosynthetic capacity.
5 A shortened crop cycle of the plant, allowing not only annual cycles but also three crops every two years in some locations.
6 The production of larger and heavier bunches. In some parts of the Canary Islands, bunch weights have increased by 62% (Galán Saúco et al., 1998), leading to average yields exceeding 80 tonnes per hectare (exceptionally 100 tonnes per hectare per year) in comparison with 60 tonnes per hectare per year average for well-managed, open-air plantations.
7 An increased in the size of banana fruits.

All these advantages reduce costs and increase yields (higher gross margin), which can compensate for the high cost of building a greenhouse (see Fig. 13) and the cost of periodically repositioning covers (see Fig. 14). This explains why, in Israel, all new banana plantings are done under greenhouse covers so that around 90% of bananas in Israel are now grown under greenhouses (see Figs. 15 and 16) (Y. Israeli, pers. comm.). This compares to around only 33% in the Canary Islands (Fig. 17) where, in some locations where tourism is very important, greenhouse cultivation is thought to have a negative aesthetic impact and is forbidden. Greenhouse banana cultivation is also forbidden on the island of Madeira for the same reason.

Different structures are used for building greenhouses in subtropical countries. In the Canary Islands, for example, a frame of galvanized steel pipes (5–10 cm in diameter and 6–7 m in height) embedded in concrete bases is used (Galán Saúco, 1992; Galán Saúco et al., 1998). Such structures last for more than 20 years, but polyethylene (PE) covers should be replaced much earlier. It is essential to select an adequate type of cover for greenhouse cultivation (see Fig. 18) since it was shown in experiments conducted in the Canary Islands that there is significant variation in respect of both the transmission of active photosynthetic

Figure 12 Banana greenhouse cultivation as a protection against sea winds. Canary Islands.

radiation (PAR) and ultraviolet (UV) radiation (Cabrera Cabrera and Galán Saúco, 2003, 2005, 2012). Some interesting observations from these studies are that:

1. UV and PAR penetrations are lower under covers than in the open-air.
2. Temperatures are higher under PE covers than under net covers.
3. The incidence of white flies (*Aleurodicus dispersus* and *Leucanoides floccissimus*) is lower under PE cover. This is explained by UV reduction, and has been reported for other crops (Antignus et al., 2001).
4. PE covers of 750 gauge (190 μm) give the most favourable conditions for banana growth and development, reducing cycle duration and increasing bunch mass.

Figure 13 Building a greenhouse. Canary Islands.

Figure 14 Renovating greenhouse covers in the Canary Islands.

5 Some PE materials, such as Celloclim®, are of special interest as they combine the best characteristics of PE with durability – a minimum of 4–5 years compared to 2–3 years for normal PE types and a minimum of 5 years for net covers.

Figure 15 Bananas under a greenhouse in Israel. Details of the structure. Courtesy Y. Israeli.

Figure 16 Young banana planting under a greenhouse in Israel. Courtesy Y. Israeli.

Figure 17 Open-air and greenhouse cultivation of banana. Canary Islands.

6 The cultivation of banana under PE covers requires careful ventilation management to avoid summer temperatures above 30°C, which may burn banana leaves. This is why many farmers prefer to use nets instead of PE to cultivate bananas in the warmer locations of the subtropics or Mediterranean areas. This is of particular importance for protected banana cultivation in Israel due to very high summer temperatures.
7 Up to now, no greenhouse cover used for banana cultivation is fully biodegradable.

Recent trials in the Canary Islands (Cabrera Cabrera and Galan Saúco, 2012) with four greenhouse covers, namely,

- *FPE*, a clear 860-gauge thermic PE cover of a three-layer film that sandwiches an expanded polymer layer with gas microbubbles between two plain polymer layers, marketed commercially as Celloclim 4S ®
- *M20*, a monofilament mesh (20 × 10 threads cm^{-2})
- *M16*, a monofilament mesh (16 × 10 threads cm^{-2})
- *MB*, a raffia mesh, marketed by Polysack ® as white net 50%,

indicate that under the environmental condition of this trial, the M16 cover, with the highest PAR and shortest UV, offers the best conditions for the growth and development of bananas, with the only inconvenience being the higher incidence of white flies. This illustrates, once more, the significant impact of the greenhouse covers over the growth, development and yield of bananas. Under this trial, all covers lasted more than eight years in good conditions, which clearly exceeds the expectations for these types of greenhouse cover.

Figure 18 Different types of greenhouse cover used for banana cultivation. Trial at the Canary Islands Agriculture Research Institute (ICIA).

Figure 19 Excessive shading under greenhouses may cause cycle delay.

Some trials have also been conducted in Israel with different coloured nets, but none of them is better than the transparent cover normally used. Furthermore, the red-coloured ones are detrimental since they induce longer and narrower leaves which have lower photosynthetic efficiency (Y. Israeli and E. Lahav, 2009, pers. comm.).

Techniques for greenhouse cultivation are similar to those for open-air cultivation, but sucker selection is easier due to the higher and quicker emission of suckers and better temperature conditions. However, the abundance of leaf surface under greenhouse cultivation requires leaf removal to avoid excessive shade and, consequently, cycle delay (see Fig. 19). As a matter of fact, during the first attempts to cultivate bananas under greenhouses at the beginning of the 1980s in the Canary Islands, the preservation of all the leaves caused a delay in the cycle of bananas cultivated under greenhouses compared with those grown in the open-air (Galán Saúco, 1992). Leaf removal (see Fig. 20) is carried out shortly after bunch emergence. A minimum of eight leaves per plant (the eight youngest leaves) are retained. Greenhouse cultivation also allows the aerial tying of bunches to the internal pipes of the roof structure, thus eliminating or reducing the need for props.

It may be logical to think that the reduction in PAR radiation occurring under greenhouses – PAR values around 73% of those obtained in the open-air, 1100 versus 1500 μmol m^{-2} s^{-1} (Galán Saúco and Robinson, 2013) – may be a problem for banana cultivation, but, on the contrary, the productivity obtained in a well-managed greenhouse in the Canary Islands exceeds 20–30% that of open-air cultivation (Galán Saúco et al., 1992, 1998). This apparent contradiction may be explained partly as a consequence of a higher temperature, and is also based on the observations of Turner (1988), which indicate that an increase of the leaf area index allows more light interception for the plants under greenhouses compared to those

Figure 20 Leaf removal under greenhouses. Canary Islands.

Figure 21 Grand Nain under greenhouse cultivation. Turkey. Courtesy H. Gubbuk.

in the open-air. It should, however, be taken into consideration that the possible excessive shading will require the removal of surplus leaves, as indicated earlier.

7 Banana cultivars for the subtropics

Grand Nain (see Fig. 21) is the most widely planted greenhouse cultivar. It is grown in Israel and Turkey (100%), Morocco (92%) and in the Canary Islands (roughly 75% of the surface under cover) (Galán Saúco et al., 2004). Dwarf Cavendish is the second most important cultivar in both Morocco and the Canary Islands. Grand Nain is now less popular in the Canary Islands where most new plantings or replantings, both under greenhouses and in the open-air, are being done with Gruesa (see Fig. 22), a locally selected Dwarf Cavendish clone (Galán Saúco, 2014). Differences in productivity and green life between Gruesa and Grand Nain are marginal (Hernández Hernández et al., 1992; Cabrera Cabrera and Galán Saúco, 2005). The smaller finger size of Gruesa and other Dwarf Cavendish clones, together with a claim about its better taste, is used as a mark of differentiation in the Spanish mainland markets compared to bananas imported from other countries.

Due to the susceptibility of Lady Finger and Cavendish cultivars to Fusarium (*Fusarium oxysporum* f.sp. *cubense*) wilt Subtropical Race 4 (FOC SR4), efforts have been made to evaluate alternative dessert cultivars with market potential, mainly tetraploid hybrids in Australia (Daniells et al., 2013; Smith et al., 2014). As a result of this experiment, interest is growing in some of them, in particular, for example High Noon (SH-3640), which is resistant to FOC SR4 but susceptible to Sigatoka (Daniells, 2010). This may be of interest for most subtropical areas that are already free of Sigatoka incidence. More recent studies conducted in Australia (Smith et al., 2014) have shown a potential for the commercial cultivation of FHIA-17, which has a level of resistance to *Fusarium* similar to Williams, a flavour similar to Gros Michel (Robinson and Galán Saúco, 2010) and has been well received by farmers in Australia (Honan, 2017). The fact that earlier studies (Walduck, 2005) indicated that Goldfinger and FHIA-18 were less susceptible to TR4 Fusarium wilt than Cavendish has recently stimulated interest in the dessert types similar to Pome. However, these and other promising hybrids need to be commercially tested in different markets. Nevertheless, the well-known consumer preference for Cavendish types is the main difficulty for any commercial changes of varieties in the future, at least for the western markets.

Figure 22 Inside a greenhouse. CV Gruesa. Canary Islands.

8 Plant protection

The cultivation of bananas under subtropical conditions has some advantages with respect to the incidence of pests and disease. The generally lower temperature and low humidity levels are equally disadvantageous for most of the important banana pests and diseases. This may explain why 'black Sigatoka' or black leaf streak (*Mycosphaerella fijiensis*), the most damaging disease in the tropics, is not generally present or has a low incidence under subtropical conditions. Consequently, the cost of controlling leaf diseases is much less or even non-existent in most subtropical areas. Even the less virulent form of Sigatoka, namely 'yellow Sigatoka', also known as 'Sigatoka disease' (*Mycosphaerella musicola*) or the burrowing nematode (*Radopholus similis*), is only present in some of the most warm and humid areas of the subtropics (Sarah et al., 1996; Galán Saúco et al., 2012).

However, *Cosmopolites sordidus,* the banana weevil (BW), causes serious problems in subtropical plantings, both in the Canary Islands and in Australia, where it is considered to be a much greater problem than in tropical Australia (Galán Saúco et al., 2012). According to Daniells, Lindsay and Smith (1996), this can be a consequence of both the slower plant growth rates, which allow BW to cause more damage to the newer part of the mat, and the fact that the more easily accessed growing point of the tissue culture plants may be more susceptible to BW during early establishment that the sucker/bit materials.

Fusarium (*F. oxysporum* f.sp. *cubense*) wilt strains such as Race 1 and Subtropical Race (SR 4) may cause more serious problems in cooler subtropical climates and is particularly damaging to Cavendish cultivars and also to Lady Finger in South Africa, Australia, Brazil and Taiwan (Daniells, 2010; Daniells and O'Keefe, 2012). However, since resistance seems to be a phenomenon related to growth rate, plants that are capable of maintaining energy reserves in excess of their immediate primary metabolic needs (for growth and reproduction) will be able to rapidly manufacture the necessary resistance compounds or secondary metabolites to meet the pathogen challenge (Daniells, 2010). This perhaps explains the relatively low incidence of Panama disease in well-managed plantings in the Canary Islands and the possibility of managing this problem with sound cultivation practices that increase plant vigour (Galán Saúco, 1992). However, since Cavendish cultivars are also susceptible to Tropical Race 4 (TR4) in the tropics in the absence of the disease-predominant conditions of low temperature occurring in the subtropics, it may be imperative to avoid the spread of TR4 all over the world. Otherwise, due to the polycyclic nature of this disease (Ploetz, 2015), Cavendish cultivars could easily be decimated and banana production in the subtropics extinguished unless resistant cultivars are planted.

Phytopathological problems under greenhouses are similar to open-air conditions, while pest exclusion measures, biological control and integrated or organic plot management are easy to apply.

9 Access to market

An important marketing advantage of banana production in subtropical areas such as Israel or South Africa is the fact that produce goes only to local or close-by markets. The requirement for a long fruit shelf life is therefore avoided and the risk of damaging fruits by bad handling or exposure to inappropriate temperatures during transport is reduced. Fruits reach the ripening rooms within two or three days after harvest; thus, bunches can be left on the plant until they are nearly mature, increasing yield and quality. In addition, local markets are more tolerant of minor quality blemishes. However, banana production in the subtropics also faces marketing difficulties due to strong competition from the tropical producing countries that are dominated by multinational companies. Tariff barriers, particularly in the European Union (EU), and phytosanitary barriers in places such as Australia or Israel have certainly helped some subtropical export producers, such as the Canary Islands or Madeira in the first case, or local producers in Australia and Israel. Yet, at present, there is a trend towards the suppression of both kinds of barriers. The future of marketing bananas produced in the subtropics may come from the advantage of lower pest and disease incidence, which will favour organic cultivation, together with the reduced costs of fungicides/insecticides. Another benefit may come from rising concerns about the carbon footprint of produce, which is much reduced in the case of subtropical production in the Mediterranean countries or subtropical European regions due to their proximity to the EU market. In addition, bananas from the subtropics must be differentiated from those of the tropics on characteristics such as taste, maturity, size, perhaps freshness, environmental credentials and so forth. Separate supply chains in supermarkets and promotional campaigns are very useful. In Spain, for example, the consumer clearly differentiates and pays higher prices for bananas from the Canary Islands, called plátanos, from 'bananas', which is the term used for fruits from other places of origin. The success of 'plátanos de Canarias', which, despite the trend towards the liberalization of the market, keeps (in the year 2016) a share of 72% of the Spanish mainland market (Nuez Yánez, 2017), is a clear example to follow.

10 Conclusion

Despite the relatively cold conditions for bananas, excellent yields can be obtained in the subtropics if the appropriate cultivation techniques are applied, not only under greenhouses but also in the open-air. Besides current cultural practices for banana cultivation elsewhere, such as the selection of good plant material and planting sites, adequate irrigation and fertilization, special care should be devoted to other cultivation practices to ensure good results. These include choosing the correct planting dates, adopting an appropriate density of planting and good planting design and ensuring the best management of sucker selection. Proximity to the market as well as the relatively low incidence of pests and diseases, compared to the tropics, contributes to the success of banana cultivation in the subtropics.

11 Where to look for further information

Besides the sources given in the reference list, the reader can find additional information at the ProMusa web page (www.promusa.org) as well as at the symposium organized by the ISHS Section for Bananas and Plantains (www.ishs.org). Of special interest are the Proceedings of the First International Symposium in the Subtropics, held in Puerto de la Cruz (the Canary Islands) in November 1997 (*Acta Horticulturae* 490), and as well as in 2010, the Proceedings of the Symposium on Citrus, Bananas and other Tropical Fruits under Subtropical Conditions, held in 2010 (*Acta Horticulturae* 928). Shortly after the publication of this book, the ISHS/ProMusa Symposium, 'Growing and Marketing Banana under Subtropical Conditions' will be held as one of the symposia of the 30th International Horticultural Congress in Istanbul, Turkey, on 12–16 August, 2018.

12 References

Antignus, Y., Nestel, D., Cohen, S. and Lapidot, M. (2001). Ultraviolet-deficient greenhouse environment affects white fly attraction and flight behavior. *Environmental Entomology* 30: 394–9.

Aubert, B. (1971). Action du climate sur le comportement du bananier en zones tropicales et subtropicales. *Fruits* 26(39: 175–88.

Cabrera Cabrera, J. and Galán Saúco, V. (2003). Crecimiento y desarrollo de la platanera. (*Musa acuminata* Colla AAA) bajo distintos tipos de cubierta. Avance de resultados. *Actas de Horticultura* 39: 257–9.

Cabrera Cabrera, J. and Galán Saúco, V. (2005). Evaluation of the banana cultivars Zelig, Grande Naine and Gruesa under different environmental conditions in the Canary Islands. *Fruits* 60(6): 1–13.

Cabrera Cabrera J. and Galán Saúco V. (2012). Evaluation of different covers used in greenhouse cultivation of Cavendish bananas (*Musa acuminata* Colla AAA) in the Canary Islands. *Acta Horticulturae* 928: 31–9.

Champion, J. (1967). *Botanique et Genetique des Bananiers. Tome 1. Notes et Documents sur les bananiers et leur Culture*. I.F.A.C., Setco, París.

Daniells, J. W. (1988). The effect of volume of root zone irrigated on yield and fruit quality of plant crop bananas. In: Guzman, J. A. (Ed.) *Proceedings of the 4th Meeting of the International Group on Horticultural Physiology of Banana, Asociación Bananera Nacional de Costa Rica (ASBANA), Costa Rica, August 1986*. ASBANA, San José, Costa Rica, pp. 37–43.

Daniells, J. W. (2010). Fusarium wilt of banana – an integrated approach to disease management. *Tree and Forestry Science and Biotechnology* 4(Special Issue 2): 50–5.

Daniells, J. W., Lindsay, S. and Smith, M. K. (1996). Realizing the full potential of tissue-cultured plants. In: *Queensland Banana Industry Protection Board annual Report 1994/1995*, p. 27.

Daniells, J. W. and O'Keefe, V. (2012). Banana production challenges in the subtropics: Are banana varieties part of the solution? *Acta Horticulturae* 928: 67–73.

Daniells, J. W., Lisle, A. T. and Bryde, A. (1994) Effect of bunch trimming and leaf removal at flowering on maturity bronzing, yield and other aspects of fruit quality of bananas in north Queensland. *Australian Journal of Experimental Agriculture* 34: 259–65.

Daniells, J.W., O'Farrell, P.J., Mulder, J.C. and Campbell, S.J. (1987). Effects of bunch covering and bunch trimming on bananas in north Queensland. *Queensland Journal of Agricultural and Animal Sciences* 44, 101–105.

Daniells, J. W., O'Keefe, V., Smyth, H., Gething, K., Fanning, K. and Telford, P. 2013. Planet of the Cavendish – understanding the domination. *Acta Horticulturae* 986: 219–22.

De Langhe, E. (1961) La phyllotaxie du bananier et ses conséquences pour la comprehension du système rejetonnant. *Fruits* 16: 429–41.

Eckstein, K. and Robinson, J. C. (1999). The influence of the mother plant on sucker growth, development and photosynthesis in banana (*Musa* AAA; Dwarf Cavendish). *Journal of Horticultural Science Biotechnology* 74(3): 347–50.

Galán Saúco, V. (1992). *Los frutales tropicales en los subtrópicos. II. El plátano*. Mundi-Prensa, Madrid.

Galán Saúco, V. (2014). Plátano, papaya y piña. In: Hueso Martín, J. J. and Cuevas González, J. (Coord.) *La fruticultura del siglo XXI en España Cajamar*. Caja Rural, Spain, pp. 381–404.

Galán Saúco, V., Ait-Oubahou, A. and Abdelhaq, H. (2004). Greenhouse cultivation of bananas. *Chronica Horticulturae* 44(2): 35–37.

Galán Saúco, V. and Damatto Junior. E. R. (2012). Chapter 1. Cultivo de bananeira em ambiente protegido. Banana. In: Chavarria, G. and Pessoas dos Santos, H. (Eds) *Fruticultura em ambiente protegido*. EMBRAPA, Brasilia, Brasil, pp. 19–48.

Galán Saúco, V. and Robinson, J. C. (2013). Fisiología, Clima y Producción de Banano. In: ACORBAT 2013. XX Reunião Internacional da Associação para a Cooperação em Pesquisa e

Desenvolvimento Integral das Musáceas (Bananas e Plátanos) 9 a 13 de setembro de 2013. Fortaleza, CE, Brazil, pp. 43–57.
Galán Saúco, V., Cabrera Cabrera, J. and Gómez Leal, P. (1996). The evaluation of different bunch covers for bananas (*Musa acuminata* Colla) in the Canary Islands. *Fruits* 51(1): 13–24.
Galán Saúco, V., Hernández Delgado, P. M. and Cabrera Cabrera. J. (1990). Preliminary evaluation of new banana cultivars for Canary Islands. *Scientia Horticulturae* 47: 169–72.
Galán Saúco, V., Cabrera Cabrera, J. and Hernández Delgado, P. M. (1992). Phenological and production differences between greenhouse and open-air bananas (*Musa* AAA, cv. Dwarf Cavendish) in the Canary Islands. *Acta Horticulturae* 296: 97–112.
Galán Saúco, V., Cabrera Cabrera, J., Hernández Delgado, P. M. and Rodríguez Pastor, M. C. (1998) Comparison of protected and open-air cultivation of Grande Naine and Dwarf Cavendish bananas. *Acta Horticulturae* 490: 247–59.
Galán Saúco, V., Robinson, J. C. Tomer. E. and Daniells, J. (2012). Current situation and challenges of cultivating banana and other tropical fruits in the subtropics. *Acta Horticulturae* 928: 19–30.
Ganry, J. (1973). Etude du développement du systéme foliare du bananier en function de la température. *Fruits* 38(7–8): 499–516.
Green, G. C. and Kuhne, F. A. (1970). The response of the banana foliar growth to widely fluctuating air temperatures. *Agroplantae* 2: 105–7.
Hernández Hernández, J., Cabrera Cabrera, J., Galán Saúco, V., Marrero Domínguez, A. and Rodríguez Pastor, M. C. (1992). Preliminary evaluation of the postharvest life of Cavendish banana cultivars. *Acta Horticulturae* 343: 86–8.
Honan, K. (2017). Banana varieties trialled in NSW to compete with Queensland's popular Cavendish, Lady finger. *ABC Rural*. Posted 14 June 2017 at 7:14 am. http://www.abc.net.au/news/rural/2017-06-14/new-disease-resistant-banana-varieties/8610844. Downloaded 02 August 2017.
Lassoudière, A. (1974). Croissance et developpement du bananier 'Poyo' en Cote d'Ivoire selon les variations climatologiques. *Fruits* 29(6): 433–55.
Nuez Yánez, J. S. (2017). Notas sobre le mercado español del plátano. *Agropalca* 35: 17–19.
Oppenheimer, Ch. (1960). *The Influence of Climatic Factors on Banana Growing in Israel*. Series 350-B. National and University Institute of Agriculture, Rehovot, Israel.
Ploetz, R. C. (2015). Fusarium wilt of banana. *Phytopathology* 105: 1512–21.
Robinson, J. R. (1990). Banana growth and development in the subtropics. *Banana Newsletter* 13: 32–3.
Robinson, J. R. and Anderson, T. (1991). Good performance of new banana cultivars at Burgershall. *CSFRI Bulletin* 227: 2–3.
Robinson, J. R. and Galán Saúco, V. (2010). *Bananas and Plantains*. 2nd Edition. CABI International, Wallingford, Oxfordshire.
Robinson, J. C. and Nel, D. J. (1984). Influence of polyethylene bunch covers on yield and fruit quality of winter-developing banana bunches. *Horticultural Science (South Africa)* 1: 26–8.
Robinson, J. C. and Nel, D. J. (1989). Plant density studies with banana (cv. 'Williams') in a subtropical climate. II. Components of yield and seasonal distribution of yield. *Journal of Horticultural Science* 64: 211–22.
Robinson, J. C. and Nel, D. J. (1990). Competitive inhibition of yield potential in a 'Williams' banana plantation due to excessive sucker growth. *Scientia Horticulturae* 43: 225–36.
Robinson, J. C., Nel, D. J. and Alberts, A. J. (1985). Parent to sucker competition within a banana mat. *ITSC Information Bulletin* 157: 11–13.
Robinson, J. C., Anderson, T. and Eckstein, K. (1992). The influence of functional leaf removal at flower emergence on components of yield and photosynthetic compensation in banana. *Journal of Horticultural Science* 67: 403–10.
Sarah, J. L., Pinochet, J. and Stanton, J. (1996). *The Burrowing Nematode of Bananas, Radopholus similis Cobb, 1913*. Musa Pest Fact Sheet 1. INIBAP, France.
Sastry, P. S. N. (1988). *Agrometeorology of the Banana Crop*. CagM Report, No. 29, Agricultural meteorology. World Meteorological Organization, Geneva, Switzerland.

Smith, M. K., Langdon, P. W., Pegg, K. G and Daniells, J. W, 2014. Growth, yield and Fusarium wilt resistance of six FHIA tetraploid bananas (*Musa* spp.) grown in the Australian subtropics, *Scientia Horticulturae* 170: 176–18.

Soto, B. M. (1991). *Bananos. Cultivo y Comercialización*. Litografía e Imprenta Lil, San José de Costa Rica.

Tai, E. (1977). Banana. In: Alvim, T. and Kozlowsky, T. (Eds) *Ecophysiology of Tropical Crops*. Academic Press, London, pp. 441–60.

Turner, D. W. (1970). Bunch covers, leaf number and yield of bananas. *Australian Journal of Experimental Agriculture and Animal Husbandry* 10: 802–5.

Turner, D. W. (1998a). The impact of environmental factors on the development and productivity of bananas and plantains. In: Hidalgo, I. (Ed.) *Proceedings XIII Meeting ACORBAT. Ecuador'98.* CONABAN, Guayaquil, pp. 635–63.

Turner, D. W. (1998b). Ecophysiology of bananas: The generation and functioning of the leaf canopy. *Acta Horticulturae* 490: 211–22.

Turner, D. W. and Fortescue, J. A. (2012). 'Bananas and mangoes at the margins - Approaches to flowering for productive management, *Acta Horticulturae* 928:113–120.

Turner, D. W. and Lahav, E. (1981). The growth of Williams banana plants in relation to temperature. In: Alstonville, D. W. (Ed.) *Tropical Fruit Research Station*. Research Report 1980–1. Department of Agriculture, New South Wales, pp. 27–8.

Turner, D. W., Fortescue, J. A. and Thomas, D. S. (2009). Chapter 2. Bananas: Environment and crop physiology. In Fabio-Dametta (ed). *Ecophysiology of Tropical tree Crops*. Nova Science Publishers, Inc. New York. pp. 1–30.

Vilela Rodrigues, M. G., Carvalho Dias, M. S., Ruggiero, C. and Lichtemberg, L. A. (2008). Planejamento, implantaçao e manejo do bananal. Informe Agropecuario. *Belo Horizonte* 29 (245): 14–24.

Walduck, G., (2005). Fusarium wilt fightback. *Australian Bananas* 20: 17–20.

Chapter 10

Nutrition and soil management in banana cultivation

Jetse J. Stoorvogel, Wageningen University, The Netherlands; and Rafael A. Segura, CORBANA S.A., Costa Rica

1 Introduction
2 Assessing nutrient requirements
3 Developing fertilizer use recommendations
4 Precision application of nutrients
5 Conclusion
6 Where to look for further information
7 References

1 Introduction

Bananas are cultivated throughout the (sub)tropics with a global production area of 12 million ha (FAO, 2017) in a wide variety of production systems. There is significant variation between and within each of the main production regions in the world (see e.g. Karamura et al. (1998) for Africa, Thiers (2017) for the Philippines and Dita et al. (2013) for Latin America). Soil management also varies considerably in these systems. On the one hand, there are subsistence systems where, after planting, soil management is almost zero. At the other extreme, there are export-oriented production systems with very intensive soil management with up to 20 applications of fertilizer, 50 applications of fungicides, 2 applications of nematicides, sometimes 2 applications of herbicides, over 2 km of 0.5–3 m deep drainage channels per ha, and very intensive tillage (mostly during the establishment of the plantation). Yields obviously vary according to the differences in management approaches and the soil natural aptitude. Countries with relatively extensive management have average yields of less than 5 t/ha (e.g. Uganda and the Democratic Republic of Congo) compared to regions with very intensive management which have yields exceeding 70 t/ha (e.g. commercial plantations in Costa Rica and the Philippines). However, at both ends of the spectrum, one can question whether these systems are sustainable. In extensive systems, production levels may be very low and soil fertility may be declining. In intensively managed systems, there is a risk of environmental contamination, including risks to human health from spraying (Marquardt, 2002), Cu contamination in soils (Thrupp, 1991), leaching of pesticides (Castillo et al., 2000, 2006; Diepens et al., 2014)

http://dx.doi.org/10.19103/AS.2017.0020.16

nitrous gas emissions contributing to global warming (Plant, 2000), imbalances in nutrient balances (Borbor-Cordova et al., 2006), acidification (Segura et al., 2015) and losses in SOM (Powers, 2004; Geissen et al., 2009).

Bananas are a special case if we talk about plant nutrition and soil management, for a number of reasons. There is simply not another crop that is comparable in terms of physiology and management. Methodologies proposed in handbooks of plant nutrition (e.g. Barker and Pilbeam, 2014) are often not applicable in case of the banana. The characteristics of the crop make it extremely difficult to carry out the typical standard fertilizer response experiment (as described by e.g. Colwell, 1994). For a perennial crop like the banana, a long-term experiment over several crop cycles with large plots would be required. Very few such fertilizer experiments have been published (e.g. Nyombi et al., 2010; Santosh and Tiwari, 2017; Nomura et al., 2017). A simple literature search in Scopus shows that fertilization studies in bananas are scarce in comparison to other crops. Crops such as coffee, rice, maize and wheat have 80%, 1643%, 2250% and 2955% more studies, respectively. Although such studies are challenging to undertake, the limited attention to fertilization studies is surprising given the fact that intensive fertilization involves significant economic (Arias et al., 2003a) as well as potential environmental costs. In the past, fertilization costs were often estimated at around 15–21% of the total production costs (López, 2000; Stoorvogel et al., 2004). However, more recently the cost of fertilization has increased up to 30% of the total production costs. These costs highlight the need to improve current fertilization schemes.

This chapter provides an overview of the general aspects of the nutrition of the banana crop in the broader context of soil management. Focusing mostly on intensively managed systems where fertilizer use is the highest, the chapter discusses a number of strategies for soil nutrition as well as advantages and disadvantages of each approach. The text is illustrated with examples from Costa Rica.

2 Assessing nutrient requirements

Different techniques have been developed to establish nutrient requirements for banana which provide the foundation for fertilization regimes and nutrition. This section looks at the various techniques that have been and are being used.

2.1 Assessing soil nutrient status

A common view of crop nutrition is that nutrient concentrations in the soil should meet a minimum, critical concentration (Arias et al., 2003b). As soon as the nutrient level falls below this threshold, it is expected that crop performance will decrease. As such, research aims to determine these critical values. This requires monitoring of the soil nutrient concentrations which are correlated with the critical values in order to determine the optimal fertilizer rate for a particular field. The results of soil analyses indicate whether the found concentrations are low, medium or high by comparison with the theoretically expected critical values. Table 1 provides the critical values and the corresponding fertilizer recommendations for Costa Rican production systems (López and Espinosa, 2000). However, it has been found that the critical values vary depending on the agro-ecological conditions, so that different studies have identified different critical values. The complexity of controlling nutrient concentrations through fertilization is illustrated by

Table 1 Critical values from the literature for soil nutrients and the resulting fertilizer recommendation (López and Espinosa, 2000; extractions with modified P-Olsen)

	Level in soil		
Nutrient	Low	Medium	High
Nitrogen			
Soil concentration		Irrespective	
Fertilizer recommendation (kg N/ha/year)		350–400	
Phosphorus			
Soil concentration (mg/kg)	<10	10–20	>20
Fertilizer recommendation (kg P_2O_5/ha/year)	100	50	0
Potassium			
Soil concentration (cmol(+)/kg)	<0.2	0.2–0.5	>0.5
Fertilizer recommendation (kg K_2O/ha/year)	700	600	500
Calcium			
Soil concentration (cmol(+)/kg)	<3	3–6	>6
Fertilizer recommendation (kg CaO/ha/year)	1160	560	0
Magnesium			
Soil concentration (cmol(+)/kg)	<1	1–3	>3
Fertilizer recommendation (kg MgO/ha/year)	200	100	0

Segura et al. (2015), indicating significant differences in nutrient concentrations in the soil and their effect on good and poor plant development.

2.2 Accounting for relationships between nutrients

Whereas the critical values discussed in Section 2.1 focus on individual nutrients, in fact the nutrients (and other soil properties) strongly interact. Most important in banana cultivation is the balance between Ca, Mg and K concentrations. Even if these concentrations are all above their critical values, any imbalance may still lead to constraints on plant growth. Similarly, nutrient availability can change with factors such as soil pH. Analytical procedures aim to correct for this using proper extraction procedures (good examples are the various extraction procedures that are available for P: P-Olsen, P-Bray, etc.; Wuenscher et al., 2015). Table 2 and Fig. 1 provide an example of equilibrium relationships between various cations in the soil, as derived for Costa Rican production systems. These values are not generic and depend on other factors so that very different values are found in other studies, as illustrated by the brief review of López and Espinosa (2000), indicating a range of values for the ratio between Mg and K of 3.3 up to 6.3.

2.3 Nutrient balances

Whilst previous approaches focus on the provision of sufficient nutrients to the plants, with increasing concern about the environmental impacts of crop cultivation, there is

Table 2 Equilibrium relationships in the soil between Ca, Mg and K (expressed in mg/L) as established by López (1983) for Costa Rica

Relation	Equilibrium values
Ca/Mg	3.5–4.0
Ca/K	17.0–25.0
Mg/K	8.0–15.0
(Ca + Mg)/K	20.0–30.0
100 K/(Ca + Mg + K)	3.0–5.0

increasing attention directed towards nutrient use efficiency (NUE) by the crop itself. In this case, nutrient use requirements are derived on the basis of the expected uptake by plants, prompting the question of how uptake can be measured. Yield monitoring may provide useful insights into the current uptake of nutrients; when combined with fertilizer use, it can be used to calculate fertilizer requirements to sustain production. However, this methodology does not provide any insights into the attainable yields. In a meta-analysis, Bertsch (2009) assessed the nutrients required to produce bananas to be 2 kg N, 0.4 kg P and 6 kg K per tonne of banana based on the nutrient extraction of the harvest. One could misinterpret these numbers as stating that an average farm in Costa Rica (producing 2950 18.14 kg boxes/ha in 2017) requires annual fertilization amounts of: 107 kg N, 21 kg P and 321 kg K, which is in fact in strong contrast to the actual amounts of fertilizer currently used in the Costa Rican banana production system (shown in Table 3). The relatively low nutrient use efficiency between 17% and 42% is the result of the perhumid conditions,

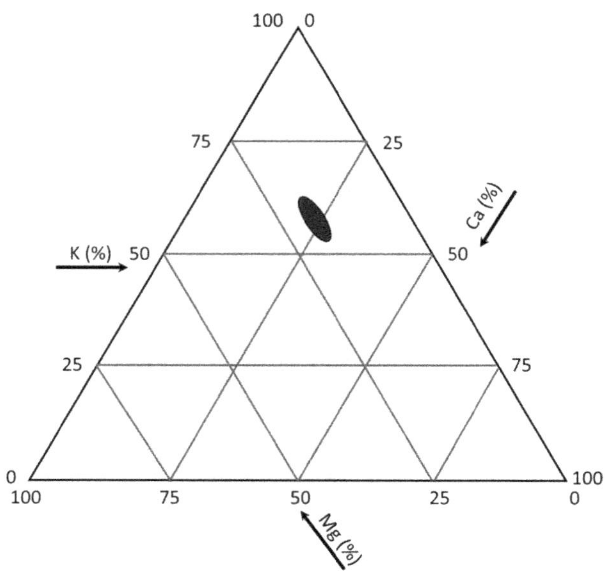

Figure 1 Triangle of balance Ca, K and Mg (Espinosa and Mite, 2008).

Table 3 Nutrient balance for a standard fertilization programme in a Costa Rican banana plantation with a 2200 boxes/ha/year yield and a 20% humidity in the banana fruit

	N	P	K
Fertilization (kg/ha/yr)	380	25	400
Nutrient concentration (%)	0.8	0.1	2.1
Nutrient removal (fruit) (kg/ha/yr)	65	8	169
Nutrient use efficiency (%)	17	31	42

soil conditions, pests, diseases and the high nutrient requirements of the banana crop. Increasing the nutrient use efficiency is the major challenge for farm management.

2.4 Assessing nutrient flows to plants

Small experiments can be carried out in existing plantations to evaluate crop response to changing fertilizer rates. Table 4 shows the results of such an experiment in a Costa Rican banana plantation in which three main soil types were identified and to which, over a period of a year, 1-ha blocks received 75%, 100% and 150% of conventional fertilization (Stoorvogel et al., 2015). In these blocks crop performance and nutrient leaching were monitored. Depending on the soil type, the different levels of fertilization resulted in changes in production and/or nutrient leaching. Results for the three main soil types are presented in Table 4. In the fertile soils, almost all additional nitrogen is taken up by the plant if we move from 75% to 100% fertilization. However, a further increase in fertilization resulted in increased leaching with the exception of the infertile soil, for which leaching already increased on moving from 75% to 100% (with a stable production), indicating that fertilization could be reduced.

Table 4 Effect of fertilization rates on a Costa Rican Banana plantation on bunch weight and nutrient concentrations in soil water on three different soil types

		Fertilization		
Soil type	Variable	75%	100%	150%
Sandy loam, fertile, well-drained soil	Weight (kg)	32.8	33	33.3
	NO_3^- (mg/L)	3.61	3.81	4.96
	P (mg/L)	0.03	0.03	0.03
	K (mg/L)	1.61	1.65	1.65
Loamy, fertile, well-drained soil	Weight (kg)	34.3	33.5	32.9
	NO_3^-	4.02	4.08	6.36
	P	0.03	0.03	0.03
	K	1.37	1.57	2.74
Clayey, infertile, well-drained soil	Weight (kg)	32.8	35	36.1
	NO_3^-	1.91	3.04	3.29
	P	0.04	0.06	0.1
	K	0.57	1.53	1.8

2.5 Assessing crop response: foliar analysis and crop NUE

As illustrated in Section 2.3, soil conditions are extremely difficult to interpret in relation to crop performance due to the complex interactions. Alternatively, crop response can be monitored in the crop itself by foliar analysis (Memon et al., 2001). Researchers have established critical values for foliar analysis (see e.g. Table 5), although it should be noted that here also the critical values depend on the agro-ecological conditions and therefore vary between different production regions. Besides the foliar rate of the nutrients should be linked with the current plant development and production. Similar to the response to the soil conditions, the balance between the cations is similarly important to the balances that are determined in the soil, suggesting that the cations are taken up in similar ratios as they are in the soil.

2.6 Other factors

The principle reason for crop nutrition is to provide the plant with nutrients in a balanced manner. As has been discussed earlier, this involves the application of nutrients through fertilization, but it may also involve other management practices such as tillage and drainage to provide the proper conditions for the development of a root system in order to facilitate the uptake of nutrients. However, nutrition is also strongly related to other aspects of crop management. Nutrition can, for example, affect the prevalence of pests and diseases, either directly, by providing the proper conditions in the soil for the development of for example fungi and nematodes (for soil born pests and diseases), or indirectly, by making the plant vulnerable to pests and diseases (for soil born pests and diseases but also for airborne pests and diseases). The interactions between crop disease and soil conditions have been investigated in a number of cases. Esnard et al. (1998), for example, demonstrated the relationship between organic amendments and nematodes. Other research, however, does not confirm such a relationship. As an example, Araya et al. (2002) showed that nematode populations are stable between years and within years and are comparable for different counties, and as such, are less sensitive to changes in soil condition.

2.7 The role of models

Crop growth simulation models are increasingly used to support crop management for a variety of crops. Van Alphen and Stoorvogel (2000) show, for example, how fertilization in winter wheat can be managed by real-time simulations. However, despite the fact that some initiatives have been taken to develop crop growth simulation models for banana,

Table 5 Critical values for foliar analysis in banana plantations (Lahav and Turner, 1992)

	Critical value (%)
N	2.6
P	0.2
K	3.0
Ca	0.5
Mg	0.3

appropriate models are still lacking for this crop. A module to represent bananas within the DSSAT framework (Jones, 2003) was added to version 4 but has not been calibrated yet. The SIMBA model was built to assess the environmental risks under a large range of cropping techniques and to help design more sustainable cropping systems (Tixier et al., 2008). SIMBA simulates the banana-cropping system over several cropping cycles using sub-models simulating soil structure, water balance, root nematode populations, yield and economic outputs; however, the model does not evaluate crop nutrition.

3 Developing fertilizer use recommendations

Given all the analytical approaches that are available and the different management options, how are fertilizer recommendations in the case of commercial banana plantations established? There is no standard framework to do so, but typically a number of steps are followed:

- Before the establishment of the farm a land evaluation is carried out to assess the suitability of the area for crop cultivation (Ansyori et al., 2010; ASBANA, 1990). Key factors that play a role in the land evaluation include the climate conditions, topography, stoniness, soil texture and soil fertility. Each factor is evaluated individually and in the end it is the most limiting factors that determine the suitability. Only in suitable areas bananas will be planted as the establishment of a banana farm requires large investment costs, although some moderately suitable areas may be planted if they fall within a more suitable area.
- No quantitative tools are available that allow us to estimate the potential production levels in a given area (e.g. crop growth simulation models). However, knowledge is available regarding the performance of the better plantations in a given region which allows a yield estimate to be made.
- By multiplying the attainable yield and the nutrient concentrations in the fruit, the expected nutrient removal is calculated (see e.g. Table 4 and López (2000)). It should be noted that, normally, all crop residues remain in the field (pseudostem and leaves) or are returned to the plantation (rachis). Although the nutrient concentrations in the fruit do vary, to form a first ballpark number concerning nutrient uptake, average nutrient concentrations can be used. This calculation is performed in particular for the most import nutrients N, P, K, Ca and Mg. Samples of fruit nutrients concentration should be performed in different seasons in the year, measuring the extraction in each of them for having a more representative data.
- A first estimate of the required amount of fertilizer is then calculated by correcting the nutrient removal for the NUE. As shown in Table 4, the nutrient use efficiencies vary between 20% and 50% depending on the mobility of the respective nutrient, climate conditions (e.g. rainfall intensity), soil properties (cation exchange capacity, clay content, and the soil organic matter content) and management (e.g. timing of applications, type of fertilizer) and the uptaking capability of the root system. The assessment of the NUE is made based on prior knowledge of nutrient use efficiencies.
- After a first soil sampling, the soil properties are compared to critical values. This may result in an adaptation of the recommendation to ensure a proper balance of

nutrients. Nutrient requirements for the banana crop are so high that the soil cannot simply supply these nutrients and as such, the crop relies almost entirely on the fertilization programme.

- As soon as the banana crop is established, leaf samples are taken once or twice a year. Together with the production, the analysis of these samples enables a determination of whether the fertilization provides a sufficient and properly balanced supply of nutrients.
- Soil sampling remains important during the course of crop production. Soil sampling does not aim to assess the optimal nutrient application, but it does allow for monitoring of soil quality in terms of, for example, soil organic matter, acidification and cations balance. Good examples are acidification due to intensive fertilizer use that needs to be compensated for by liming, the decline in soil organic matter that needs to be corrected by organic fertilizers and high addition of calcium amendments for increasing the lost Ca by lixiviation.
- If nutrients are not supplied in sufficient quantities, it may lead to specific physiological conditions that can be observed by the farmer, although it is unwise to wait until, for example nutrient deficiencies or nutritional imbalances become visible in the crop.

4 Precision application of nutrients

Whilst intensive fertilization is required to maintain yields, there should be a focus on increasing fertilizer use efficiency to reduce the environmental impact of cultivation, including side effects such as reduced organic matter contents and acidification (Booltink et al., 2001; Stoorvogel et al., 2015). Many banana production systems in humid regions also suffer a high risk of nutrient losses through leaching. To increase NUE, nutrient concentrations should be kept as low as possible (although still adequate for plant nutrition). Producers are using a range of options to reach this lowest possible level including split fertilizer application, and the use of slow release or organic fertilizers. If the banana crop is irrigated, fertigation can control the release of nutrients precisely. A good example is the application of the fertilizer in a fertilizer band or broadcast application. Application in the fertilizer band has the advantage that the nutrients are directly applied to the soil, but it does lead to a differentiation in the applied nutrient levels across the plantation (Segura et al., 2015). Although there are many options available, knowledge of their impacts is still very limited. Less is known about the application of micronutrients, despite the fact that these are viewed as increasingly important in crop nutrition (López, 2000).

Research has also highlighted the benefits of targeting fertilizer more precisely, for example timing nitrogen application to coincide with the establishment of the root system to maximise its effectiveness, and splitting application of nitrogen and potassium to target particular growth stages (Nyombi, 2014; Rahman et al., 2017). There has also been a range of recent research highlighting the advantages of fertigation techniques in optimising fertilizer use and nutrient uptake efficiency in banana cultivation (Mustaffa and Kumar, 2012; Mahendran et al., 2013; Pawar and Dingre, 2013; Pramanik et al., 2014, 2016; Pawar et al., 2017; Senthilkumar et al., 2017). Research by the Department of Agriculture in Queensland, Australia, for example, showed that fortnightly fertigation produced similar yield, fruit quality, improved bunch weight and sucker growth as compared to conventional fertilization using monthly broadcast applications of fertilizer, with up to 40%

saving in nitrogen fertilizer (QWP, 2011). Targeted broadcast applications of solid fertilizer were used during extended periods of wet weather, when flooded irrigation pumps and saturated soil prevented fertigation. This research has also shown that nitrogen leaching can be significantly reduced in systems which use fortnightly nitrogen applications and a grassed interrow compared to a conventional system with monthly applications.

Research on banana nutrition has highlighted the broader importance of improving overall soil health, for example by reduced tillage and applying compost to act as a mulch to improve water and nutrient retention as well as improve organic matter, soil biological activity and nutrient composition, thus reducing the need for supplementary fertilizer application (Bananuka et al., 2000; Pattison and Lindsay, 2006; QWP, 2011; Baijukya, 2004; Hazarika et al., 2015). Recent research has also started to look at ways of enhancing rhizobacteria in the soil, both to improve crop yields and enhance resistance to diseases such as *Fusarium* wilt (Fu et al., 2016). These issues are discussed in more detail in other chapters in the book.

5 Conclusion

Whereas in the past decades major steps have been made in formalizing fertilizer use recommendations in a range of different crops, research on the banana crop has been lagging behind. Various approaches have been elaborated, but formal, well-described methods are not available. As a result, many producers rely on general recommendations followed by a period of trial and error in which the fertilization is optimized in which expert knowledge in combination with soil and foliar analysis form the foundation. Large steps can still be made, and are necessary, to increase the fertilizer use efficiency through the implementation of precision agriculture and the further development of, for example, crop growth simulation models. However, more holistic approaches to soil and nutrient management will pose new questions on research to better understand the relation between for example crop diseases and fertilization. Despite the options to increase nutrient use efficiency, it should be recognized that extremely high production levels are obtained and sustained. Due to intensive monitoring, problems that arose like declining soil organic matter contents and acidification are managed in a timely manner through organic fertilizer and liming.

6 Where to look for further information

There have been several attempts in the past to develop guidelines for banana nutrition (e.g. Lahav and Kalmar, 1988; Lahav, 1995; López and Espinosa, 2000; Rosales et al., 2008; Robinson and Galán Saucó, 2010). There is also a range of handbooks on banana (e.g. Gowen, 1995; Soto, 1992). However, these guidelines had a very general character and did not provide insight in the actual fertilizer recommendation at a particular location. As shown in this chapter, key elements of the banana production vary between the different production regions. As a result the fertilizer recommendations vary between countries. Although generic guidelines are missing there is considerable knowledge in the various countries available in terms of grey literature and expert knowledge. Research institutes where this knowledge can be found include CORBANA in Costa Rica, INIBAP in Ecuador, Embrapa in Brazil and NARO in Uganda.

7 References

Ansyori, Sudarsono, Poerwanto, R. and Darmawan (2010), 'Land suitability criteria for intensively managed Cavendish banana crop in Way Kambas East Lampung, Indonesia', *Journal of Tropical Soils*, 15, 159–67.

Araya, M., De Waele, D. and Vargas, R. (2002), 'Occurrence and population densities of nematode parasites of banana (*Musa* AAA) roots in Costa Rica', *Nematropica*, 32(1), 21–33.

Arias, P., Dankers, C., Liu, P. and Pilkauskas, P. (2003a), *The World Banana Economy 1985–2002*, FAO, Rome.

Arias, F., Segura, R., Serrano, E., Bertsch, F., López, A. and Soto, E. (2003b), 'Niveles críticos de Calcio, Magnesio, Potasio y Fósforo en suelos dedicados al cultivo de banano (*Musa* AAA) en Costa Rica', *CORBANA*, 29, 69–81.

ASBANA (1990), *Manual de procedimientos para presentación y realización de estudios detallados de suelos y clasificación de tierras para el cultivo del banano*, ASBANA, San José, Costa Rica.

Baijukya, F. P. (2004), 'Adapting to change in banana-based farming systems of northern Tanzania. The potential role of herbaceous legumes', PhD thesis, Wageningen University, Wageningen, the Netherlands.

Bananuka, J., Rubaihayo, P. and Zake, J. (2000), 'Effect of organic mulches on growth, yield components and yield of East African highland bananas', *Acta Horticulturae (ISHS)*, 540, 293–300.

Barker, A. V. and Pilbeam, D. J. (2014), *Handbook of Plant Nutrition*, 2nd Edition, CRC Press.

Bertsch, F. (2009), *Absorbción de nutrimentos por los cultivos*, ACCS, San José, Costa Rica.

Booltink, H. W. G., Van Alphen, B. J., Batchelor, W. D., Paz, J. O., Stoorvogel, J. J. and Vargas, R. (2001), 'Tools for optimizing management of spatially-variable fields', *Agricultural Systems*, 70, 445–76.

Borbor-Cordova, M. J., Boyer, E. W., McDowell, W. H. and Hall, C. A. (2006), 'Nitrogen and phosphorus budgets for a tropical watershed impacted by agricultural land use: Guayas, Ecuador', *Biogeochemistry*, 79, 135–61.

Castillo, L. E., Ruepert, C. and Solis, E. (2000), 'Pesticide residues in the aquatic environment of banana plantation areas in the north Atlantic zone of Costa Rica', *Environmental Toxicology and Chemistry*, 19(8), 1942–50.

Castillo, L. E., Martínez, E., Ruepert, C., Savage, C., Gilek, M., Pinnock, M. and Solis, E. (2006), 'Water quality and macroinvertebrate community response following pesticide applications in a banana plantation, Limon, Costa Rica', *Science of the Total Environment*, 367, 418–32.

Colwell, J. D. (1994), *Estimating Fertilizer Requirements. A Quantitative Approach*, CAB International, Wallingford, UK.

Diepens, N. J., Pfennig, S., Van den Brink, P. J., Gunnarsson, J. S., Ruepert, C. and Castillo, L. E. (2014), 'Effect of pesticides used in banana and pineapple plantations on aquatic ecosystems in Costa Rica', *Journal of Environmental Biology*, 35, 73–84.

Dita, M. A., Garming, H., Van Den Bergh, I., Staver, C. and Lescot, T. (2013), 'Banana in Latin America and the Caribbean: Current state challenges and perspectives', *Acta Horticulturae*, 986, 365–80.

Esnard, J., Marban-Mendoza, N. and Zuckerman, B. M. (1998), 'Effects of three microbial broth cultures and an organic amendment on growth and populations of free living and plant-parasitic nematodes on banana', *European Journal of Plant Pathology*, 104, 457–63.

Espinosa, J. and Mite, F. (2008), 'Estado actual y futuro de la nutrición y fertilización del banano', *Informaciones Agronomicas*, 48, 4–10.

FAO (2017). FAOstat. http://www.fao.org/faostat/en/#home/ (accessed 15 December 2017).

Fu, L., Ruan, Y., Tao, C., Li, R. and Shen, Q. (2016), 'Continuous application of bioorganic fertilizer induced resilient culturable bacteria community associated with Fusarium wilt suppression', *Nature Scientific Reports*, 6, 27731. (DOI: 10.1038/srep27731).

Geissen, V., Peña-Peña, K. and Huerta, E. (2009), 'Effects of different land use on soil chemical properties, decomposition rate and earthworm communities in tropical Mexico', *Pedobiologia*, 53, 75—86.

Gowen, S. (1995), *Bananas and Plantains*, Chapman & Hall, London, UK.
Hazarika, T. K., Nautiyal, B. P. and Bhattacharyya, R. K. (2015), 'Conjunctive use of bio-fertilizers and organics for improving growth, yield and quality of banana cv. Grand Naine', *Indian Journal of Horticulture*, 72, 461–5. doi:10.5958/0974-0112.2016.00003.7.
Jones, J. W., Hoogenboom, G., Porter, C. H., Boote, K. J., Batchelor, W. D., Hunt, L. A., Wilkens, P. W., Singh, U., Gijsman, A. J. and Ritchie, J. T. (2003), 'The DSSAT cropping system model', *European Journal of Agronomy*, 18, 235–65.
Karamura, E., Frison, E., Karamura, D. A. and Sharrock, S. (1998), 'Banana production systems in eastern and southern Africa', *in* C. Picq, E. Fouré and E. A. Frison (Eds), *Bananas and Food Security*, INIBAP, Montpellier, France, pp. 401–12.
Lahav, E. (1995), 'Banana nutrition', *in* S. Gowen (Ed.), *Bananas and Plantains*, Chapman & Hall, London, UK, pp. 258–316.
Lahav, E. and Turner, D. W. (1992), 'Fertilización del banano para rendimientos altos', Segunda edición, Boletín No. 7, Instituto de la Potasa y el Fósforo, Quito, Ecuador, 71p.
López, A. (1983), 'Diagnóstico del estado nutricional de plantaciones bananeras', *ASBANA*, 6(19), 13–18.
López, A. (2000), 'Fertilización convencional del cultivo de banana en Costa Rica y su relación con la producción sostenible', *in* F. E. Rosales, S. C. Tripon and J. Cerna (Eds), *Producción de banano orgánico y/o ambientalmente amigable. Memorias del taller internacional sobre producción de banano orgánico y, o, ambientalmente amigable*, INIBAP, pp. 63–82.
López, A and Espinosa, J. (2000), *Manual on the Nutrition and Fertilization of Banana*, Potash & Phosphate Institute (Atlanta, Ga.), Potash & Phosphate Institute of Canada, 57p.
Mahendran, P., Yuvaraj, M., Paramewari, C., Gurusamy, A. and Krishnasamy, S. (2013), 'Enhancing growth, yield and quality of banana through subsurface drip fertigation', *International Journal of Chemical, Environmental and Biological Sciences*, 1(2), 391–402.
Marquardt, S. (2002), 'Pesticides, parakeets, and unions in the Costa Rican banana industry, 1938–1962', *Latin American Research Review*, 37(2), 3–36.
Memon, N., Zia-Ul-Hassan and Cad, F. C. (2001), 'Banana nutrition management through plant analysis', *Pakistan Journal of Applied Sciences*, 1, 563–74.
Mustaffa, M. and Kumar, V. (2012), 'Banana production and productivity enhancement through spatial, water and nutrient management', *Journal of Horticultural Science*, 7(1), 1–28.
Nomura, E. S., Cuquel, F. L., Damatto Junior, E. R., Fuzitani, E. J. and Borges, A. L. (2017), 'Fertilization with nitrogen and potassium in banana cultivars "grand Naine", "FHIA 17" and "Nanicão iac 2001" cultivated in Ribeira Valley, São Paulo state, Brazil', *Acta Scientiarum – Agronomy*, 39, 505–13. doi:10.4025/actasciagron.v39i4.32919.
Nyombi, K. (2014), 'Fertilizer management of highland bananas in East Africa', *Better Crops with Plant Food (IPNI)*, 98(1), 29–31.
Nyombi, K., van Asten, P. J. A., Corbeels, M., Taulya, G., Leffelaar, P. A. and Giller, K. E. (2010), 'Mineral fertilizer response and nutrient use efficiencies of East African highland banana (*Musa* spp., AAA-EAHB, cv. Kisansa)', *Field Crops Research*, 117, 38–50.
Pattison, T. and Lindsay, S. (2006), *Banana Root and Soil Health User's Manual: FR02023*, Department of Primary Industries and Fisheries, Queensland, Australia.
Pawar, D. D. and Dingre, S. K. (2013), 'Influence of fertigation scheduling through drip on growth and yield of banana in western Maharashtra', *Indian Journal of Horticulture*, 70, 200–5.
Pawar, D., Dingre, S. and Bhoi, P. (2017), 'Productivity and economics of drip-irrigated banana under different planting and fertigation techniques in subtropical India', *Communications in Soil Science and Plant Analysis*, 48, 449–58.
Plant, R. A. J. (2000), 'Regional analysis of soil-atmosphere nitrous oxide emissions in the northern Atlantic Zone of Costa Rica', *Global Change Biology*, 6, 639–53.
Powers, J. S. (2004), 'Changes in soil carbon and nitrogen after contrasting land-use transitions in northeastern Costa Rica', *Ecosystems*, 7, 134–46.

Pramanik, S., Tripathi, S., Ray, R. and Banerjee, H. (2014), 'Economic evaluation of drip-fertigation system in banana cv. Martaman (AAB, silk) cultivation in new alluvium zone of West Bengal', *Agricultural Economics Research Review*, 27(1), 103–9.

Pramanik, S., Lai, S., Ray, R. and Patra, S. K. (2016), 'Effect of drip fertigation on yield, water use efficiency, and nutrients availability in banana in West Bengal, India', *Communications in Soil Science and Plant Analysis*, 47, 13–14.

Queensland Wetlands Program (QWP) (2011), *Case study: Bananas – Nurturing the Soil and Neighbouring Wetlands on a Banana Farm in the Wet Tropics*, Department of Environment and Resource Management, Queensland, Australia.

Rahman, H., Akter, A. and Rahman, J. (2017), 'Effect of split application of N and K fertilizer and bagging on the growth, yield and post-harvest quality of banana', *Research and Reviews: Journal of Crop Science and Technology*, 6(1), 12–22.

Robinson, J. C. and Galán Saucó, V. (2010), *Bananas and Plantains*, 2nd Edition, CAB International, Wallingford, UK.

Rosales, F. E., Pocasangre, L. E., Trejos, J, Serrano, E. and Peña, W. (2008), *Guía de diagnostico de la calidad y salud de suelos bananeros*, Bioversity International, Montpellier, France.

Santosh, D. T. and Tiwari, K. N. (2017), 'Response of tissue cultured banana (*Musa acuminate* L.) cv. grand naine to different levels of nutrients under drip fertigation and black plastic mulch', *Applied Ecology and Environmental Research*, 15, 1473–88. doi:10.15666/aeer/1504_14731488.

Segura, R., Serrano, E., Pocasangre, L., Acuna, O., Bertsch, F., Stoorvogel, J. J. and Sandoval, J. A. (2015), 'Chemical and microbiological interactions between soils and roots in commercial banana plantations (*Musa* AAA, cv. Cavendish)', *Scientia Horticulturae*, 197, 66–71.

Senthilkumar, M., Ganesh, S., Srinivas, K., Panneerselvam, P., Nagaraja, A. and Kasinath, B. L. (2017), 'Fertigation for effective nutrition and higher productivity in banana – a review', *International Journal of Current Microbiology and Applied Sciences*, 6(7), 2104–22.

Soto, M. (1992), *Banano cultivo y comercialización*, Litografia e imprentat LIL SA, San José, Costa Rica.

Stoorvogel, J. J., Bouma, J. and Orlich, R. A. (2004), 'Participatory research for systems analysis: Prototyping for a Costa Rican banana plantation', *Agronomy Journal*, 96(2), 323–36.

Stoorvogel, J. J., Kooistra, L. and Bouma, J. (2015), 'Managing soil variability at different spatial scales as a basis for precision agriculture', *in* R. Lal and B. A. Stewart (Eds), *Soil-Specific Farming: Precision Agriculture. Advances in Soil Science*, CRC Press, Boca Raton, FL, pp. 37–72.

Thiers, R. (2017), 'Flying bananas: Small producer tactics and the (un)making of Philippine banana export chains', *Journal of Peasant Studies*, 1–21 (online). doi:10.1080/03066150.2017.1376659.

Thrupp, L. A. (1991), 'Long-term losses from accumulation of pesticide residues: A case of persistent Copper toxicity in soils of Costa Rica', *Geoforum*, 22(1), 1–15.

Tixier, P., Malezieux, E., Dorel, M. and Wery, J. (2008), 'SIMBA, a model for designing sustainable banana-based cropping systems', *Agricultural Systems*, 97(3), 139–50.

Van Alphen, B. J. and Stoorvogel, J. J. (2000), 'A methodology for precision Nitrogen fertilization in high-input farming systems', *Precision Agriculture*, 2, 319–32.

Wuenscher, R., Unterfrauner, H., Peticzka, R. and Zehetner, F. (2015), 'A comparison of 14 soil phosphorus extraction methods applied to 50 agricultural soils from Central Europe', *Plant Soil Environ.*, 61 (2), 86–96.

Chapter 11

Harvesting and packaging of bananas

Juan José Aycart, Dole, Ecuador

1 Introduction
2 Criteria determining when to harvest bunches
3 Harvesting techniques
4 Bunch movement from the fields
5 Packing house operations
6 Fruit selection processes
7 Packing bananas
8 Future trends
9 Where to look for further information
10 References

1 Introduction

Banana and plantain belong to the family Musaceae. Although they are not seasonal like other tropical crops, they are classified as perennial and annual crops (Karunarate and Niroshini, 2010). Whilst bananas originated in Southeast Asia, banana cultivation is now widely distributed around the world in more than 120 countries (Robinson and Galan Sauco, 2012). The wide distribution of banana cultivation is related to the fact that they are both one of the most widely traded crops globally and, at the same time, a staple food in the tropical countries where they are produced (Sharrock, 1997). This duality of bananas and plantains, as both a global cash crop and local staple food, can be observed in some mixed-crop-producing areas where export and locally consumed varieties coexist (Fig. 1).

The cycle of banana production begins when a shoot (sucker) appears alongside the main (mother) plant. The shoot grows, producing leaves until flowering, after which the banana bunch is formed. Each banana tree produces a single bunch consisting of individual fruits (known as 'fingers') clustered together in 'hands'. Lassoudière (1978) outlined three different growth phases of fruit fingers. The first phase is marked by accelerated cell division and elongation, taking up to 28 days after flower emergence. This phase is followed by a transition phase (18–38 days) determined by the rate fruit cells fill with starch. The last phase extends from day 38 up to ripening (Lassoudière, 1978).

After the bunch is harvested, the pseudostem dies. The next sucker (shoot) replaces the dying pseudostem, thus ensuring continuous production. Growers select suckers which will

http://dx.doi.org/10.19103/AS.2017.0020.20

Figure 1 Plantain bunches ready for sale in local markets in Ecuador.

act as followers for the next round of production. This combination of a large pseudostem-producing fruit, a smaller sucker (called a follower) growing to take over growth and fruit production after the initial one is harvested, and a small sucker for the next cycle forms the basic unit of production. The process of selecting the follower per unit of production, based on the position and vigour of the suckers, and then removing all other suckers is known as desuckering. This practice ensures the continuity of production by managing the succession of productive pseudostems (Sierra, 1993).

With appropriate management of different production units, it is possible to produce bunches ready for harvest on a continuous basis (Goncalves and Kernaghan, 2014). Key aspects of management include fertilization to control plant growth, regular monitoring of suckers and selection of followers and control of population density to maximize yield whilst maintaining individual plant vigour (Sierra, 1993). Monitoring of suckers makes it possible to assess future production and labour requirements, as well as estimate when bunches can be harvested to meet market requirements. Generally, triploid banana varieties reach maturity under ideal growing conditions 13–15 weeks after bunch emergence. Diploid bananas have a shorter fruit fill period of approximately 6 weeks. Bunches should be harvested when the fruit is still green but already succulent (i.e. when the banana fingers have filled) if the right sensory quality is to be achieved (Soto, 1985).

Establishing when bunches can be harvested thus relies both on tracking the age of shoots as well as on specific techniques to measure fruit quality. When bananas can be harvested also depends on market requirements such as final destination and transit time which determine how early or late harvesting needs to be to ensure optimal quality at the point of consumption. The age of fruit at harvest determines the ripening of (green) packed fruit during transport and storage through to the point of sale. This means that harvesting and packing are both critical and challenging stages in banana production (Nuevo and Maunahan, 2001). This chapter discusses key aspects of harvesting and postharvest operations, from the identification of the age of the fruit in the field to packaging for the fruit's final retail destination. Postharvest losses can be reduced, and quality characteristics such as appearance, flavour and shelf-life optimized, by adopting the right harvest and postharvest practices (Moreno et al., 2009).

2 Criteria determining when to harvest bunches

Various criteria have been used to determine when to harvest bunches. A traditional measure is the degree of fruit fill (i.e. the proportion of edible pulp in the fruit). One of the major international companies producing bananas for the export market, United Fruits (now Chiquita), established four categories of ripeness: ¾ thin, ¾ normal, ¾ full and full (United Brands, 1975). The physiological grade can be estimated in practice using a cross-sectional cut of the fruit. A full fruit shows a higher proportion of pulp with a rounded shape compared to the more angular shape of pulp fill shown by a ¾ grade fruit (Thompson, 2003).

A second technique to assess fruit fill is to measure finger diameter using calipers. A typical sampling technique is to measure the central finger of the second hand on a bunch. Measurements are typically in units of 1/32 of an inch. This system was adopted, for example, by Standard Fruit (now Dole) with quality parameters of 46–48 for North America markets and 43–45 for European markets. These basic parameters are modified depending on the requirements of individual countries or markets (e.g. conventional or organic).

A major problem with both these measures is that fruit fill is very sensitive to environmental conditions and is not always a reliable indicator of bunch age and thus harvest date. An alternative approach is to measure bunch age by using coloured tags (Soto, 1985). The initial tag indicates the week when the flower emerges from the pseudostem, providing a benchmark for assessing the age of the bunch. Subsequent tags can be used to mark later stages of development. This system has, for example, been standard in Ecuador since 1976, using a four-colour tag code. Subsequently, an eight-colour tag code was developed at the end of the 1980s, using different colour tags for specific weeks. This code has been adopted by most international banana producers and is typically combined with fruit caliper grading to decide bunch harvest dates.

Keeping track of the age of fruit is a major requirement in banana production. This is typically done on a weekly basis to both determine when to harvest bunches and, more generally, maintain a steady flow of fruit moving to maturity as well as to avoid potential postharvest losses (Woltering, 2017a). As has been noted, harvest dates are also determined by the distance fruit needs to travel to reach its market. Transit times depend on the final destination and mode of transport. Typical shipping times to European markets, for example, are 14 days for Colombia and 21 days for Ecuador. Some producers recommend that fruits should be harvested at 90% maturity if transit times to market are short, and at 75% maturity for longer transit times (Mohapatra et al., 2010). As an example, in Ecuador, bunches are harvested 11–12 weeks after bunch emergence for distant export markets (such as Europe), subject to caliper grading to measure fruit fill. It is important to harvest bananas by 13–14 weeks after bunch emergence to avoid bananas becoming over ripe and developing texture problems at the point of consumption.

3 Harvesting techniques

Harvesting bunches is one of the most delicate activities in banana production. The process of cutting the bunch free makes it vulnerable to postharvest diseases as does subsequent damage such as bruising from poor handling (Toivonen et al., 2017). At large-scale commercial export farms, harvesting is a group activity, with teams of 3–4 people, depending on farm size and number of plants/ha (Goncalves and Kernaghan, 2014). Small farms growing mixed crops will rely on one person.

All harvest methods start with identification of hanging bunches for harvesting. Bunches are identified initially by age and then by caliper grade. The selected stems are marked with sidelong cuts into the pseudostem to make identification easier (Fig. 2). The choice of harvesting technique is based on bunch dimensions and conformation (Champion, 1968). Different cutting tools include a big pruning blade called a 'Podón', modified machete-type blades or, in the case of harvesting smaller varieties of banana, an ordinary machete. Other harvest techniques for taller varieties involve the use of ladders, sometimes with pulley systems to help harvest bunches (Fig. 3). Metal support devices called 'caimana' or 'cóndor' are used in Ecuador and Colombia (Fig. 4). Other devices include the 'float-up' or 'flora' which can be used by one person and are used commercially on some farms in Peru, Ecuador and Colombia.

With medium to tall varieties (such as Cavendish), two persons are required to harvest a bunch. One cuts into one side of the pseudostem in the shape of a seven ('siete'), which allows the bunch to be lowered safely. The second operator holds the bunch to avoid excessive mechanical damage. The first operator will usually need to cut away leaves to expose the stalk (rachis or raquis). The bunch is secured from the upper part of the raquis to prevent it from falling before the stalk is cut to release the bunch. The second operator helps to ensure that the bunch is lowered slowly to avoid bruising. Additional protection can include the incorporation of plastic coverings between finger rows (daypas) to reduce potential bruising as the bunch is removed. A padded shoulder support (called a 'cuna') is used to protect both the bunch and the operator when carrying the bunch away from the plant (Figs. 5 and 6). A variation of the 'cuna' uses an inflated pneumatic tube to provide even more cushioning (Fonsah and Chindebelu, 2012). It is important to note that very little energy is needed to bruise a banana finger: only 20–200 mJ are needed to produce visible bruising of the epidermal layers after a few minutes (Bugaud et al., 2013). It has been reported that the compression resistance of the fruit against bruising is related to latex exudation at harvest and the moisture content/turgidity of the fruit (Banks and Joseph,

Figure 2 Plant selected for harvest, showing a sidelong cut in the pseudostem to facilitate identification.

Figure 3 Harvest ladder and padded shoulder support for bunch placement.

Figure 4 The use of a 'caimana' during harvesting.

1991). This means that harvesting operations require significant levels of skill and teamwork if bunches are to be harvested both efficiently and in a way that preserves quality.

4 Bunch movement from the fields

There are a range of methods to transport bunches from the field to the packing plant. These fruit transport systems have a common objective to avoid transport bruises and

Figure 5 Padded support ('cuna') used in harvesting.

other quality problems. Common problems are broken necks (the stems linking fingers to a hand), bruises, friction between rows of fingers and damage to finger tips, usually when fruit hands have excessive curvature. Polyethylene universal protectors between hands are commonly used, as well as jersey fabric circular protectors placed between finger rows of hands. The polyethylene protectors are placed on the bunch before harvesting, whilst the jersey protectors are usually placed at the point bunches are placed on a cableway.

Cableways are frequently used in larger, export-focused farms in the Americas (Figs. 7 and 8). A typical design consists of a 11-mm steel cord supported by galvanized steel arches. The usual weight of the steel cord is 0.746 kg per linear meter, and the support arches are sited at distances of 7–8 m. The cableway design can have a slight slope towards the packing plant of no more than 3/1000, but preferably the slope must trend to zero. The height of the cableway steel arches will determine the height of the fruit rail at the packing plant. The cableway infrastructure network must ensure the capacity to move fruit from the field whilst avoiding excessive bunch movement. The system usually consists of a principal cableway connected to more distant parts of the farm by secondary cableways (Moreno et al., 2009). In countries such as the Dominican Republic, initial grading by size is carried out before bunches are loaded onto cableways (Goncalves and Kernaghan, 2014).

Up to 25 bunches on average can be loaded as a single group (convoy) pulled by an operator. Each bunch hangs from a chain supported by a trolley ('garrucha'). This consists

Figure 6 Cuna pad support during harvesting.

Figure 7 Bunch movement to the cableway.

Figure 8 'Caimana' harvested fruit delivered to the cableway.

of two steel wheels (typically 2.25 inch wide) that move over the cableway (Fig. 9). The trolleys are separated from each other by metal bars to help ensure bunches are carried smoothly. The number of bunches in a convoy can be increased to up to 80 if the cableway is motorized. Some systems use mules to provide the power to move larger convoys of bunches. An important design issue is to fit a cableway system around a farm's irrigation/drainage system which typically includes a network of ditches to supply water to the fields. Tractor-driven trailers are commonly used to move batches in Africa (Fonsah and Chindebelu, 2012). In Australia, plantations are designed with double rows which allow a tractor and trailer to be used in harvesting (Horticulture Innovation Australia, 2017). Other farms harvest the fruit and transport it to modified bunch trailers to move the fruit to the packing houses (Queensland Government, 2010).

5 Packing house operations

Once bunches arrive at the packing house, they are transferred to an area known as the bunch patio (Fig. 10). At this point, bunches are still in the plastic bags used to protect them during cultivation. These are rolled up and tied to the stalk (raquis) so that bunches can be grouped for the packing process. The fruit is inspected and production data such as bunch weight, bunch age, fruit grade caliper and finger length compiled (Table 1). An

Figure 9 Transport of banana convoy by cableway system to the packing plant. (a) Fruit convoy transported by the cableway. (b) Trolley chain supporting the bunch. (c) Banana trolley supporting bunches at arrival at the bunch patio (with close-up of trolley mechanism shown right).

Figure 10 Bunch protection plastic bags are removed at the bunch patio.

Table 1 Fruit reception criteria at the bunch patio

Issue	Description	Risk
Pulp colour	If the pulp colour tends to be more 'cream', especially at the central axis, it is rejected	Fruit has begun the ripening cascade prior to harvest
Creamy pulp	There is a significant loss of turgor in the pulp. There is no pulp resistance to thumb mechanical pressure	Fruit has begun the ripening cascade prior to harvest
Full grade fruit	This fruit is probably old or was not harvested young	A physiologically 'old' fruit can turn yellow during transit to market
Under/over specification	The inspected fruit is under/over specification when graded by caliper	Problems with identification of fruit ready for harvest, or poor agronomic conditions at the plantation

important additional test at this stage is a pulp firmness test. An experienced operator will slice a banana to expose the pulp and test its firmness using a finger and thumb to squeeze the pulp. The degree of firmness, extent of latex exudation, colour and odour may all be taken into account to decide bunch quality (Andrade and Zurita, 2009).

Bunches are also washed with pressurized water to remove field dirt, flower residues and insect pests (Mohapatra et al., 2010). It is particularly important to remove insect pests hidden between hands and fingers since these are both a disease risk and, depending on the requirements of export markets, a quarantine risk which may lead to impounding of imported bananas. Major pests in this respect include mealybugs and scales (Fig. 11) (Agrocalidad, 2013). Since they may harbour disease, flower residues may also be removed at this stage to reduce the risk of mould or rot. It is important to do this carefully without turning the bunch unnecessarily or causing bruising which can reduce quality. However, removing flower residues inevitably causes superficial wounds and exudation of latex from the wound which requires subsequent washing (Fig. 12). Best practice is for all workers to wear protective clothing and latex gloves to minimize the spread of potential infection.

Bunches at the patio are also dehanded (i.e. individual hands are separated from the stalk or raquis) using a curved-neck knife or bent bladed spatula (cuchareta) (Fig. 13). The decision over which dehand tool should be used depends on issues such as final

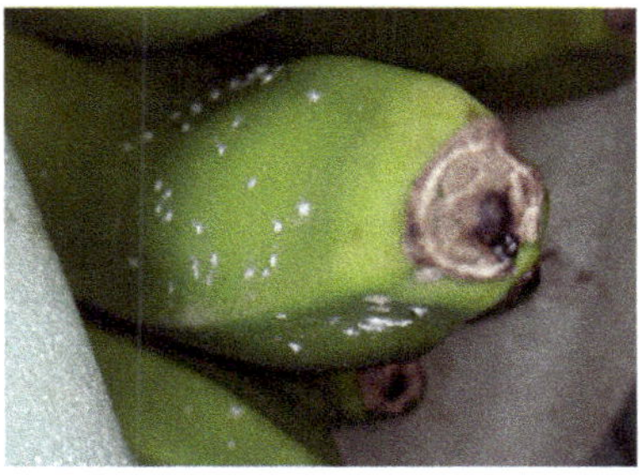

Figure 11 Mealybug infestation of banana bunch prior to pack house patio inspection.

Figure 12 Fresh flowers removed at the bunch patio, showing latex exudation from the flower scar.

box specifications and crown conformation requirements (discussed later). The raquis are discarded after hand removal. However, a set number of raquis may be weighed to establish their average contribution to the overall bunch weight.

Two water baths are typically used to clean banana hands as well as sort them by size and class. A typical layout of a pack house showing the position of these baths as well as other stages in the process is shown in Fig. 14. After cleaning, the fruit is sorted out and classified into small, medium, large and extra-large hands, to facilitate the final fruit arrangement over the plastic trays (Table 2). Additional factors in classification include finger length and curvature. Typical baths are constructed from lined concrete. Traditionally, the lining was ceramic, but due to modern food security standards, the use of epoxy food grade coatings is required. The size of the sorting and packing facility needs to be large enough to cope with the number of boxes of bananas a farm produces on a daily basis. A farm producing 1920 boxes per day will, for example, need a 9 m × 6 m bath, compared to a farm producing 480 boxes per day which will need a 7 m × 2.5 m bath.

Figure 13 Bunch dehanding using a cuchareta.

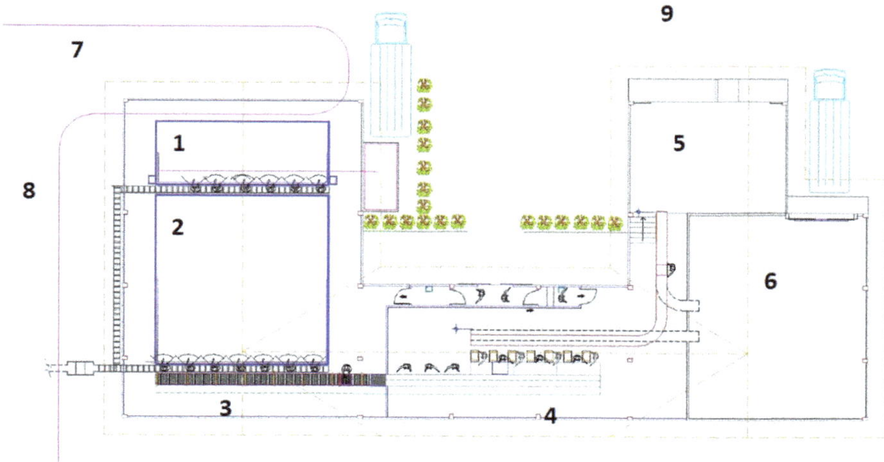

Figure 14 Banana linear pack house design. 1: Dehanding tank; 2: Delatexing/selection tank; 3: Clustered tray conveyor; 4: Packing zone; 5: Pallet assembly area; 6: Carton warehouse; 7: Cableway; 8: Bunch patio.

After washing in the first bath, the hands are trimmed into predetermined finger clusters joined by the crown, using a curved-neck knife, according to cluster unit specifications determined by the final market customer (Fig. 15). The required number of fingers per cluster can vary from nine to four (Andrade and Zurita, 2009); however, certain markets require complete hands. When wounded by cutting, crowns exude a milky sap known as latex (Milburn et al., 1990; Lagenheim, 2003). As a result, they need to be washed in a second delatexing bath to remove the latex which would otherwise stain the fruit peel (Fig. 16). 'Washing time' to remove latex varies from 3 to 15 min (Fonsah and Chindebelu, 2012). Even after washing, some latex will continue to exude (Mohapatra et al., 2010). Postharvest latex exudation and staining remain potential quality problems throughout the rest of the supply chain (Akpoebi and Chukwumah, 2012).

Table 2 General hand classification

Hand classification	Description	Principal use
Extra large	More than 10 inches in length	These hands are separated for a special three-row packing pattern
Large	Between 10 and 9.5 inches in length	Mainly used for the three-row packing pattern, or the final four-row packing pattern
Medium	Between 8 and 9.5 inches in length	A curved medium hand is used for the second row of the packing pattern. The straight medium hands are used for the third row
Small	Between 7.75 and 8 inches in length	Fruit used for the first row, of the four-row packing pattern. The smaller fruit can be used for single finger boxes

Crown conformation examples	Comments
	Bad crown conformation. The crown faces are irregular, increasing the fresh wound area.
	Bad crown conformation. There are regular crown faces, but there is very little crown left. This increases dehydration and the risk of postharvest diseases.
	Acceptable crown conformation. The crown faces are regular. Enough crown is left to avoid transit dehydration issues.
	Excellent crown conformation. The crown faces are very regular. Enough crown its left to avoid transit dehydration issues.

Figure 15 Common problems in crown conformation.

Figure 16 Fruit cluster selection water bath.

6 Fruit selection processes

Selection staff are often the most experienced personnel in the packing plant. Located at the end of the delatexing bath stage, they select and pre-assemble the fruit required for the final packing. Since weight is a key criterion, a key item of equipment is an electronic or conventional industrial scale with a 50 kg capacity. As an example, 18-kg banana and 22.64-kg plantain boxes are prepared in operations in Ecuador. Criteria for selection include categories such as small curved fruit, medium curved fruit, medium straight fruit and large straight fruit. Clusters are sorted into trays to facilitate later packing (Fig. 17). To avoid bruises, a plastic foam protector is placed over the tray (Fig. 18). The polyethylene foam must be impermeable with a density of 30–40 kg/m^3, with no impurities that might cause contamination. The plastic trays with the fruit are moved through the packing line using a plastic roller conveyor table (Fig. 19). The rollers are placed every 13–15 cm and

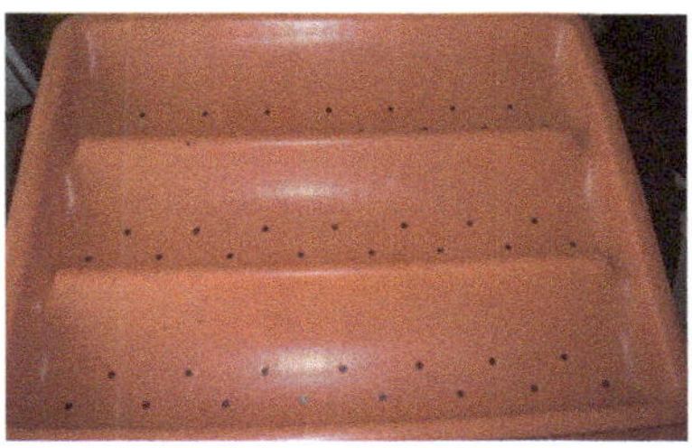

Figure 17 Plastic tray for cluster selection.

Figure 18 Polyethylene foam tray protection.

are made of 35-mm external diameter plastic tubes and typically 60–80 cm long. This type of design reduces vibration and therefore bruising. There are two main types of the roller conveyor design: linear conveyors and circular conveyors. Circular conveyors make it possible to work more efficiently, shifting from one type of package to other simultaneously. Small farms usually follow the same roller conveyor designs of the bigger farm pack houses, but with a predominantly linear configuration.

As the fruit moves along the conveyor, operatives apply a postharvest treatment to the exposed crowns of the fruit placed in the plastic tray. Whilst banana skin provides a natural barrier to disease, postharvest cutting operations create wounds which risk contamination from fungal disease (Dadzie and Orchard, 1997). Protective antifungal solutions include citrus seed extract, wax-based adjuvant and fungicides like thiabendazole and imazalil. Solutions can include ammonium sulphate, potassium sorbate, sodium benzoate, oxalic and maleic acids, essential oils, chlorine water and chitosan solution (Mohapatra et al., 2010). Other used fungicides are myclobutanil and azoxystrobin which give similar results.

There are three main ways of applying a postharvest treatment. The most popular method is a 18L knapsack pump diaphragm sprayer. Small farms with production of fewer than 200 boxes/day use a 5L version with a similar nozzle configuration. Another application technique is a fogging chamber that impregnates the fruit with a mist containing the

Figure 19 Plastic roller conveyor to move the fruit from the delatexing tank to the fruit packing area.

postharvest treatment. The third method is to brush the application directly on the crown. The postharvest treatment drips onto the brushes to keep them impregnated. This method is the most efficient in terms of volume of treatment needed with the fogging chamber requiring the most solution.

7 Packing bananas

Since they affect quality further along the supply chain, some authors suggest that packing operations are the most important operation in farms focused on export bananas (Fonsah and Chindebelu, 2012). The packaging system must ensure that fruit will reach the final market in a safe and sound condition at an affordable cost. Poor packing creates quality problems such as bruising and latex staining which may lead to boxes being rejected by the retailer. Typical quality problems which may become evident at this stage, and may lead to fruit being rejected, are problems related to poor harvesting and bunch patio operations such as poor crown trimming, broken necks where the finger connects to the crowns, knife cuts to the fruit neck or body, as well as problems from poor packing such as crown or tip bruises, box 'burn' or latex staining. These problems reduce visual and texture quality and may lead to other problems such as water loss and microbial growth during packaging and transport (Woltering, 2017b).

The fruit arranged over the plastic trays is delivered to the packing area where operators use different kinds of packing materials to meet differing market specifications. These market requirements relate to factors such as material quality, packing pattern, postharvest treatment, certification, labelling, fruit weight, fruit length, cluster numbers and so forth (Table 3). Labelling needs to show market information such as exporter brand, country of origin and any certification logos (e.g. bananas produced under the Rainforest Alliance certification scheme). Some boxes for gourmet bananas may need to be designed to be used for retail display.

In the past, producers used wooden boxes to move fruit to markets. More recently, producers in countries such as Australia and many Latin American countries have used telescopic corrugated carton boxes (Queensland Government, 2010; APEDA, 2002). The cardboard paper has different specifications (60–120 gr/m^2) with a burst parameter between 130 and 250 kPa. The box design needs to take into account ripening requirements later in the supply chain. These include holes to allow easy handling at either end of the box, as well as further holes along each side to facilitate opening of packing bags to allow the flow of ethylene gas used to manage ripening. Plastics are used to provide further protection

Table 3 Different box specifications and markets

Box type	22XU	22XU	22XU	22XU plantain (47 lbs)	BB Orito	BM Morado
Packing	Vacuum	Polytube	Pack bag	Vacuum/linear plastic	Vacuum/pack bag	Vacuum/ pack bag
Length	8 inch	8 inch	8 inch	9 inch	3.5 inch minimum	5–9 inch
Grade range	42–46	42–46	42–46	52–60	33–38	40–47
Cluster fingers	16–19	16–20	16–19	65–70 fingers	12	12–16
Market	ME	USA	EU	USA/EU	USA/EU	USA/EU

(e.g. to reduce bruising) and prevent fruit moisture loss. Packing plastics use different resin combinations of HDPE, LDPE and LLDPE. As an example, pack tube and vacuum bags (0.7–0.8 mm) are made of 80% HDPE and 20% LLD butene, whilst pack bags (0.8 mm) are made only using LLDPE. Absorbent paper can be also used at the bottom of boxes to reduce latex staining of the crowns and tips of fruit on the bottom row. Plastic LLDPE strips can also be placed between finger tips in clusters to avoid tip injuries from compression during packing and transport.

There are various methods for placing bananas in boxes to group different sizes together. The right packing pattern is also important to prevent bruising, minimize problems from latex exudation and optimize the use of space. A four-row banana packing pattern is obtained with curved clusters placed in the first row since curved bananas are more resistant to compression (Fig. 20). The second row typically contains medium-curved clusters with the crowns placed opposite the first row. The following clusters use the similar crown-opposed position but use less-curved clusters. The last row usually contains straighter/longer fruit which is more vulnerable to bruising. A three-row packing patterns is used with long, plain fruit. This pattern uses medium-curved bottom row clusters, with crown-opposed subsequent rows using long plain fruit. Packing operators and selection staff need the appropriate training to be able to select and then pack four-row and three-row packing boxes. Organic bananas are sometimes placed using packing pattern based on complete hands.

Different varieties of banana may have different packing requirements. As an example, small sweet diploid bananas called Oritos (also known as Sucriers or Pisang Mas) are packed in telescopic display boxes in complete hands and clusters. Eight hands are placed vertically in a 20-lb box, separated by polyethylene foam. Since varieties such as Oritos are particularly susceptible to bruises, polyethylene foam protection is placed at the bottom of the box. Packs of Manzano (silk) bananas use the same pattern of two parallel rows of fruit, with the crown facing the longer side of the carton. Six to eight clusters are packed in 20-lb carton box. Red variety bananas (Red Dacca) use a similar packing pattern with 12–15 clusters.

Figure 20 'Four row' banana clusters packing, polybag and carton bard lamina protection.

A single fruit packing pattern is common for plantains and some varieties of banana (Fig. 21). This pattern typically consists of two baseline rows placed side by side. Curved fruit is used on the next level, with further layers of fruit placed in opposite directions with the crowns toward the bottom of the box. The top fruit row usually uses straight fruit. Banana single finger boxes (40–50 lbs) contain 100 to 160 fingers compared to the plantain of 50-lb box with 50–67 fingers (IICA, 2006). The fruit is also labelled according to market requirements. Two adhesive labels are typically placed per cluster. The labels provide key information on issues such as: country of origin, product and farm codes (for traceability), certification schemes (e.g. for organic bananas), marketing promotions, final market branding and so forth.

Processed banana boxes are then weighed before they can then be palletized (Fig. 22). Pallets are usually hard wood or plastic. Typically 48 conventional (≈40 lb) boxes are placed in eight layers. They are secured with plastic or cardboard corner supporters and held together using a plastic strip bandage (zuncho) attached using a metal stapler. If a pallet is well constructed, it reduces vibration and thus bruising during transport as well as making logistics operations such as storage, loading and unloading easier. As a result, pallets have largely replaced the use of bulk boxes. However, more bulk boxes (1200)

Figure 21 Banana singles packing pattern, with fruit placed sideways at the bottom to provide sufficient support for the following rows.

Figure 22 Final fruit box weight control.

can be placed inside a container compared to the palletized system (1080 boxes). Pallet operations are often conducted on a raised platform to make it easier to transfer the pallet into lorries transporting the bananas.

8 Future trends

Consumers have shown a growing interest in the way crops such as bananas are produced, from safety and environmental concerns (such as concerns about pesticide or other chemical residues, the environmental impact of agricultural production and the problem of food losses and waste in the supply chain) to fair trade issues related to incomes and working conditions for agricultural workers and smallholders. However, most customers are still used to bananas as a cheap commodity which places continued pressure on production costs.

These broader concerns mean that there will be more pressure on producers to maintain product quality whilst making production more efficient and reducing losses in the supply chain (Woltering, 2017a,b). This involves the use of certification schemes in areas such as environmental performance and social responsibility as a way for producers to differentiate their product in the market (Schouten, 2017). Areas for improvement include better measurement and reduction in water use in production. Whilst crop irrigation consumes most water, there is also a focus on more efficient use of water in packing operations (Clercx et al., 2016). To maintain product quality and reduce spoilage, there will also be greater interest in more novel preservation techniques such as the use of modified atmosphere packaging or ultraviolet light in packing, transport and storage (Turtoi, 2013; Kerbell, 2017).

There will be a continuing focus on safety (Beuschat et al., 1998). An increase in outbreaks of food poisoning related to horticultural as well as other agricultural products, as well as broader concerns about bioterrorism led the USA, for example, to enact The Food Safety Modernization Act in 2011. This has led to an increase in safety management regulation at the farm level as well as to the development of Foreign Supplier Verification Programs to ensure a consistent level of safety in imported produce (Kennedy, 2016; FDA, 2017a,b; National Sustainable Agriculture Coalition, 2017). The Act empowers the Food and Drug Administration (FDA) to initiate a preventive approach by requiring food firms to adapt traceability and hazard analysis critical control point systems along their supply chains. The FDA is also required to maintain strict border measures to ensure that unsafe foods of international origin are not allowed into domestic supply chains (Nakuja, 2012). The implementation of these new requirements begins with a risk analysis that identifies the processes that must be modified. Common areas for improvement relate to process sanitation methods and monitoring.

These improvements to food safety systems are based on international safety standards set down by the Codex Alimentarius Commission (CAC), jointly established by the FAO and the World Health Organization. The CAC includes regional committees such as the Coordinating Committee for Latin America and the Caribbean (CCLAC). CCLAC includes banana-producing countries such as Colombia, Ecuador and Peru and has moved to develop risk-based food inspection systems in harmony with developments in other parts of the world such as the USA and Europe (Leake, 2015). These requirements will increase the need for improved and more standardized processes for the safe cultivation, harvesting and packing of fruits such as bananas.

9 Where to look for further information

The trend in tropical fruits postharvest can be sought out on crops with similar origin, distribution channels and markets on the fresh fruit side. Any advance obtained in pineapples and mango could be developed for bananas since the fruit must be peeled for consumption and transportation is performed using carton boxes. All postharvest mould-rot control active ingredients of fresh fruits must be reviewed for banana in case market legislation will limit the use of the accepted fungicides. Any postharvest treatment must ensure the satisfaction of the Maximum Residue Limit (MRL) accepted at the market.

For further information on new trends that will affect banana harvest and postharvest, please review:

1. Food Safety Modernization Act (FSMA) at https://www.fda.gov/Food/GuidanceRegulation/FSMA/
2. Standards for the Growing, Harvesting, Packing, and Holding of Produce for Human Consumption. https://www.fda.gov/Food/GuidanceRegulation/FSMA/ucm334114.htm
3. General information related to banana agronomy at https://www.cabi.org/bookshop/book/9781845936587
4. MRL database at https://www.fas.usda.gov/maximum-residue-limits-mrl-database

10 References

Agrocalidad. (2013). http://www.agrocalidad.gob.ec. Retrieved from http://www.agrocalidad.gob.ec/inspeccion-fitosanitaria-para-productos-de-exportacion/productos-de-exportacion/banano/normativa:http://www.agrocalidad.gob.ec/wp-content/uploads/pdf/sanidad-vegetal/2-certificacion-fitosanitaria/inspeccion-fitosanitaria-para-productos-de-exportacion/productos-de-exportacion/banano/normativa/manual-de-procedimientos-de-inspeccion-banano.pdf

Akpoebi, D. and Chukwumah, C. (2012). Determination of the characteristic of plant exudates from selected tropical trees and unripe fruits. *African Journal of Plant Science*, 6(4), 150–6. Retrieved 1 March 2017, from http://www.academicjournals.org/journal/AJPS/article-full-text-pdf/1BCCCBD11792

Andrade, A. and Zurita, M. (2009). *Reduccion de defectos y control en una empacadora de Banano*. Guayaquil: ESPOL/FIMCP. Retrieved 13 February 2016, from www.Espol.edu.ec: https://www.dspace.espol.edu.ec/handle/123456789/1733

APEDA. (2002). Report on updation of technical specifications of packaging for fresh fruits and vegetables for export. Retrieved from APEDA.gov.in: http://apeda.gov.in/apedawebsite/trade_promotion/study_report/F-and-V-Reprt-revised.pdf

Banks, N. and Joseph, M. (1991). Factors affecting resistance of banana fruit to compression and impact bruising. *Journal of the Science of Food and Agriculture*, 56, 315–23.

Beuschat, L., Nail, B., Adler, B. and Clavero, M. (1998). Efficacy of spray application of chlorinated water in killing pathogenic bacteria on raw apples, tomatoes, lettuce. *Journal of Food Protection*, 61(10), 1305–11.

Bugaud, C., Ocrisse, G., Salmon, F. and Rinaldo, D. (2013). Bruise susceptibility of banana peel in relation to genotype and post-climacteric storage conditions. *Postharvest Biology and Technology*, 87, 113–19.

Champion, J. (1968). *El Platano, Tecnicas agricolas y produciones tropicales*. Paris: Blume, Barcelona.

Clercx, L., Torres, E. Z. and Kuiper, J. D. (2016). Water footprint assessment of bananas produced by small banana farmers in Peru and Ecuador. *Acta Hortic.* 1112, 21–28. doi:10.17660/ActaHortic.2016.1112.1

Dadzie, B. and Orchard, J. (1997). *Routine Post-Harvest Screening of Banana/Plantain Hybrids: Criteria and Methods*. Rome, Italy: INIBAP.

FDA. (2017a). FSMA final rule on Foreign Supplier Verification Programs (FSVP) for importers of food for humans and animals. Retrieved from U.S. Department of Health and Human Services: https://www.fda.gov/food/guidanceregulation/fsma/ucm361902.htm

FDA. (2017b). Standards for the growing, harvesting, packing, and holding of produce for human consumption. Retrieved from FSMA Final Rule on Produce Safety: https://www.fda.gov/Food/GuidanceRegulation/FSMA/ucm334114.htm

Fonsah, G. and Chindebelu, A. (2012). *Economics of Banana Production and Marketing in the Tropics. A Case Study of Cameroon*. Bamenda, Cameroon: Langaa Research & Publishing CIG.

Goncalves, A. and Kernaghan, J. (2014). Banana production methods a comparative study. Retrieved from https://www.naturskyddsforeningen.se: https://www.naturskyddsforeningen.se/sites/default/files/dokument-media/banana_report_final_version.pdf

Horticulture Innovation Australia. (2017). Australian bananas.com.au. Retrieved from Australian banana production: https://australianbananas.com.au/Pages/all-about-bananas/production

IICA. (2006). *Guia práctica para la exportación a EE.UU Platano*. Managua, Nicaragua: Instituto interamericano de Cooperación para la agricutura. Representacion del IICA en Nicaragua. Retrieved 1 March 2017, from http://www.bio-nica.info/biblioteca/IICA2006Platano.pdf

Karunarate, A. and Niroshini, W. (2010). Agrochemical free treatment to improve postharvest shelf-life of banana. In: A. Cohen (Ed.), *Bananas Nutrition Disease and Trade*. New York, NY: Nova Science, pp. 1–32).

Kennedy, P. (2016). FSMA and imported food – Will the FDA's new strategy improve food safety? Retrieved from Food Safety & Quality Blog: http://foodsafety.merieuxnutrisciences.com/fsma-and-imported-food-will-the-fdas-new-strategy-improve-food-safety

Kerbell, E. (2017). Como lograr que el nanano llegue con optima calidad consistente todo el año a los mercados de destino. *XIV Foro Internacional del Banano*, Guayaquil, Guayas, Ecuador.

Langenheim, J. (2003). *Plant Resins*. Portland, OR: Timber Press.

Lassoudière, A. (1978). Quelques aspects de la croissance et du développement du bananier 'POYO' en Côte d'Ivoire. *Fruits*, 33, 457–91.

Leake, L. (2015). Food safety in South America. Retrieved 8 March 2018, from Food quality and safety web site: http://www.foodqualityandsafety.com/article/food-safety-in-south-america/

Milburn, J., Kallarackal, J. and Baker, D. (1990). Water relations of the banana. I. Predicting the water relations of the field grown banana using the exuding latex. *Australian Journal of Plant Physiology*, 17, 57–68.

Mohapatra, D., Mishra, S. and Namrata, S. (2010). Banana post harvest practices: Current status and future prospects- a review. *Agricultural Reviews*, 31(1), 56–62.

Moreno, J., Blanco, C. and Mendoza, R. (2009). *Buenas prácticas agricolas en el cultivo de banano en la región de Magdalena.* Medellin: AUGURA. Retrieved 16 January 2017, from http://cep.unep.org/repcar/proyectos-demostrativos/colombia-1/publicaciones-colombia/cartilla-banano-definitiva.pdf

Nakuja, T. (2012). Political influence and the implications of the US Food Safety Modernization Act for fruit and vegetable trade. Thesis, Canada: University of Saskatchewan. doi:S7N 5A8.

National Sustainable Agriculture Coalition. (2017). http://sustainableagriculture.net. Retrieved from http://sustainableagriculture.net/fsma/overview-and-background/: http://sustainableagriculture.net/fsma/overview-and-background/

Nuevo, P. and Maunahan, M. (2001). AGRIS: Postharvest handling systems of banana in the Philippines. Retrieved from Food and Agriculture Organization of the United Nations: http://agris.fao.org/agris-search/search.do?recordID=PH2003001697

Queensland Government, AU. (2010). Harvesting and post-harvest handling of bananas. Retrieved from Queensland Government: https://www.daf.qld.gov.au/plants/fruit-and-vegetables/fruit-and-nuts/bananas/harvesting-and-post-harvest-handling

Robinson, J. and Galan Sauco, V. (2012). *Banana and Plantains*. (C. International, Ed. and E. M. Prensa, Trans.) Wallingford, UK: CAB International. doi:ISBN-13: 978-1845936587.
Schouten, R. (2017). Quality measurements and quality modeling. Post harvest technology course, Wageningen, the Netherlands: Wageningen University.
Sharrock, S. (1997) Uses of Musa, INIBAP Annual Report 1996, INIBAP, Montpellier, France 42–4.
Soto, M. (1985). Cosecha y empaque de fruta. In: M. Soto (ed.), *Bananos: Cultivo y comercializacion*. San José, Costa Rica: Litografía e Imprenta LIL, p. 345.
Thompson, K. (2003). *Fruit and Vegetables: Harvesting, Handling and Storage*. Oxford, UK: Blackwell Publishing.
Toivonen, P., Mitchman, E. and Terry, L. (2017). Postharvest care and treatment of fruits and vegetables. In: G. Dixon and D. Aldous (Eds), *Horticulture: Plants for the People and Places*, vol. 1. Dordrecht: Springer Science Business Media, pp. 465–77.
Turtoi, M. (2013). Ultraviolet light treatment of fresh fruits and vegetables surfaces: A Review. *Journal of Agroalimentary Processes and Technologies*, 9(3), 325–37.
United Brands. (1975). *Guía práctica para el cultivo del banano*. La Lima, Honduras: Departamento de investigaciones agricolas tropicales.
Woltering, E. (2017a). Global loses and health aspects of fresh produce. Postharvest workshop, Wageningen University, Wageningen, the Netherlands.
Woltering, E. (2017b). Factors determining quality of fruits and vegetables. Postharvest technology course, Wageningen University, Wageningen, the Netherlands.

Chapter 12

Ripening systems for bananas

Frits Popma, Popma Fruit Expertise, The Netherlands

1 Introduction
2 Factors affecting banana ripening
3 Banana ripening cycles
4 The banana supply chain
5 Ripening technologies and rooms
6 Quality control
7 Summary
8 Future trends
9 About the author

1 Introduction

Banana ripening begins in the country of origin. A grower who is ignorant of the stages of the ripening process will not be successful in ripening aesthetically pleasing bananas with a good taste and shelf life. Once the banana stem is cut from the parent plant, the supply of nourishment is interrupted. This changes the chemistry of the fruit. An irreversible process begins as the starch converts into sugar. Unripe, green bananas contain 80% water and about 20% starch, vitamins and so on. Starch does not provide a pleasing taste for consumers who generally prefer sweet-tasting bananas. The ripening process is critical in converting the starch into sugar. Ripening is also important to produce bananas with an acceptable colour and a shelf life of at least four days.

Retailers require a consistent product quality. Bananas travelling from Latin America to Europe, for example, are ready to harvest between 9 and 13 weeks, with 14 weeks typical where transit distances are shorter. The smell of unripe bananas resembles the smell of cucumber and the absence of this odour indicates that the fruit is older. Freshness may be tested by breaking a banana in two and pushing the pieces back together. If latex drips from the banana, it indicates freshness. Many retail purchasers and consumers mistakenly believe that green bananas have a longer shelf life, leading them to prefer ordering green bananas.

http://dx.doi.org/10.19103/AS.2017.0020.22

2 Factors affecting banana ripening

2.1 Temperature

When bananas arrive in the ripening facility, the product temperature is below 14.5°C. If it falls below 13.0°C, the bananas start to show chilling damage. This is referred to as UPD, or under peel discolouration. If the temperature at which the bananas are stored is too cold, the fruit will remain brownish grey and will be difficult to ripen. Attempting to ripen these bananas is inadvisable.

2.2 Ethylene

Raising the product temperature will result in activation of the enzyme, ACC, or 1-aminocyclopropane-1-carboxylic acid, which is responsible for the production of internal ethylene. Although bananas produce ethylene naturally, it is added by ripeners to achieve a consistent rate of ripening, and acceptable colour. All the bananas in a single box will not come from the same plant and will have a different background and history. By gassing the bananas or adding ethylene (C_2H_4) in ripening room, all the fruit will start to develop the colour break at the same time. It is important that the pulp temperature between the bananas does not differ by more than 0.3°C. It is recommended that ethylene is applied at 15.0°C for a seven- or eight-day ripening cycle (Table 1). The recommended quantity or dosage may vary per banana ripening room. Whether there is a single carton in the room, or 1152 boxes in a standard 24-pallet room, the commercial value is 1000 ppm. The recommended quantity for modern, airtight banana ripening rooms is between 300 and 400 ppm ethylene.

2.3 Humidity

Many ripeners have difficulty with the ripening process during winter or spring. During this period, bananas do not change colour, and not all ripeners are aware of this phenomenon. One explanation is that when humidity levels drop below 50%, pores in the banana peel close and ethylene cannot enter the peel and the pulp. Humidifiers may be used to rectify this. However, the simplest method is to moisten the floor with water. Ideal ripening conditions require a humidity level of roughly 95%. High humidity guarantees better taste and a longer shelf life. The bananas will appear more shiny and attractive, and have 50% less scarring when compared to bananas ripened in low levels of humidity.

2.4 Oxygen

Even after being cut from the plant, bananas still require oxygen. When pulp temperatures are low, the need for oxygen is limited. This changes during the 24-h period in which bananas are exposed to heat. Heat exposure encourages the fruit to turn from colour stage

Table 1 Ethylene and ripening (Source: UC Davis Dr. Adel A. Kader)

Temperature	13°C (56°F)	15°C (59°F)	18°C (65°F)	20°C (68°F)
µL C_2H_4/kg·hr[1]	0.1–2	0.2–5	0.2–8	0.3–10

[1] Low end for mature green bananas and high end for ripening bananas.

3 (more green) to colour stage 4 (more yellow then green). During this period of heat exposure, the bananas need a lot of oxygen. A banana with a shiny bright colour indicates it has been exposed to a fresh environment and will have a real banana taste. For the safety of the ripener, and to produce a bright yellow colour, the air in the ripening room must be replenished with oxygen twice daily.

2.5 Carbon dioxide and respiration

The rate of respiration directly correlates with the temperature at which bananas are ripening (Table 2). Low temperatures will result in low respiration rates, high temperatures in high respiration rates. The same applies to the internal ethylene production. It is important to refresh the air in the banana ripening room as high CO_2 levels will slow down the ripening process, resulting in a dull, blotchy appearance and an unappetising flavour. When the level of CO_2 exceeds 1.5% in a ripening room, it presents a danger to staff.

2.6 Enzymes

Various enzymes are responsible for the many chemical reactions which occur during banana ripening. Enzymes act as catalysts to the chemical reactions, increasing the ripening rate. Without these reactions, a banana will not become yellow and edible or have a pleasing flavour. Bananas have about 21–22% starch and only 1% sugar when harvested. The conversion process is very complex.

As previously mentioned, the enzyme ACC activates the internal ethylene production. ACC not only changes the fruit colour, but also activates other enzymes to make a banana edible and attractive. The enzyme hydrolyse is responsible for the breakdown of chlorophyll. Chlorophyll creates the green colour of a banana skin. Hydrolyse unmasks the carotids and creates the yellow colour of the banana skin. Amylase converts the complex structure of the starch into simple sugars such as glucose and fructose. This enzyme is responsible for creating fructose, thus making the banana sweeter. Pectinase is a group of enzymes that break down a central part of the plant cell walls. Hydrolase changes the structure of the molecules, thus altering the taste.

3 Banana ripening cycles

The ripening cycles range from a period of four to eight days. Bananas usually change from colour stage 3 to colour stage 4 over a period of roughly 24 h. During a four-day ripening cycle, bananas are ready to be exposed to heat 60 h after the ripening process starts as ethylene gas enters the ripening room. Ripening takes place at temperatures between 14.5 and 20° C.

Table 2 Carbon dioxide and ripening (amount of CO_2 produced per kg of bananas per hour at different temperatures)

Temperature	13°C (56°F)	15°C (59°F)	18°C (65°F)	20°C (68°F)
mL CO_2/kg·hr[1, 2]	10–30	12–40	15–60	20–70

[1] Low end for mature green bananas and high end for ripening bananas.
[2] To calculate heat production multiply ml CO_2/kg·h by 440 to get Btu/ton/day or by 122 to get kcal/metric ton/day.

Should the ripener choose to prolong the ripening process, heat exposure should begin after 72 h. The decision to prolong the process depends on the product quality at arrival.

In summary, bananas convert starch into sugar. During this process, the banana needs a specific environment. The pulp temperature of a single banana increases by 8°C when the fruit is ripened using a four- to five-day programme. An extended ripening programme results in a pulp temperature increase of 4°C. It is essential to manage humidity as well as ethylene and CO_2 production during the banana ripening process. Without controlled heat exposure, the bananas will ripen too quickly.

Figure 1 displays an ideal six-day ripening cycle. The actual results may vary slightly. Temperature adjustments can be made while bananas are in the ripening room. After 144 h, the pulp temperature will be 15°C and the bananas will have reached colour 4.

Whenever a buyer requires a greener banana, the fruit will be removed from the ripening room earlier than the ripening period. Without proper cooling, the pulp temperature may easily increase to 24°C or higher. Higher pulp temperatures result in weight loss. One kilo of evaporated water during the banana ripening process is equal to the price of one kilo of bananas. It is perhaps of greater concern that a higher pulp temperature will reduce the shelf life of the fruit.

A common-sense approach to measuring shelf life has been adopted in Europe, where it is measured from the time when the bananas leave the ripening facility. The shelf life is then 100 h, the period necessary for the bananas to reach colour stage 7 (Yellow bananas with sugar spots). After stage 7, the quality declines. The taste will become mealy, resulting in fewer sales and increased waste.

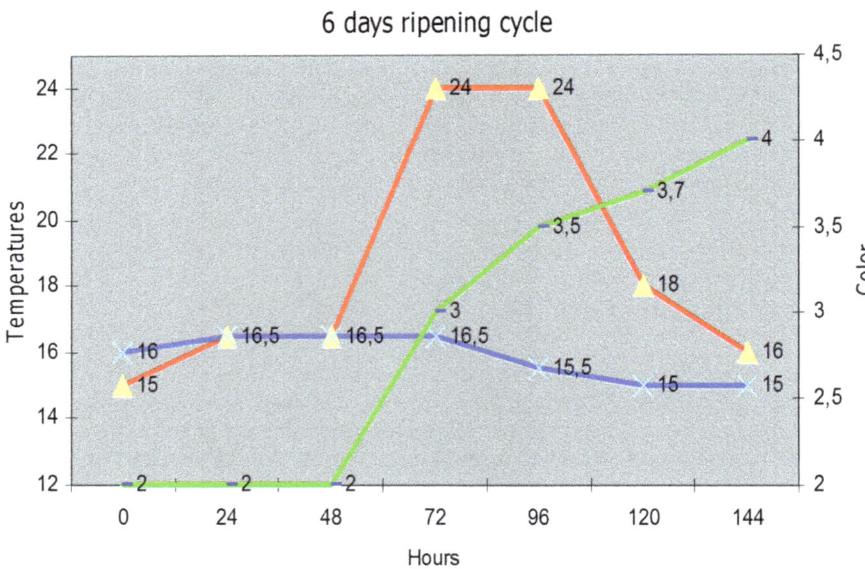

Figure 1 Banana ripening cycle. Notes: The green line represents the colour of the bananas during ripening. The blue line represents the temperature setting for a six-day ripening programme. The red line represents the banana pulp temperature. The *x*-axis shows the hours during the ripening process, the *y*1-axis shows the pulp temperatures and the *y*2-axis shows the fruit colour.

4 The banana supply chain

4.1 Growing

The growing stage in the banana supply chain is illustrated in Fig. 2. During the growing period, specifications are implemented to benefit the grower and the banana ripener. These specifications are designed to produce the best possible product in the carton or plastic crate and on the retail shelf. One of the specifications is age control. As soon as a banana bunch forms fingers, it is protected with a bag to avoid scarring or damage from insects and birds. The bag also provides protection against dust and creates a higher temperature for the fruit. After bagging, the bunch is colour-coded. Each week has a specific colour. In Europe and the Middle East, bananas are harvested between 9 and 12 weeks. Younger bananas will not reach the minimum grade. Bananas harvested at 13 or 14 weeks will display ripening or texture issues at the point of arrival (ripe and turning and/or soft green).

Black sigatoka causes significant problems for growers and ripeners. Bananas affected by this disease will start to ripen during transit and damage is caused to the leaves. It is essential that bananas are not harvested from a plant which has fewer than seven healthy leaves. Tropical Race 4, or TR4, is a bacterium which destroys the root system of banana plants and causes plant death. Plantations in the Philippines and parts of China, Australia, Mozambique and Jordan are infected by this disease. Combating the TR4 bacteria is almost impossible. Severe weather conditions, such as excess rain or periods of drought, also create stress for the banana plant. Ripeners often blame weather conditions for producing poor results, although poor ripening cannot always be attributed to the weather.

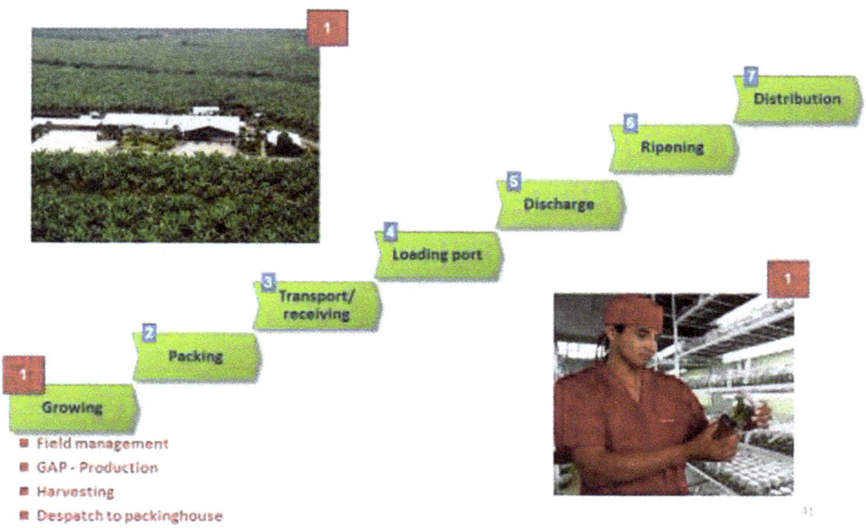

Figure 2 The banana supply chain: growing.

Organic bananas are not treated with fungicide or other chemical products, as are conventional bananas. These bananas are packed in Banavac plastic bag. (Banavac is a word combined from the words, bananas and vacuum.) The plastic is thicker than that of a normal Polypack liner and differs in being without holes. When the bananas are packed, air is removed by a vacuum device. Because there are no holes, free water comes from the banana surface and from cooling in a vessel or container. The free water will drip to the bottom of the carton, creating an ideal environment for bacteria which cause rot and moulds during the ripening process.

4.2 Packing

Grading the bananas is an important stage in the supply chain process (Fig. 3). Grade and length are measured before the bunches are cut into hands or clusters. These are Class Extra, Class I and Class II. As Class I and Class II are usually smaller hands or clusters, the ripening may become more difficult. It is not advisable to ripen these bananas in the same ripening room. Generally, the Class I and Class II bananas are packed under a different label. Bananas which are not within these specifications are sorted and removed. This may be due to defects which include scarring, red rust, maturity stain and others. A simple grading technique is to cut a banana into two equal parts. This shows, for example, if it is angled or full (Fig. 4). The wrong selection may lead to uneven ripening.

Not every hand or cluster will have the same size or shape. Selectors sort the clusters into curved, medium curved or straight and place them on a tray for packing. The net weight of the bananas when ripened should usually be more than 18.14 kilos, as during pre-cooling, transport and ripening, the fruit will lose weight.

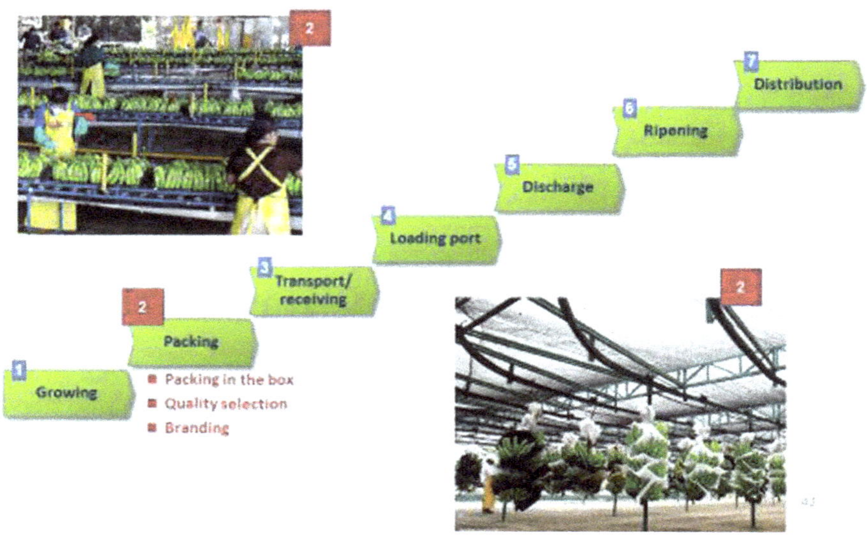

Figure 3 The banana supply chain: packing.

Figure 4 Full bananas left and angled right.

The bananas are packed in cardboard boxes or plastic crates. Together with Mr. Willem Kokkeel of Eurasia Connection (www.eurasiaconnection.com) we carried out a feasibility study in 2015 on the use of plastic collapsible crates in India. This was sponsored by the Dutch Government (www.rvo.nl). A carton banana box was initially developed to reduce the field heat to 13.7°C. During transit in a banana vessel, the airflow is vertical. During ripening, the airflow is horizontal. The carton must be sufficiently strong to protect the bananas during transport overseas. A plastic crate provides an ideal alternative to cartons made from paper materials.

4.3 Pre-cooling

Bananas should be pre-cooled as soon as possible: within 24–48 h. Once harvested, they must be placed in a refrigerated container or a banana reefer vessel. A pre-cooling facility, where the bananas are pre-cooled within 24 h to a temperature of 13.7°C, is ideal. One option is to use a forced-air system. It is essential to follow proper procedures. Extreme cooling causes weight loss, dehydration of the fruit and will not result in a properly ripened product.

4.4 Transport

The transport stage is illustrated in Fig. 5. Boxes stacked on pallets are better protected against handling defects such as bruising as there is less manual handling during loading and unloading. However, transport rates per box are higher. Break-bulk-loaded containers accommodate more boxes. Loading and unloading the containers or vessels are faster when the bananas are palletised.

If pre-cooling takes place in a reefer container, it is important to ensure that air is forced from bottom to top through the pallets (using holes in the boxes to facilitate airflow). When pre-cooling takes place in a banana vessel, the fruit is loaded within 24 h and the ideal set point (13.7°C) is reached within that period. When bananas are transported under controlled atmosphere, it is important that the set point of the bananas is reached before the atmosphere is changed.

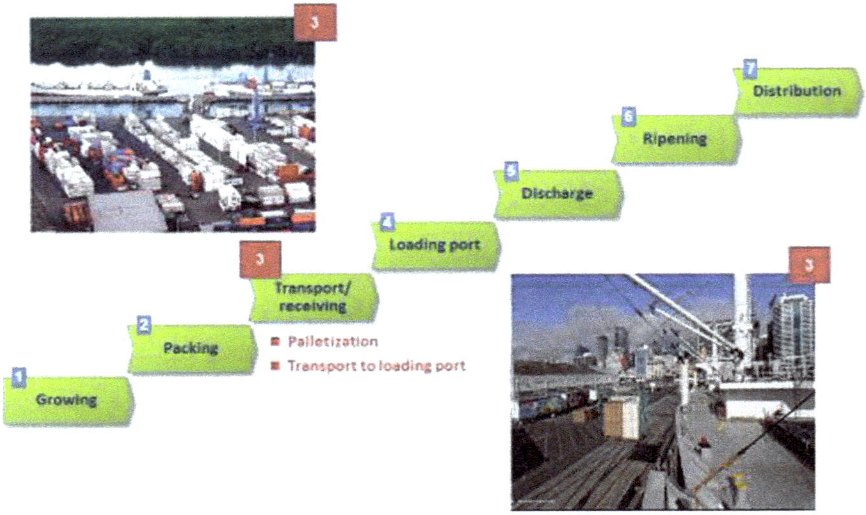

Figure 5 The banana supply chain: transport/receiving.

4.5 Loading

On arrival, the bananas are checked by banana company quality inspectors, EU inspectors or independent surveyors. The quality of the bananas must meet EU or country-specific specifications. In addition to EU inspections, banana companies have their own specifications. Usually, these do not differ greatly from each other. It is advisable to collect the bananas as soon as possible rather than to leave them in port warehouses as the ripener will have no control over bananas stored with other brands in the same warehouse. In some ports, the bananas still arrive as 'break bulk'. The cartons have to be palletised, involving increased manual labour which may cause defects such as scarring and bruising.

4.6 Discharge

As soon as the vessels or containers arrive, an initial quality check on the transport conditions must be carried out (Fig. 6). Fruit in damaged cartons and pallets will not receive an ideal airflow in the ripening room and will therefore ripen unevenly. These boxes should be removed.

Where temperatures are too low (<13.7°C), UPD of the bananas will occur. If temperatures are too high (>17°C) the bananas will ripen and when they have reached colour 3 (turning) and colour 4 (ripe), the boxes should be removed. (Colour 3 is more green than yellow, colour 4 is more yellowish than green.) Soft green condition is mainly found when bananas have been packed in Banavac bags. If there is insufficient oxygen in the bag, the fruit starts to ripen and remains green but soft (see Section 4.1).

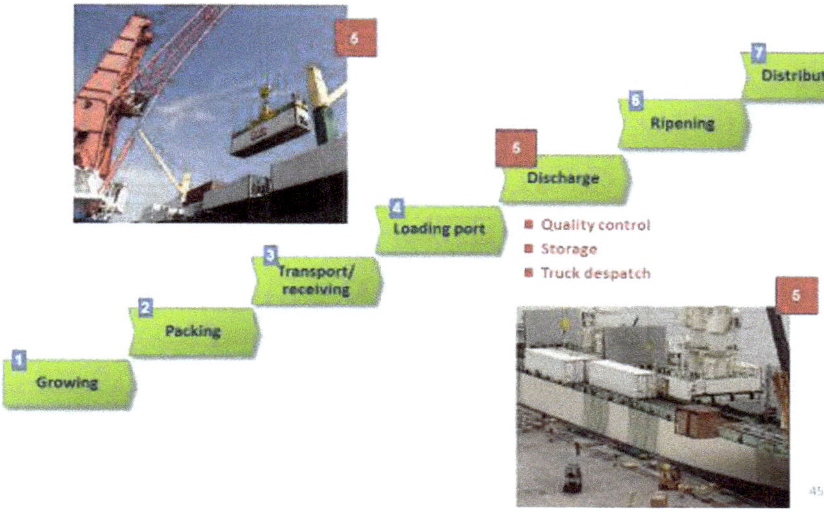

Figure 6 The banana supply chain: discharge.

4.7 Ripening

It is advisable for the ripener to make quality checks as soon as the bananas arrive in the ripening facility. This enables the placing of a claim against the supplier or transport company within 24 h. When the bananas leave the ripening facility, a quality control check should also be made. When there are discrepancies, a plan of action is necessary. The ripener can introduce a simple shelf life test for good communication with customers.

Consumers usually buy colour 4 bananas and consume them when the colour is 5 (yellow with green tips and necks). These colours are more attractive for the consumer and lead to higher sales. However, most banana buyers believe green bananas have a longer shelf life. A banana ripener has to deal with these requirements. Colour preferences will differ from country to country.

Before the bananas are ready for dispatch to the retail distribution centre, they can be stored in the ripening room, typically 14.5–15.5° C. If this is not a temperature-controlled environment, it will lead to high banana pulp temperatures and shorter shelf life.

4.8 Distribution

Figure 7 illustrates the distribution stage. Quality control for distribution is not usually complicated. Inspection determines the correct colour for the bananas and the temperature of the pulp. Some retailers will have their own quality specifications. Bananas normally have a shelf life of four days. This is defined as being from the time they leave the ripening facility to the time the first sugar spots appear. Sugar spots are graded as colour 7. Bananas should be transported to the retail point as soon as possible. Colour 4 and a pulp temperature between

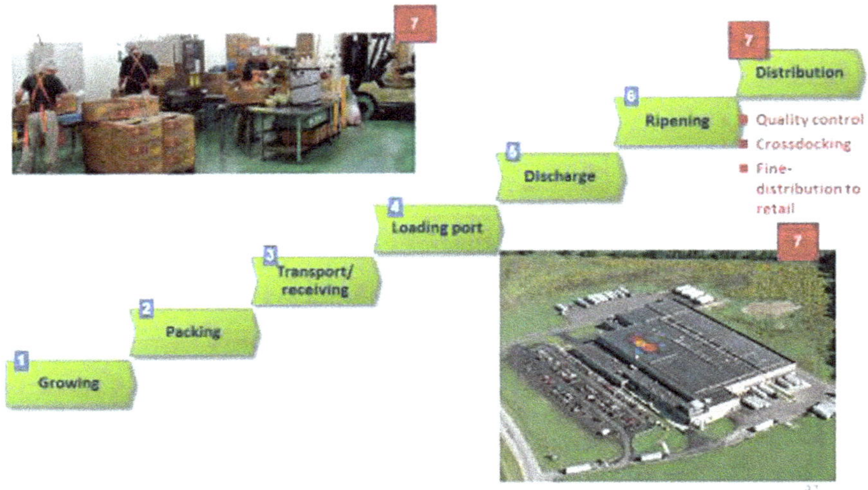

Figure 7 The banana supply chain: distribution.

14.5 and 15.5°C are considered to be the optimum condition. Pallets should be placed with a minimum gap of 10 cm to avoid hot spots and the first in should be the first out.

5 Ripening technologies and rooms

In the past, bananas arrived in Europe, the United States and the Far East in bunches. These bunches were then hung from ceilings. To ripen the bananas, natural gas was ignited. As a by-product, ethylene was released and the banana turned yellow. The ripener then cut the yellow bunches and packed them in crates. The author has recently observed a similar process in countries such as Egypt, where carbide is used instead of natural gas (Figs. 8–10).

Since the introduction of the banana box in 1968, several companies have developed different technologies, including pressurised ripening systems. Examples include a block stacked system, composed of five boxes layered with spaces between them to simulate pressurised ripening rooms. The first pressurised rooms were built in Sweden. An evaporator is placed in the middle of the room, which blows air to the walls, creating resistance due to the boxes. The ventilators of the evaporator suck in the air from the middle of the room, creating a lower pressure which forces the air through the boxes of bananas. This is known as a TH system.

The Del Monte or tarp system is an alternative system in which the room is filled with pallets of bananas and the tops of the boxes are covered with a tarpaulin (Figs. 11 and 13). Air is blown from an evaporator at the back of the room. The pressure increases at the front, creating an airflow across the room. At the back of the room there is a suction wall, behind which the air returns. All the air will be mixed in this zone, the warmer air rising and passing through the evaporator. The advantage of this system is that all the air comes

Figure 8 Banana ripening room in Egypt.

Figure 9 Banana ripening cellar in India cooled with ice.

Figure 10 So-called 'pyramid' stacking (Syria).

Figure 11 Modern tarp ripening system (Romania).

from one evaporator and is at a uniform temperature. The length of the room is important, and is limited to a maximum of 28 pallets in a single tier. An evaporator which covers the full length of the room is ideal. A by-pass, a space between the evaporator and the wall, negatively impacts the even airflow.

Another type of system is the Interko or TH system. Interko rooms, designed by a leading company in the manufacture of high-tech banana coolers, may be built as single-tier to four-tier ripening rooms. Air distribution is from the sides (Figs. 12 and 14). Interko

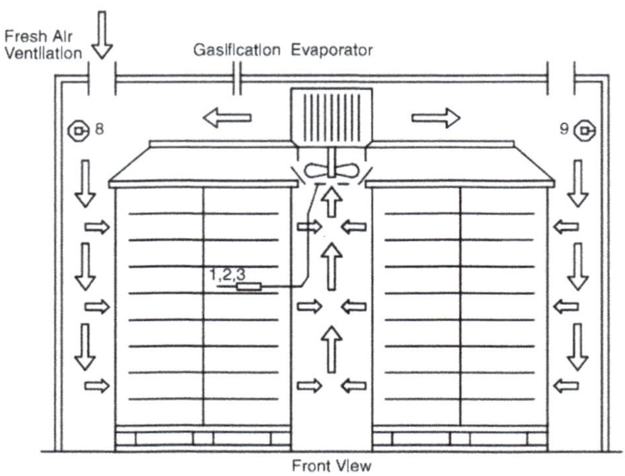

Figure 12 Schematic display of a modern ripening room and its essential elements, showing how air is forced though the banana boxes for optimal ventilation.

Figure 13 Tarp system.

Figure 14 Interko or TH system.

Figure 15 Ripening room in Europe (left); Indian ripening room (right).

Figure 16 Building insulation panels, local panels and labour.

also has a reversible system in which the airflow changes every few hours. This banana ripening system is sold in more than 50 countries. In total, 6000 banana ripening rooms of this type have been installed globally. The author has added the company Interko BV to this chapter as their ripening system is now in use in several countries.

The development of new software has been one of the key factors prompting companies to develop new ripening systems. Before 1995 there was little development, and block stack systems were replaced by pressurised ripening rooms until Chiquita International Services Group in Antwerp developed a new system: Chiquita 2000. The major difference was the use of computer software. After initial scepticism, cooling and ripening companies realised change was necessary. Cooling companies invested in developing a different type of ripening room. The result was a banana ripening facility with reversible airflow. Reversible airflow creates an even colour, longer shelf life and better taste and in some cases, reduces energy consumption.

There are a number of general considerations to take into account in the design and construction of banana ripening rooms. Their construction is not complex; however, 'the devil is in the detail'. The author's experience is that banana companies wishing to construct ripening rooms have the knowledge of bananas but are without the technical skills necessary to build ripening rooms.

In India, various types of ripening rooms are to be found (Figs. 15 and 16). In the author's opinion, these are simply cold stores. India is the biggest producer of bananas in the world but considers the banana to be 'the poor man's food'. There is therefore little investment in expensive equipment as the bananas are cheap, of low quality and subject to an imperfect supply chain.

A company constructing cooling equipment may have a reputation in this field but construction of banana ripening rooms is considered to be a different enterprise. As mentioned above, 'the devil is in the detail'. It is for this reason that only a few countries and companies are able to build modern banana ripening rooms. They tend to rely upon copying existing systems and lack the knowledge for building effective ripening rooms.

The experience in developing countries is that the installation of ripening rooms increases rental costs. Money spent on this technology, which cannot move with a company, is not seen as an efficient investment.

At least six rooms each able to fit 20 pallets are necessary for a viable commercial ripening enterprise. Cooling may come from a central cooling system or on an individual room basis. For a cooling company to be viable, it must be possible for it to lease a banana ripening room to its customers for five years.

In the Netherlands, De Laat Koudetechniek BV in Ammerzoden together with BG Door International BV in Barendrecht have developed an 'ECO-TOP' ripening system which is reliable, giving good ripening results and, when compared to other systems, fairly priced (Figs. 17 and 18).

The components of ECO-TOP Start are assembled and tested in the factory. Insulation panels can be built with local materials and labour. The equipment is placed at the back of the room and connected to the electricity supply. The gas-tight door and the tarp may be manually or electrically operated. The advantage, besides good ripening results, depending on banana quality, is that this equipment is more portable, making it easier if a company needs to relocate.

Figure 17 Installing the pre-fab kit into the banana ripening room.

Figure 18 Loading the room.

6 Quality control

The objective of ripening procedures is to guarantee a consistent product for customers in terms of shelf life, taste and appearance. For effective communication with the advisor, customer or supplier, it is necessary for ripeners to use a ripening process control sheet (RPCS) (Fig. 19). This form is utilised as soon as the bananas arrive in the premises. Information on this document can be used to communicate with the supplier, clients and an external advisor. Any claim must be sent to the grower or supplier within 24 h of the bananas arriving at the ripening facility. A quality inspector describes any problems or defects and the number of cartons affected, along with the station numbers. Wherever possible, the quality inspector provides photographs of the problems, showing the defect and station number.

A minimum of six boxes need to be inspected before the delivery truck leaves. The first step is to check quality at arrival. The ripener takes a random sample of six boxes to measure pulp temperatures:

1 The top box.
2 Box number 4 from the top of the pallet.
3 The bottom box.
4 From the second pallet in the truck.
5 From the last pallet in the truck.
6 Measure the pulp temperature of the banana in the fourth line (coldest) and the second line, second cluster (warmest). In 60 × 40 boxes the second cluster in the third line (bottom line).

The ripener also inspects the:

1 Latex flow (Yes/no).
2 UPD (Trace-Light-Medium-Severe).
3 Grade (Thin-Middle-Thick).
4 Rots and moulds.
5 Freshness (Fresh-shiny).

The first inspection:

- Determines the temperatures, freshness, major defects and conditions of the boxes and whether materials other than polyethylene have been used for packaging.
- Measures bananas in the second line, the second cluster of 40 lb and the third line boxes measuring 60 × 40.
- Warms the thermometer needle and measures again. Records the temperature.
- Looks at the grade of the bananas and takes the average.
- Breaks a banana into two equal pieces and examines at the latex flow. (A plentiful flow indicates that the bananas are fresh and will therefore ripen more slowly: no latex flow shows the bananas to be old or to have become dehydrated during transport.)

NADER & EBRAHIM S/O HASSAN PHILS., INC.
Ladislawa Building, Ladislawa Avenue, Buhangin, Davao City 8000 Philippines
Tel. No. +63.82.224.5637 to 38 Fax No. +63.82.224.5636

POPMA Fruit Expertise
EXPERTISE IN RIPENING TROPICAL PRODUCTS

RIPENING PROCESS CONTROL SHEET

ROOM:

Name ripener:

Facility:

Ripening Information

Fruit arrival in depot:

Ripening program:

Delivery date:

Expected color:

Logistics information

Transport:

Brand:

Arrival date: (vessel)

Port:

Vessel:

Division:

# of boxes:

Pallet & Box information

Pallet condition:	Tight	Small gaps	Not closed	
Box type:	40 pounds	60/40	Smalls	Others

Fruit information

Fruit temperature:					
Grade:	Low	Middle	High		
Defects:	UD	R&T	Scarring	R&M	
Latex flow:	Yes	No	Appearance:	Dull	Shiny

Date	Time	Pulp temp should be	Temp sensors			Temp. adjusted	Pulptemperatures			Correction	Gassing time	Colour	Dull or shiny
			1	2	3		1	2	3				

RM-03 4-Rev00-RPCS

Figure 19 Ripening process control chart.

Issues to be considered in loading a ripening room:

- Before the ripening room is loaded with bananas, it must be cleaned. It should be free of dust, empty pallets or old bananas.
- The ripener checks to ensure every fan runs in the same direction.
- The boxes are stacked to ensure the correct airflow.
- Plastic packaging or expandable foam is forced into gaps to prevent pressure loss.
- Poorly stacked pallets should be re-stacked.
- Damaged boxes should be removed and the bananas repacked.

UPD or low temperatures at arrival:

- A box of which the pulp temperature is lower than 13.5°C should be marked.
- The bananas should be placed in a heated room as soon as possible, until the pulp temperature of the marked box is above 14.0°C. The equalisation programme can then be started.
- Bananas which are severely chilled cannot be ripened.

Equalisation period:

- During the equalisation period, the ripener uses the RPCS (Ripening Process Control Sheet) form to write down the pulp temperatures, the set points, the readings of the sensors (if available) and any alterations in the programme.
- The equalisation period is necessary when pulp temperatures differ by more than 0.3°C.
- The set point is always 0.5°C below the gassing temperature.
- As soon as the temperatures are within a range of 0–3°C the gassing can start.
- The equalisation period is no longer than 24 h.
- During the equalisation period, the product sensors are not active.
- Before gassing the bananas, the ripening room is ventilated for 30 min.

Product sensors:

- The product sensors are built into the fourth box between the clusters of lines 1 and 2.
- The product sensor should not be in contact with the tunnel path and/or the plastic.
- The tunnel path must be open.
- The sensor has to be free of latex or banana pulp.
- The product sensor must not be placed in a box without first inspecting the bananas.
- The sensor must be placed in a box with an average grade.

Gassing the bananas:

- The room must be airtight.
- The doors must be inspected for leakage on a regular basis.
- The pulp temperature should be above 14.5°C, preferably 15°C.
- Prior to gassing, ventilate the room for 30 min.
- Gas the bananas as soon as possible and no longer than 24 h after arrival.
- The ripening programme begins when the bananas are gassed.
- After 48–60 h, depending on the ripening programme, the bananas start to break colour and become softer.

Procedure for gassing:

- Enter the room after equalisation, and check the sample boxes to ensure the temperature variation within the banana is at an acceptable level, and the fruit has reached the required gassing temperature.
- Make a final check to ensure that all pallets are tightly sealed together and that the airbags are properly positioned and secured.

- Record all the temperatures of the fruit on the ripening record, the day and time of gassing, and any other relevant information on RPCS.
- Ensure that there is sufficient gas pressure in the system to gas the room in question. If not, connect a new full gas cylinder. Each gassing requires a minimum of 70 bar pressure in the system. The pressure is shown on the pressure gauge above the gas station.
- Check the ripening plan to ensure that the correct programme has been selected for that room.
- Start procedure to gas the room.
- The room will be automatically closed off for a period of 24 h during the gassing cycle, or by locking the door by using the control panel near the door.

Ventilation:

- The room must be ventilated directly after the gassing period. A minimum of two times per day is required.
- Every ventilation cycle lasts for 30 min.
- Every cycle is 2–3 times the volume of the room.
- Insufficient oxygen will produce dull, grey bananas.
- Concentrations of ethylene and carbon dioxide which are too high will delay the colour development of the banana, and are harmful for human health.

Daily temperatures inspection, RPCS:

It is important to conduct a daily temperature inspection. Before entering a banana ripening room, open the door and wait for two minutes. A high concentration of carbon dioxide is dangerous. Good inspection practice includes the following:

- The banana's pulp temperature must to be measured twice daily.
- This should always be done in the box opposite the product sensor. (This is always the fourth box.)
- Mark the boxes with an X on the side and the top.
- Always measure the same box as these are the reference boxes of the ripening programme
- Measure the temperature in the second cluster in the second line, in the case of 60 × 40 boxes in the third line second cluster.
- Place the thermometer in a banana and wait until the temperature has been determined.
- Measure the temperature of another banana.
- In subsequent inspections, always warm the sensor up in a measured banana and then measure the temperature of a new banana.
- Always test three bananas, one from the second pallet, one from the middle and one from the second pallet at the end of the room.
- In two-tier rooms, measure at least one temperature from the second tier.
- Measure three boxes with the largest quantity. The ripening programme is judged by the majority of boxes in the room.
- Measure the temperature of the minority number of boxes.
- Leave damaged clusters in the box until the end of the ripening cycle.
- A pod/finger is removed from the top row of the box and broken up to examine the consistency of the fruit pulp and to ensure that the fruit is ripened at the required rate.

- The ripener records the temperature found in the test boxes on the ripening record RPCS, ensuring that the same box is recorded in the same column on the ripening record.
- The ripener records any adjustments made on the ripening record on RPCS.
- The ripener records the actual stock at the end of the day on the RPCS.
- If the fruit, for whatever reason, is too warm, the ripener cools it down in steps of 0.5°C. If the program is set to 'air ripening' the steps are one degree per eight hours.

Calibration and maintenance:

- The sensors should be calibrated with a gauged thermometer every second month.
- The calibration must be recorded in a logbook.
- It is necessary for the ripener to keep the rooms tidy.
- It is necessary that a maintenance contract is signed.
- Maintenance should be available within 24 h, in case of a breakdown.
- The ripening company has a written procedure on how to cope with full system failure.

7 Summary

There are various practical ways to improve banana ripening in developing countries:

The process:

1. Investigate what equipment is available for banana ripening in a facility.
2. Identify the largest decrease of quality in the banana supply chain.
3. Write a simple training program for improvements.
4. Reducing the claims will lead to higher sales figures.

The banana ripening room:

1. Information regarding packaging, cartons or crates, weight of the carton boxes with bananas, type of packaging, the cooling capacity of the evaporator, temperature and/or air sensors available is basic to design a banana ripening room.
2. Create overpressure by using a tarpaulin, and build two rows of cartons/crates with adequate space between the two rows. A corridor which is too narrow will not give a good ripening result.
3. Ripen the bananas in six days and prevent large temperature shocks.

8 Future trends

China imports a huge quantity of bananas, mainly from Ecuador and the Philippines, during some periods of the year. China has its own production which is not year round. The Ecuadorian government set the price for a 18-kg box at $6.50 for 2016. In India, bananas are sold for a higher price in the morning and a lower price at the end of the day, by which time, the quality is much reduced. In China, waste could be greatly reduced. The author's observations were that 15–20% waste level is not abnormal in Chinese supermarkets. As mentioned before, there are no university courses on banana ripening. How might we share our knowledge with the Chinese and Indian banana ripening industry? These countries account for almost 75% of the world's population.

9 About the author

I am from the Netherlands and I have been active in the banana supply chain for more than 27 years. In 1990 I started working for Chiquita Banana Company BV in Zwijndrecht, the Netherlands, as 'Quality Inspector for Tropical Fruit'. In 1992, I worked in quality systems in Honduras, where I learned a great deal about pre- and post-harvest bananas.

In 1995 I became the manager of Ripening and Technical Services for the Chiquita International Service Group in Antwerp, Belgium. I was responsible for the banana ripening process and project management. I advised customers on building new banana ripening rooms. When I accepted this position my first question was: 'Why is the banana industry so conservative, why isn't anyone investing in new ripening technologies?'

I was very fortunate to work among a team of experts in developing new ripening systems. I also worked with Dutch and Belgian universities to improve the quality of ripened bananas. In 2001 a Belgian professor I contacted for a project called me and said: 'Frits all I need to learn about banana growing would fill the whole room with books; all I want to know about banana ripening will not even cover my desk'. Inspired by this encounter I founded Popma Fruit Expertise, a consulting company specialising in the banana supply chain (www.popmafruitexpertise.nl). I believe this to be a great opportunity to contribute and share my experience with 'bananeros' around the world.

Part 3

Assessing and improving sustainability

Chapter 13

Life cycle assessment and carbon footprint of banana cultivation

Louis Bockel, Laure-Sophie Schiettecatte and Orane Debrune, Food and Agriculture Organization (FAO), Italy

1 Introduction
2 Assessing the environmental impact of banana production
3 An integrated socio-economic carbon footprint banana value chain tool
4 Future trends and conclusion
5 Where to look for further information
6 Abbreviations
7 References

1 Introduction

1.1 Increasing awareness of environmental issues in agriculture

The 1980s and 1990s witnessed increasing public awareness of environmental issues. This development can be seen in the 1992 United Nations Conference on Environment and Development in Rio de Janeiro, when governments recognized the importance of good stewardship of natural resources in achieving sustainable development. As consumers became more aware of environmental issues relating to the food they eat, intensive methods of agricultural production attracted growing attention (FAO, 2009a; FAO, 2011). Increasing environmental awareness among consumers in developed countries created a demand for food suppliers and retailers to measure the environmental impact of the production and distribution of their products. This has led to the introduction of environmental labelling of products; for example in France different groups of retailers have developed their own approaches (Lescot, 2012). In particular, consumer concern about reducing greenhouse gas (GHG) emissions has led to the introduction of several standards designed to measure a product's carbon footprint (CFP). These describe the quantity of GHG generated by an item or activity, generally expressed in CO_2 equivalent (CO_2-e), and are designed to inform consumers, when making choices, about what they purchase (Craig et al., 2012). Governments have now built on these labelling initiatives by developing a common methodology based on life cycle assessment (LCA) (Lescot, 2012).

Over the past ten years, there has been a general shift in food LCA from final products to developing preventative measures earlier in the supply chain, and from only considering particular stages in production (e.g. packaging) to studying entire systems (Svane and

http://dx.doi.org/10.19103/AS.2017.0020.23

Aronsson, 2013). Improving the efficiency of food production and consumption, including reducing food waste which represents a third of overall food production (FAO, 2011), is at the heart of debates about sustainable development (Notarnicola et al., 2015). Strategic objectives for improving food safety, reducing rural poverty, and developing sustainable and resilient agro-ecosystems are becoming increasingly important in the debate about sustainability. These priorities can be seen, for example, in countries involved in the European Joint Programming Initiatives prioritizing:

1 sustainable intensification of European agriculture,
2 maintaining agriculture within limits for GHGs, energy, biodiversity and contaminants and
3 increased resilience to climate change in agriculture and food systems (Soussana, 2014; Notarnicola et al., 2017).

The Food and Agriculture Organization of the United Nations (FAO) has provided guidance to help achieve the international goal of a sustainable global food value chain through numerous local initiatives (FAO, 2014a).

Even though environmental assessment has existed for decades and numerous methodologies have been developed, their use in the food and agricultural sector is relatively recent (Lescot, 2012). The environmental assessment of agricultural and food products is now a strategic global issue. Value chain analysis is a strategic starting point to understand the socio-economic aspects of food production systems such as markets and their relationships, as well as the environmental challenges in the production chain from agricultural production to consumers (Rota and Sperandini 2010, FAO, 2011). The distributed nature of food value chains presents both a challenge and an opportunity to make meaningful progress in de-carbonizing our global economy by developing relevant and precise methods for environmental assessment such as CFP analysis and LCA. The aim of this chapter is to set out in detail the importance of developing environmental assessment methodologies for the banana value chain, in particular LCA and CFP which are two complementary models used increasingly to assess the environmental impact of production.

1.2 The banana value chain

Bananas are the world's most exported fresh fruit both in volume and in value. This fruit is exported primarily from developing countries to industrialized countries which account for 90% of net global banana imports. In 2013, over 110.7 million tonnes of banana were produced in more than 130 countries around the world corresponding to US$ 9.7 billion (FAO, 2017). This global trade relies mostly on one banana variety, the Cavendish, selected for its high yield, resistance to Fusarium wilt race 1, ability to withstand long-distance transport and consistent quality of appearance (FAO, 2014b, BASIC, 2015). In 2013, the volume of global gross banana exports was estimated to be 20.1 million tonnes, or about 15–20% of world production, with the main exporter countries concentrated in Latin America, the so-called 'dollar zone' (including Ecuador, Colombia, Costa Rica and Guatemala), the Philippines, and African and Caribbean countries (the ACP group) (FAO, 2017, UN Comtrade Database, 2017). The five leading banana-exporting countries (Ecuador, the Philippines, Guatemala, Costa Rica and Colombia) account for almost 77% of global banana exports (Fig. 1).

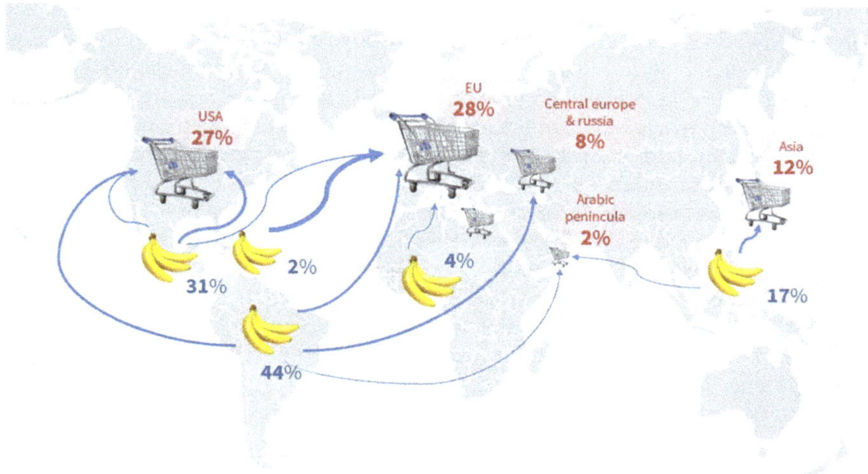

Figure 1 Share of banana world trade. Source: BASIC (2015).

As the banana market has always been heavily globalized, buyers shift happily from one country to another and from one supplier to another to keep consistent quality bananas. In addition, exporters in banana-producing countries distribute their sales and risks as much as possible between clients and consumer countries to maximize their gains and/or reduce their losses (BASIC, 2014). As a result of this highly concentrated production and trade in large transnational companies, the banana sector came under close scrutiny in the 1990s, particularly in relation to the environmental and social impact of its production, since it is still a key crop for millions of households in developing countries in providing food, nutrition and income.

1.3 Climate change and the impact of value chains on greenhouse gases

The agricultural and food sectors now face the challenge of increasing food production by more than 60% by 2050 (FAO, 2013) without intensifying pressure on the environment. The requirement to feed a growing population wanting an improved diet has led to an intensification of agricultural production. At the same time, environmental threats such as climate change, loss of biodiversity, as well as degradation of soil and water resources have spurred public concern about agriculture's environmental footprint, resilience of agricultural production and the effect on agricultural earning, food prices, quality and safety (Gornall et al., 2010, IPCC, 2007, Thornton et Cramer, 2012, Vermeulen et al., 2012, Notarnicola et al., 2017).

Whereas the agricultural sector is the second largest producer of GHGs after the energy sector, representing about 10–12% of global anthropogenic emissions (Smith et al., 2014), the full supply chain, from manufacture of inputs such as fertilizers, agriculture production and processing, transport retail, household consumption to waste disposal, could contribute to19–29% of global anthropogenic GHG emissions (Vermeulen et al.,

2012). To achieve food security and agricultural development goals, adaptation to climate change and lower emission intensities of food produced will be necessary. Indeed as the FAO has said, 'Enhancing food security while contributing to mitigate climate change and preserving the natural resources and vital ecosystems services will require the transition to agricultural production systems that are more productive, use inputs more efficiently, have less variability and greater stability in their outputs and are more resilient to risks, shocks and long-term climate variability' (FAO, 2013).

In the case of the banana value chain, different environmental issues arise between production and consumption, from agrochemical-intensive production on large-scale plantations to post-production processes. Improving sustainability, and in particular steps to climate mitigation, includes better agronomic practices (reduced use of fertilizer) and a shift towards more sustainable production (e.g. intercropping, crop rotation and biological control). It also involves upgrading of transport, shipping and ripening technologies. To enable these changes at every stage of the banana chain, those involved in achieving a more sustainable system need detailed environmental analysis of every stage of the banana value chain to enable effective changes to take place.

Some countries, such as the United Kingdom tried, with the participation of both public authorities and private companies, to voluntarily establish a system based around the Carbon Trust and the Publicly Available Specification (PAS 2005). Some other methodologies centred more on the LCA have been developed and used in agriculture and other industries to estimate the environmental impact of production, and identify the resource use and GHG emission processes within the life cycle of the product (FAO, 2010, Medeiros DuBois et al., 2012). LCA can be applied at every stage of a value chain to target sustainability goals and improve environmental and social food standards.

1.4 Life cycle assessment (LCA)

LCA is a quantitative method defined by international standards ISO 14040 and 14044 (ISO 14040:2006) to analyse the environmental impact of product systems (Lescot, 2012; Curhran, 2014) over its entire life cycle from production and transport of inputs (e.g. fertilizer, pesticide and feed) to transportation, processing, packaging and distribution to retailers, that is, from 'cradle to gate'. This analysis can be extended to the use and disposal phase of the product post consumption, that is, 'cradle to grave'. In both cases, LCA has the scope to highlight the environmental impacts of the value chain at each stage (FAO, 2010, Medeiros DuBois et al. 2012, Svanes and Aronsson, 2013).

By rigorously examining each step in the product life cycle, from cradle to grave, LCA takes into account how raw materials are extracted; the resources involved in planning or designing the product; the materials and energy used during manufacture, packaging and distribution; and finally the impacts from using the product as well as the waste and pollution created throughout the process and at the end of life (Deloitte, 2012) (Box 1 and Fig. 2). The methodology behind LCA can, to some extent, be associated with the term 'footprinting'. Often used interchangeably, this concept is a subset of LCA, only taking into account one environmental measurement (e.g. the carbon or water impact of a product), whereas LCA involves analysing a host of complex environmental metrics such as ozone depletion potential and eutrophication to understand the relative trade-offs involved in a specific activity.

Of particular interest in the present chapter is the term 'carbon footprint', which is used to describe the quantity of GHG generated by an item or activity. It is expressed in CO_2 equivalent (amount of carbon dioxide that would have the same impact on climate change). Hence, the CFP of a food/beverage product or service is the total amount of carbon dioxide

Box 1 Definition of LCA by the international standards (ISO 14044)

The international standards defined LCA as follows: 'LCA studies the environmental aspects and potential impacts throughout a product's life (i.e. cradle-to-grave) from raw material acquisitions through production, use and disposal. The general categories of environmental impacts needing consideration, include resource use, human health and ecological consequences' (Curran, 2014).

(CO_2) and other GHGs emitted over the life cycle of that product or service, expressed as kilograms of CO_2 equivalents (kgCO_2-e) per kilogram of product. This includes all GHGs generated in the agricultural phase – including emissions from the production and transport of all inputs such as fertilizers, as well as emissions due to on-farm energy use and non-energy-related emissions (such as methane and nitrous oxide) from soils and livestock. However, it also includes those generated after the agricultural phase from the processing and packaging to delivery. For the food service industry – including restaurants and cafeterias – the CFP generally extends to GHG emissions from on-site meal preparation. CFP is one of the environmental impact indicators that also include eutrophication, toxicity, acidification and abiotic resource consumption, which are included in an LCA (Lescot, 2012). The use of

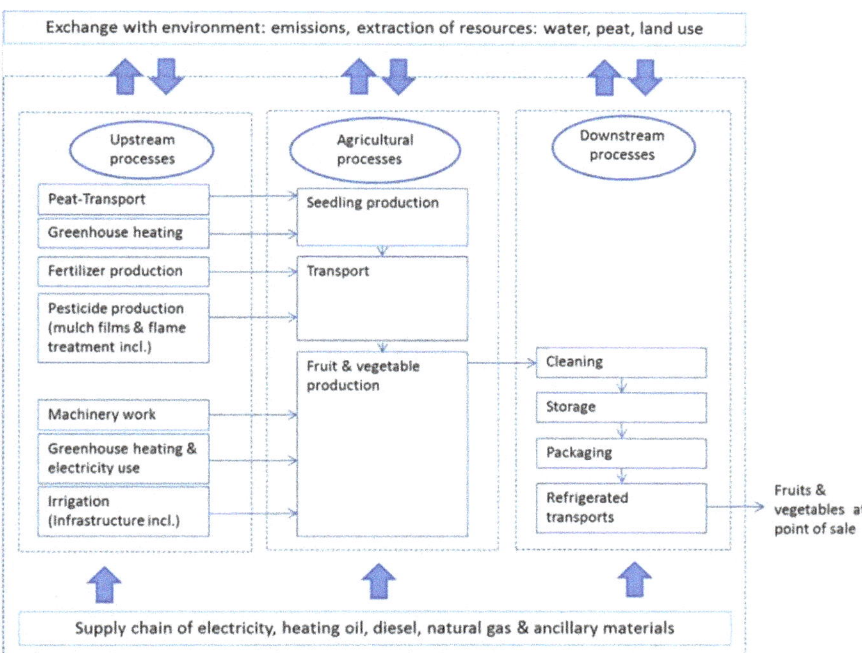

Figure 2 System boundaries for cradle-to-gate fruit and vegetable production. From Stoessel et al. (2012).

CFP in assessing the environmental impact of the food value chain has increased widely over the past decade as it is easily understood by non-experts, facilitating the diffusion of results (Roibás et al., 2016). In the banana industry, a CFP is created across the entire value chain, which can be divided into three broad steps: production, transport and ripening.

By considering all the raw materials, transport, production processes, and use and disposal of the product, the CFP provides an overview of emissions for individual products and services, and allows a comparison of products in terms of eco-friendliness. This enables the development of climate-friendly and/or environmentally friendly products, such as the climatop label (http://www.myclimate.org/) securing a climate-neutral product or service by comparing CO_2 emissions.

Another environmental footprint is the water footprint which is the volume of water used to produce the product (water volumes consumed and or polluted per unit of time) calculated over the various steps of the production chain, for example, production (or growing phase) and processing/packing facilities for the banana. The water footprint for the banana has been reported to range between 200 and 230 L per tonne of banana (Sikirica 2011).

The environmental LCA (CFP, water footprint, energy footprint) can also be complemented with social and socio-economic aspects of a product, from cradle to gate. A socio-economic LCA focus is a social impact (and potential impact) assessment technique that aims to assess the social and socio-economic aspects of the products and their impacts (positive or negative) during the product's life cycle, that is, from the extraction and processing of raw materials to manufacture, use, reuse, maintenance, recycling and final disposal. Social LCA makes use of generic and site-specific data, which can be quantitative, semi-quantitative or qualitative. 'Social and socio-economic aspects assessed in social LCA are those that may directly affect stakeholders positively or negatively during the life cycle of a product. They may be linked to the behaviors of enterprises, to socio-economic processes, or to impacts on social capital. Depending on the scope of the study, indirect impacts on stakeholders may also be considered' (UNEP, 2009).

By considering all the above, LCA is used to help people working in the food sector to make business decisions and improve performance. LCA may help reduce costs by recycling components used in the production process, enhance the environmental and/or societal aspect(s) of the brand for competitive differentiation, and improve design and other decisions that can affect sustainability. LCA can also be used to communicate the positive environmental attributes generated by products and to develop product labelling that communicates a sustainability commitment to retailers and consumers. On a larger scale, LCA applied in policies and regulation can help improve the environmental awareness of decision-makers and develop environment-oriented strategies at local, national and international levels (Deloitte, 2012).

2 Assessing the environmental impact of banana production

2.1 Environmental impact and carbon footprint in the banana sector

After coffee, cocoa and cotton, bananas have received increasing interest due to their social and environmental impacts including loss of biodiversity, their high water consumption, as well as contamination of air, water and soil (Svane and Aronsson, 2013). Use of unsustainable production and trade methods have not only contributed to the depletion of natural resources but are also responsible for negative impacts in terms of income and health, particularly for farmers, workers, their families and local communities. Banana production is the origin of a significant amount of pollution caused by the intensive use of agrochemicals in monoculture production. This intensive use remains an important challenge affecting productivity, the environment and the socio-economic performance. Moreover, banana production may evolve globally with climate change, especially with the increase in temperature and change in the rainfall pattern (FAO, 2016). More favourable conditions may appear in different parts of the world, creating new considerations in terms of competitiveness throughout the global market.

This study does not aim to provide a comprehensive review of the environmental impact issues in the banana sector, but to open a debate about the progress and limitations of this type of study. Indeed, it will be seen that the methodologies used are multiple or are still being developed. Furthermore, they do not factor in local production conditions: the emissions of a particular fertilizer in a tropical environment or transport of producers. Without this geolocation, the results could well be misleading. This raises a fundamental research question about the development of research specific to tropical environments.

The increasing awareness of the role of consumers in reducing GHG emissions from banana production, and the environmental impact of transoceanic transportation and their potential contribution to global warming within the food sector, has led to numerous programmes designed to measure emissions embedded in products, that is, the products' CFP (Craig et al., 2012, Roibás et al., 2016). Table 1 lists the main GHG sources at each step of the banana value chain as well as possible mitigation options.

So far, only a few published CFP studies of banana production exist with varying results and methodologies. Published LCA or CFP results vary considerably from one study to another from 0.249 to 1.37 kg CO_2-e per kg of banana produced (Lescot, 2012; Craig and Blanco, 2009; Svanes and Aronsson, 2013, Iriarte et al., 2014; Roibás et al., 2015). The results usually vary according to the boundary of CFP, that is, from production to packaging or from production to consumption (Mordini et al., 2009). As shown in Fig. 3 the major sources of GHGs are transportation (mainly from transoceanic transportation) and distribution to the destination market, followed by usage of chemicals during cultivation (Ingwersen, 2012, Craig and Blanco, 2009). Existing corporate as well as scientific studies have also evaluated the CFP of the transnational crop along longer segments of the value chain and come to the result that roughly 1 tCO_2-e is emitted per tonne of exported banana, which is consistent with the estimated range mentioned above, that is 1–1.3 kg CO_2-e (Craig and Blanco, 2009).

However, recent initiatives from both international bodies and local communities have promoted the importance of developing a sustainable banana value chain not only to mitigate the negative environmental effect of banana production but also to ensure socio-economic security for the millions of households depending on this trade (World Banana Forum[1], Rainforest Alliance, Dole, BananaLink 2016). Therefore, quantitative and robust methodologies have to be developed and improved to be able to compare different systems of banana production and different studies and to assist the decision-making process.

1. http://www.fao.org/world-banana-forum/en/.

Table 1 Key activities generating GHG emissions along the banana supply chain and non-exhaustive possible mitigation options, based on Svanes et al., 2013, Lescot, 2012 and Lukse, 2010

Value chain step	Major CO_2-e sources	CO_2-e mitigation options
Production	Indirect energy consumption for the production and transport of plant materials, plastic tools, field preparation and transport to the farm	Reduced transport distance Reduced tillage Use of local rough materials (straw, wood)
	Energy consumption for the preparation and transport of fertilizers and herbicide	Increased use of natural local biofertilizers Use of green coverage
	Application of nitrogen fertilizers	Soil conservation intercropping, cover cropping, weed management, soil conservation, water conservation, precision agriculture (soil analysis and monitoring) and fertilization application
Harvesting and upstream transport	Fuel	Reduction of fuel consumption, alternative fuels
Packaging	Energy consumption for the manufacture and preparation of cardboard boxes	
Downstream transport and shipping	Fuel consumption for downstream road transport of bananas	Idle reduction; improved aerodynamics; improved freight logistics; automatic tire inflation systems; wide-based (and single-wide) tires; low-viscosity lubricants; weight reduction; hybrid vehicle technologies and alternative fuels; driver training; intermodal transport (road plus rail) Additional measures adopted by truck companies include: use of compressed natural gas (CNG) tractors in port haulage; use of liquefied natural gas (LNG) tractors for heavy loads and mountainous routes; evaluation of 'only natural gas' equipment, as the network of CNG and LNG improves; use of re-refined oils and re-tread tires that require less energy and emit less GHGs in the production processes; establishment of programmes to reduce gas, energy and water consumption; environmental self-inspections conducted annually; speed reduction to 100 km/h resulting in a significant increase in miles per gallon
	Refrigerant leakage from maritime containers and energy consumption	Replacement of older containers or improved refrigerated containers (improved designs, temperature monitoring by software); R12 replaced with R134a or R141b replaced with other lower-emission refrigerants; Implementation of improved monitoring and control equipment

Value chain step	Major CO_2-e sources	CO_2-e mitigation options
Ripening	Energy consumption for temperature and illumination, and refrigerant (ethylene)	Alternative natural refrigerant (NH_3 and H_2O); energy-efficient illumination; patented reversed air technology; recovery of heat during ripening for heating warehouse and offices; optimal use of daylight through roof tunnels on the top of the buildings, use of electric cars and scooters for low-emission mobility

2.2 The banana value chain and sustainability

There is a growing demand within the banana sector (from producer to buyer) to be able to link LCA with a sustainability footprint at every stage of the chain (from seed provision to consumer). Sustainability indicators such as CFP and LCA have become key factors for growth and value creation (FAO, 2013), and there is no doubt that sustainability has to be integrated much more closely into the decision-making process (Frank et al., 2014). Increasingly seen as a key concept to ensure more sustainable production and consumption patterns, LCA is still some way off from including and addressing all possible environmental questions at once (Ekvall and Weidema, 2004; Mordini et al., 2009). In theory, with unlimited time, unlimited resources and instantaneous access to all data, this could be possible, but in practice it is not. There are still huge issues in data availability, database inter-comparability, modelling approaches, validity and relevance of some impact assessment methods, as well as normalization and weighting (Notarnicola et al., 2016).

As an example, several methodological issues are not clearly addressed in ISO 14067 or in the LCA standards ISO 14040 and leave a lot of flexibility in methodologies and assumptions as well as a serious lack of transparency. As a result, LCA continues to suffer from variations in practice which can have a real impact on the end result. Moreover, the

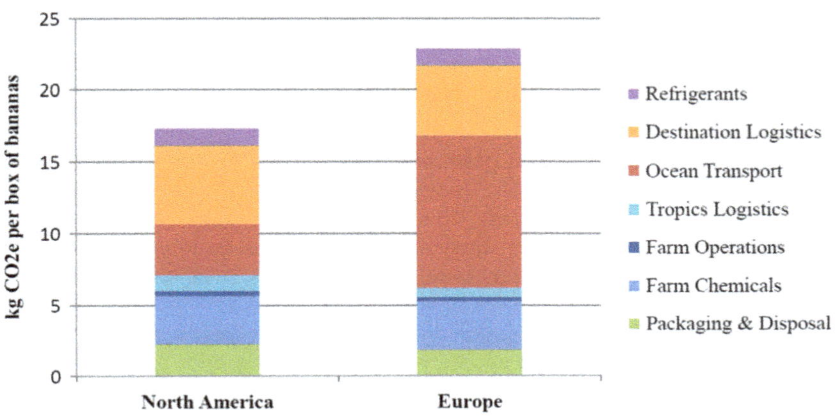

Figure 3 Comparison of banana carbon footprints (in kg CO_2-e) per box of banana. From Craig and Blanco (2009).

quantity of data needed is high and there are no rules concerning the emission factors that have to be used. These factors may lead to inaccurate end results and poor decision-making (Svanes and Aronsson, 2013). The constant development of the methodology can also generate issues over the lifetime of the different studies already completed. For instance, CFP results are generally computed for a given year, but can be greatly affected by database updates, especially for emission factors in tropical environments, or the arrival of new impact measurement methodologies.

At another level, system borders of the analysis are also often unclear or defined differently (Mordini et al., 2009). Indeed, LCA boundaries are difficult to define properly and can produce very different results, depending on the life cycle stages included and inclusion or exclusion of the unit processes. An important issue to consider is whether or not production of infrastructure or capital goods should be included in the analysis (Svanes and Aronsson, 2013), which might affect the accuracy and comparability of results.

Even though LCA seems to be a meaningful tool for assessing global warming potential, there is a clear need for an internationally agreed methodology, in order to have clear and strong indicators that can help develop a sustainable banana value chain. As already discussed above, this can have an important impact on the results, the analysis and the decision-making process. Some limitations, such as modelling missing impact indicators and the availability of the life cycle inventory, should be addressed through continued research and development.

If methods are improved, LCA may then become a robust and accurate tool for decision-making both in business and in policy development. Indeed, progress towards environmentally sustainable systems requires refining the methods for quantitative, integrated assessment and promoting the use of these methods in different areas such as improving current production and consumption. LCA, and so CFP, can then be used to compare different production methods (conventional production of banana versus agroforestry plantation for instance) to assess which could be most effective locally and with the lowest environmental impact. Since incentives for farmers to shift from conventional systems to more environmentally friendly systems are related to the global economic policy framework which remains principally focused on monoculture (Bellamy, 2013), these kinds of appraisals help decision-makers develop more environmentally friendly strategies and help farmers increase the sustainability and economic efficiency of production.

2.3 Banana labelling and certification

There are a number of certification programmes that apply to fruit and vegetables produced for export, including bananas (Box 2). The bulk of bananas that enter international trade are certified to one or more standards. This chapter covers those environmental and/or social standards which are the most significant to the banana trade in terms of quantities certified, namely organic agriculture, fair-trade, Rainforest Alliance, SA8000 and ISO 14001.

As the volume and value of certification and labelling sales grow across the world, there is an increasing demand to measure and demonstrate the difference that engagement has been made into participating producers, workers and other stakeholders in developing countries. This in turn increases the need to develop common and standardized methodologies to assess the environmental impact of food production from cradle to grave. This demand comes from a variety of stakeholders including consumers, media, political authorities, donors and supply chain actors (traders and retailers), who have a

Box 2 Basic principles of certification

Certification is a written guarantee by an independent certification agency that a production process or a product meets the criteria or requirements contained in a certain standard. The basic elements of a certification system (also called certification 'program' or 'scheme') are the standard and the system to control compliance of the certified entity with the standard. Environmental and social standards are generally aimed at the production process (and sometimes also the trading process, as in fair-trade standards). Environmental standards focus on issues such as soil conservation, water protection, pesticide use or waste management, whereas social standards focus on for example worker rights, occupational health and safety. Some other standards like food safety may be part of other standards. The improvements can result in protection of local resources, healthier workers, and other benefits for producers, consumers and local communities. The certification is defined as voluntary when producers are free to decide whether or not they want to certify their production methods (FAO, 2009).

Box 3 Stop climate change seals

In Germany, the stop climate change seals shows products that are carbon neutral. It implies that efforts have been made both to reduce GHG emissions across the value chain and to offset measures taken to reach carbon neutrality; that is, emission rights have been purchased. The company Biotropic GmbH was the first one to be awarded the label for its organic bananas from Ecuador and the Dominican Republic. As they found that they were working with a partner who used a lot of energy for ripening (about three times more than that with modern techniques), they decided to construct their own ripening facility. They therefore offset the remaining emissions by buying credits from a power plant in Russia using wood.

Source: Stop Climate Change: http://www.stop-climate-change.de/en/Siegel.html

legitimate interest in learning whether the labelling system is meeting its aims and objectives as well as improving the situation of smallholder producers and plantation workers (Smith et al., 2010).

CFP labelling helps consumers to choose more environmentally friendly products by comparing products or by labelling superior products as in the German example (Box 3).

Numerous banana producers have expended considerable effort improving plantation management, packing plants, transport and storage to reduce the environmental impact of production (Svanes and Aronsson, 2013). Strong pressure from NGOs, negative media coverage and a shift in consumer preference towards 'eco-friendly' products led some companies to take measures to reduce the adverse impacts of banana cultivation on the environment.

As an example, the year 2016 has seen a spate of announcements by supermarkets expressing their renewed commitment to sustainability. The flagship product chosen by many to symbolize this commitment was bananas and to a lesser extent pineapples. Several chains (e.g. Lidl and Aldi) are promising 100% fair-trade-certified bananas (BananaLink, 2016). Some evidence for this is the growing quantity of bananas sold under certification such as 'organic[2]' or 'fair trade[3]'. The annual quantity of exported fair trade bananas, 468 200 metric tonnes in 2014 (out of a total of 110 million tonnes), is growing by 40–50% per year (45.5% in 2013) representing US$ 183 million of exports in 2014. However the global expansion of organic banana areas which increased by 19.5% between 2005 and 2009 (from 30 183 ha in 2006 to 61 593 ha in 2009) did slow down more recently to 6.6% per year between 2009 and 2013. It gives a global trend of 11.8% annual growth between 2004 and 2014 (FAO, 2014b; FiBL-IFOAM[4]).

However, this rapid expansion has invited a growing suspicion that much of its success can be attributed to the laxity of the standards themselves and the undemanding nature of the certification process (BananaLink, 2016). This raises the importance of the methodological approach in such a process. Green labelling and CFP are frequently used to differentiate value chains. This is often an initiative by an individual company (retailers or industry) or launched by governments with the voluntary collaboration of the private sector. If accurate, labelling could help consumers choose more environmentally friendly products according to the CFP results. At another level of the value chain, CFP is also a support to management and decision-making in the value chain, giving information and orientation on better management practices, cost saving and performance, and can strengthen relationships with suppliers. It can help enhance brand image and reputation by greening their image and products, ensuring certain credibility to their environmental declarations. It can ensure product differentiation and access to a niche segment. However, no harmonized methodology exists for CFP computation while there is a multiplication of methodologies according to the user and type of tool developed. It is therefore much more difficult to compare the footprint of the same category of product and more specifically across countries.

3 An integrated socio-economic carbon footprint banana value chain tool

Banana environmental assessment cannot be addressed solely with LCA or CFP since banana production and trade is highly dependent on human resources. It also requires a social and economic approach from producers to retailers to provide a global interpretation of upgrading action for environmental mitigation and adaptation. Use of impact categories that most decision-makers can easily understand and communicate is also required. Ideally, all impact categories should have the simplicity and accuracy of the GHG impact category (Notarnicola et al., 2016). An integrated approach to value chain development encompassing agro-processing, storage and market development can be effective in terms of market self-regulation and stability (IFAD, 2009). An understanding of the existing value chains is therefore useful to identify new models that would be more efficient, profitable and sustainable, while at the same time promoting producer and consumer welfare (Ouma and Jagwe, 2010).

2. Briefly, organic refers to a production where soil fertility is maintained and enhanced by the use of organic fertilizers (e.g. coffee husks, animal dung, guano and fish bones), weed control is manual and/or with machetes, and pest problems are treated using a variety of non-chemical methods.
3. Fair trade standards are designed to support the sustainable development of small producers' organizations and agricultural workers in the poorest countries. It includes common principles such as social, economic and environmental (organic production) developments and prohibits forced and child labour.
4. http://www.organic-world.net/statistics/statistics-data-sources.html.

3.1 FAO EX-ACT banana value chain tool

The World Banana Forum (WBF) is an international multi-stakeholder forum on sustainable production and trade implemented in 2009 by FAO to meet the multiple and complex international challenges of the banana sector and to promote the involvement of all stakeholders within the chain in adopting best practices for sustainable banana production and trade worldwide. The main focus is on promoting sustainable management of natural resources, mitigation of climate change, labour conditions and the fair distribution of value along the value chain. To simultaneously tackle those different issues, FAO and WBF decided to develop an easy-to-use tool to measure carbon and socio-economic performance that could be specifically adapted to the banana sector, the EX-Ante Carbon-Balance Tool for Banana Value Chain (EX-ACT BVC[5]).

The EX-ACT BVC was developed by the FAO EX-ACT team in 2016 to help current and future stakeholders in the banana sector appraise the socio-economic and environmental performance of the banana chain at different levels using quantitative and qualitative indicators (FAO-ESA, 2016). In particular, the EX-ACT B-VC tool allows

1. farm level and value chain appraisals,
2. encouragement of value chain improvements,
3. decision making on value chain upgrading options and
4. promotion of carbon labelling for some value chains (FAO, 2016).

The EX-ACT BVC aims to help design effective and sustainable banana value chains. It is a freely accessible tool downloadable at http://www.fao.org/tc/exact/ex-act-tool-for-value-chains/en/. The methodology provides both a quantified socio-economic appraisal of the value chain at micro and meso levels (by agent, group or the whole chain) and an environmental carbon balance appraisal of the value chain impact, in terms of climate mitigation, adaptation and value chain resilience (Bockel et al., 2016). Thus,

- The impact on climate mitigation is reflected through quantitative indicators, derived directly from the EX-ACT tool. These indicators are used to obtain and analyse the mitigation impacts in terms of tCO_2-e of the current value chain and in an upgrading project scenario. The CFP of the product is calculated both for the whole value chain and at different required stages, aiming to analyse the environmental performance of the chain. The equivalent economic return is also determined as it could be an important aspect to be considered when attempting, for example, to access payments for environmental services.
- The value chain resilience is assessed using simple quantitative and qualitative indicators. Adaptation indicators measure the reduction in the vulnerability of people, livelihoods, ecosystems and markets to climate change.
- The socio-economic impact of the value chain is assessed in terms of value added, income and job generation using a socio-economic appraisal.

Such a multiple co-benefit appraisal value chain tool is suited to perform ex-ante scenario analysis on the environmental and economic effects of any fluctuation in demand for goods and services. EX-ACT BVC requires current situation data from the banana value chain (production, agricultural inputs, economic information) and assumptions about

5. http://www.fao.org/fileadmin/templates/ex_act/images/EX-ACT_VC/Guideline_EX-ACT_B-VC__10082016.pdf.

socio-economic factors related to changes in product consumption or production. The business mantra 'you cannot manage what you cannot measure' applies here as much to carbon emissions, income generated and other impact performances as to resources and costs. In other words, EX-ACT BVC allows the orientation of public/private decisions and comparison/simulation of future development scenarios at the scale of a value chain or a segment of it (Bockel et al., 2011).

3.2 Case studies

Banana LCA-CFP analysis through a value chain has been implemented within EX-ACT BVC. The first analysis is based on a Peruvian banana value chain from a micro investment of the Agrobanco National Bank. The second case study targets the whole Ecuadorian banana value chain with a scenario simulation of progressive expansion of the organic banana over a period of ten years. Data here are mostly issued from a FAO case study report from 2016 'Ecuador Banana Sector under climate change' (FAO, 2016).

3.2.1 Micro investment in a Peruvian banana value chain

The Peruvian banana sector is changing dramatically both in terms of value and type of production because Western consumption is becoming increasingly oriented towards more organic and Fairtrade products. This case study appraises an investment of US$ 17 000 in a group of four household farmers aiming to increase organic banana production through:

1. an additional plantation of organic banana on 4 ha of unused fallow land (defined as 'other land' in the EX-ACT BVC terminology) and
2. improvement of the current 4 ha of conventionally produced banana into organically produced banana, that is, improved in terms of agronomic practices and manure application with reduced fertilizer and pesticide application.

The investment only targeted farm production. The packaging and transport profiles were assumed unchanged between the current and upgraded scenario.

On an annual basis, the conventional banana production had a yield of 28 tonne per hectare using 250 kg of urea, 100 kg of Haifa MAP and 400 kg of NPK (13-0-46%) and for the upgraded organic production the project produced a yield of 30 tonne of banana per hectare, with changes in agricultural inputs, that is, 1000 kg of lime, 200 kg of urea and 2000 kg of compost.

The mitigation potential of the investment, that is, an increase in banana production area (8 ha) and a shift from conventional to organic banana production, could lead to a GHG reduction of about 50 tCO_2-e per year as a result of better management practices and improvement of degraded land (Fig. 4). This is equivalent to a mitigation impact of US$ 62 per ha per year with an assumed price of US$ 10/tCO_2-e (in between market price of CO_2 from US$ 1 to US$ 137, World Bank 2016[6]) as a consequence of adopting organic agricultural practices.

At the production level, the investment helps decrease the CFP to −0.204 tCO_2-e per tonne of banana produced compared to 0.123 tCO_2-e per tonne of product with

6. https://openknowledge.worldbank.org/bitstream/handle/10986/24288/CarbonPricingWatch2016.pdf?sequence=4&isAllowed=y.

Price tCO_2 10 US$

Climate Mitigation dimension of the value chain (s)	Current situation	Value chain upgrading project	Mitigation impact
1 GHG impact in tCO_2-e per year	14	-36	-50
2 GHG impact in tCO_2-e per year per hectare	2	-4.50	-6
3 Carbon footprint per tonne of production, in tCO_2-e per tonne of product	0.1	-0.20	0
4 Incremental in tCO_2-e [emitted (+) / reduced or avoided (-)]		-5	
5 Equivalent project cost per tonne of CO_2 reduced or avoided, in US$ per tCO_2-e		346	
6 Equivalent value of mitigation impact per year , in US$ /tCO_2-e		50	
7 Equivalent value of mitigation impact per year per ha, in US$ /tCO_2-e per year per hectare		62	

Carbon footprint at the different levels of the Value Chain	Emissions (tCO2/t product)		Mitigation
	Current VC	Improved VC	impact
PRODUCTION	0.123	-0.204	-0.328
PROCESSING	0.107	0.107	0.000
TRANSPORT	0.852	0.852	0.000
PRODUCT LOSS	0.032	0.007	-0.025
RETAIL	0.000	0.000	0.000
TOTAL	1.115	0.762	-0.352

Figure 4 Climate mitigation dimension of the value chain and carbon footprint at each step of the value chain, for the current scenario (conventional banana on 4 ha) and the upgraded scenario (organic banana on 8 ha).

traditional methods, due to soil carbon improvement after adopting organic agricultural practices and converting degraded land.

A detailed CFP gives a better understanding of the main source of emissions within the value chain. The CFP of the whole organic banana value chain, integrating GHG emissions/sources from and beyond farm production, is about 0.762 tCO_2-e per tonne of banana against 1.115 tCO_2-e per tonne of banana in the conventional banana. Processing and transport (transoceanic shipping), which remain unchanged in the current and upgraded scenario, are the main GHG sources in the analysed value chain, contributing 0.852 and 0.107 tCO_2-e respectively per tonne of banana. The mitigation potential of the investment project is therefore −0.352 tCO_2-e per tonne of banana through good agricultural practices and an increased area managed under climate-resilient practices and increased soil carbon[7].

Considering the whole chain, EX-ACT BVC values are in agreement with those mentioned earlier in this chapter, in particular from Craig and Blanco, 2009, that is, 1 tCO_2-e per tonne of banana exported to the United States.

The present investment, which focuses on the production level of the banana value chain, brings to the fore the strong climate mitigation impact of measures such as adopting organic practices (less agricultural inputs such as synthetic fertilizers) and the conversion of degraded land into banana production. The conversion of 4 ha of degraded land to agricultural production and the shift to organic banana results in a decrease of 0.328 tCO_2-e per tonne of banana of the CFP.

EX-ACT B-VC also gives access to an economic value of the mitigation dimension due to the carbon market. In the present case, improving banana production practices corresponded to an increase of US$ 62 per tCO_2-e per year per ha (Fig. 4), which could eventually be used towards payment for environmental services.

7. Increase in soil carbon originates from the land use change from degraded land converted into organic plantations of banana trees and from changes in agronomic practices (improved agronomic practices and manure application), which boost annual plant systems, residues and soil carbon sequestration, see computations and methodology in Bernoux et al., 2014.

Nb of operator eq	3	7	4	jobs
Gross production Value (GPV)	33	78	44	000 US$
Value Added (VA)	27	72	45	000 US$
Gross Income (GI)	24	65	41	000 US$
VA / tonnes of product	266	316	50	US$
VA / ha	6847	9013	2166	US$
Gross income / farmer	5976	16212	10236	US$

Figure 5 Socio-economic performances (in US$) of the value chain at the production level for the current scenario (conventional banana) and the upgraded scenario (organic banana).

A more detailed analysis is also available concerning economic performance. When focusing on the socio-economic analysis of the production level (Fig. 5), the impact of the investment project resulted in a benefit of US$ 44 000 from conventional to organic banana production. Benefits are also translated to the value added (defined as the gross production – inputs) which more than doubled, from US$ 27 000 to 72 000, owing to a reduction in the cost of agricultural inputs in the organic banana production. This is also reflected in the value added per tonne of product and the value added per hectare. The gross annual income per farmer could increase by about US$ 5976–16 212 from the current to the upgraded scenario. The increase in production area and the associated increase in production also resulted in four additional operator-equivalent employments.

3.2.2 Upgrading a banana value chain in Ecuador

In the case of Ecuador's banana sector, biophysical and economic studies were performed simultaneously to evaluate the scope for increasing productivity and sustainability. A key concern in terms of banana production is the extensive use of pesticides to maximize yield and the social impact on the value chain. Hence this case study was selected to simulate a green scenario of a progressive transformation of the value chain through an increase in the organic banana areas. This qualitative upgrading scenario neither provided for any expansion of the whole banana area nor increased production.

The current banana value chain with 147 692 ha of conventional banana plantations and 17 308 ha of organic banana[8] is anticipated to see an annual growth of 20% in organic banana (based on current trends in Ecuador) reaching 104 630 ha of organic banana in ten years' time. Therefore the value chain should benefit from a progressive price increase linked to a 23% surplus price currently paid for organic banana (US$ 8 per 18 kg versus US$ 6.5 for conventional). The scenario is based on a constant number of 7120 banana farmers.

In terms of climate mitigation impact, the shift from the conventional cultivation of banana (with high fertilizers and pesticide inputs) to organic production allows a decrease in GHG emissions from 9.3 to 6.5 tCO_2-e per ha (data not shown) giving a potential mitigation from the investment project of about 3 tCO_2-e per ha.

A detailed CFP gives better understanding of the main sources of emissions within the value chain (Fig. 6).

In terms of CFP as calculated by the North American Market, such upgrading would allow a 7% reduction in the CFP from 1.155 to 1.076 tCO_2-e per tonne of banana traded.

8. http://www.organic-world.net/statistics-data-sources.html.

Carbon footprint at the different levels of the Value Chain	Emissions (tCO2/t product) Current VC	Improved VC	Mitigation impact
PRODUCTION	0.254	0.182	-0.072
PROCESSING	0.112	0.112	0.000
TRANSPORT	0.762	0.762	0.000
PRODUCT LOSS	0.027	0.020	-0.006
RETAIL	0.000	0.000	0.000
TOTAL	1.155	1.076	-0.078

Figure 6 Carbon footprint at each step of the value chain and mitigation impact in tCO_2-e per tonne of banana for the current value chain (conventional banana) and the upgraded scenario (organic banana).

Values from the present study also echo CFP values from the studies cited above (Lescot, 2012; Craig and Blanco, 2009; Svanes and Aronsson, 2013, Roibas et al., 2015). Along the different steps of the value chain, the main reductions arise from the shift from conventional to organic banana (assuming less agricultural inputs) and to a lesser extent a reduction in lost production, −0.072 and −0.006 tCO_2-e per tonne of banana respectively, consistent with the results from the Peruvian Banana Value Chain (see previous analysis).

With an average area around 23 ha per household, farmers have an average gross income of around US$ 17 000 per year which could increase by 75% over ten years in the upgrading scenario, that is, US$ 30 000 (Fig. 7). The impact of the upgrading scenario is stronger for the smaller farmers with around 10 ha – their annual income increases by an order of magnitude, from US$ 7500 for conventional banana to US$ 13 000 for organic banana (data not shown).

Upgrading the value chain results in better performance in terms of creating wealth and distribution of income at every stage of the banana chain. Indeed, organic installation creates an incremental value added of more than US$ 190 million with a gross income reaching US$ 214 million for the whole banana value chain. As seen in the previous analysis, because there are proportionally fewer costs at the production level, value added increases by US$ 36 per tonne of production, leading to a higher gross income for the different beneficiaries.

These two analyses illustrate the range of practical scenarios, which could be implemented either at micro level or at global value chain level using more sustainable banana farming systems. The results could be considered by every developer (public agency, decision makers, authorities) to target the real needs of the population and demonstrate their ability to consider every component in the banana sector to improve both environmental and socio-economic performances.

Socio-economic performances of the value chain	Current situation	Value chain upgrading project	Balance	
Production level				
Nb of operator eq	643500	726000	82500	jobs
Gross production Value (GPV)	1320022	1458576	138554	000 US$
Value Added (VA)	936504	1126725	190220	000 US$
Gross Income (GI)	123054	214275	91220	000 US$
VA / tonnes of product	172	209	36	US$
VA / ha	5676	6829	1153	US$
Gross income / farmer	17283	30095	12812	US$

Figure 7 Socio-economic performances of the value chain at the production level, considering 23 ha per household.

4 Future trends and conclusion

4.1 Systematic carbon footprint appraisal of the banana value chain

From the analysis of LCA and CFP methodologies applied to banana production, it is clear that developing accurate and harmonized methods for every participant remains an important but difficult challenge. Increasingly seen as key for ensuring the transition towards more sustainable production and consumption patterns, LCA and CFP still do not take into consideration every aspect of sustainability that should be included to access global performance analysis. A new trend is emerging in international discussions, highlighting the importance of considering the value chain in a more systemic way. Developing a sustainable value chain means targeting the three pillars of sustainability: economic, social and environment. The new agenda of sustainable development launched in 2015 does not simply want to consider poverty, hunger and inequality but also takes into account the impact of climate change, to mitigate and adapt to its effect. However, none of these components is obtained separately but in a way that they can simultaneously benefit the whole population. Value chain performance analysis should not just be addressed for environmental impact but also according to every stage that has an influence on economic and social performances. In the light of this, it seems that environmental assessment will probably move towards an integration of the different impacts from both an environmental and socio-economic point of view. Consequently, numerous projects are being led throughout the world by different organizations and institutions to integrate these three components within one project or programme. The WBF considers not only GHG emissions generated from intensive agrochemical production, but also the health and productivity of workers. A systemic approach can focus on all these problems and highlight the main issues that can be solved. Integrating every component of sustainability in the analysis could help investors who want to support value chain development towards the achievable aim of meeting the global demand for more sustainable production.

4.2 Strategic research and development orientations

Analysing value chain performances allows access to quantitative and/or qualitative indicators that could be used to improve sustainability in the different sectors. Value chain analysis can be used for assessing either ex-ante or the current impact of a specific value chain. A combination of tools such as LCA with cost–benefit analysis could for instance assess whether strategic objectives for developing sustainable value chain are reached, monitor the different progresses achieved and predict the different impacts that could be generated. Being able to use such tools could make it easier to take decisions concerning the strategic orientation of the different operators of the chain, but more importantly it promotes what can be done within Research and Development. Finally, every indicator, measure and analysis gives the opportunity to highlight weak points in the value chain and helps research the best solution to develop better strategies, and to direct investors and public policy to finance development.

5 Where to look for further information

The WBF, created in 2009 by the FAO's Trade and Markets Division, is a permanent platform that promotes collaboration and open dialogue on challenges faced by the banana industry by bringing together key stakeholders, including representatives of the global supply chain. Specifically, the WBF and its specialized working groups and partners work together to promote: the sustainable management of natural resources, mitigation of climate change and uncontrolled environmental impact, respect for human rights (labour rights, health and safety at work), promotion of gender equity, fair distribution of value along the supply chain, and knowledge sharing as a tool for development (sharing information on best practices). The EX-ACT BVC, developed on request by the WBF, is used in the above context to assess mitigation impact and socio-economic performance of various types of banana value chains.

More information on the WBF can be found here: http://www.fao.org/world-banana-forum/en/

For EX-ACT BVC: http://www.fao.org/tc/exact/ex-act-tool-for-value-chains/en/

6 Abbreviations

The following abbreviations are used in the text:

AFOLU: Agriculture, forestry and other land use
CFP: Carbon footprint
EX-ACT B-VC: EX-Ante-Carbon-balance Tool for Banana Value Chain
FAO: Food and Agriculture Organization of the United Nations
GHG: Greenhouse gases
ISO: International Organization of Standardization
LCA: Life cycle assessment
PAS: Publicly available specification
tCO_2-e: Tonne of CO_2-equivalent

7 References

BananaLink. (2016). *Rainforest Alliance and the Discount Supermarkets : Low Prices and Easy Standards?* (September).

BASIC. (2014). *Analysis of German Banana Value Chains and Impacts on Small Farmers & Workers.* Bureau d'Analyse Sociétale pour une Information Citoyenne, June, pp. 1–41.

BASIC. (2015). *Banana Value Chains in Europe and the Consequences of Unfair Trading Practices.* Bureau d'Analyse Sociétale pour une Information Citoyenne, October, pp. 1–51.

Bernoux, M., Bockel, L., Branca, G., Colomb, V., Gentien, A. and Tinlot, M. (2014). EX-Ante Carbon-balance Tool (EX-ACT) technical guidelines for Version 4.. *EASYPol module 101*, FAO, Rome.

Bellamy, A. S. (2013). Banana production systems: Identification of alternative systems for more sustainable production. *Ambio*, 42(3), 334–43.

Bockel, L., Touchemoulin, O., Jonsson, M. and Sanz Cortes, R. (2011). Carbon footprinting across the food value chain: A new profitable low carbon initiative? FAO EASYPOL resources for policy making.

Bockel, L., Debrune, O., Toudert, A. and Schiettecatte, L.-S. (2016). Practical guidance: EX-ACT Value Chain Tool (EX-ACT VC) Co- benefit appraisal of Value Chain on GHG emissions, climate resilience and income, http://www.fao.org/fileadmin/templates/ex_act/images/EX-ACT_VC/Working_document_EX-ACT_VC_29082016_english.pdf.

Craig, A. and Blanco, E. (2009) The banana carbon footprint case study, MIT Center for Transportation & Logistic, September 2009, http://CTL.MIT.EDU/RESEARCH.

Craig, A. J., Blanco, E. E. and Sheffi, Y. (2012). ESD Working Paper Series A Supply Chain View of Product Carbon Footprints: Results from the Banana Supply Chain, Massachusetts Institute of Technology. Engineering Systems Division. http://hdl.handle.net/1721.1/102939.

Curran, M. A. (2014). *Strengths and Limitations of Life Cycle Assessment* (pp. 189–206). Springer, Netherlands.

Deloitte (2012). *Enhancing the Value of Life Cycle Assessment*. Deloitte Consulting LLP, 8p. http://www.deloitte.com.

Ekvall, T. and Weidema, B. P. (2004). System boundaries and input data in consequential life cycle inventory analysis. *The International Journal of Life Cycle Assessment*, 9(3), 161–71.

FAO. (2009a). *Certification in the Value Chain for Fresh Fruits: The Example of Banana Industry*. Trade Policy Service, Trade and markets Division. Food and Agriculture Organization of the United Nations, Rome.

FAO. (2009b). *The State of Food and Agriculture*. FAO, Rome, 180p.

FAO. (2010). *Greenhouse Gas Emissions from the Dairy Sector: A Life Cycle Assessment*. Animal Production and Health Division, Rome, 98p.

FAO. (2011). *Global Food Losses and Waste. Extent, Causes and Prevention* (available at http://www.fao.org/docrep/014/mb060e/mb060e00.pdf).

FAO. (2013). *Climate Smart Agriculture, Sourcebook*. FAO, Rome, 570p.

FAO. (2014a). *Developing Sustainable Food Value Chains. Guiding Principles*. FAO, Rome, Italy.

FAO. (2014b). *Banana Market Review and Banana Statistics 2012–2013*. Market and Policy Analysis of raw Materials, Horticulture and Tropical (RAMHOT) Products team, Rome, Italy, 39p.

FAO. (2016). *Ecuador's Banana Sector Under Climate Change: An Economic and Biophysical Assessment to Promote a Sustainable and Climate-compatible Strategy*, by A. Elbehri, G. Calberto, C. Staver, A. Hospido, L. Roibas, D. Skully, P. Siles, J. Arguello, I. Sotomayor and A. Bustamante. FAO, Rome, Italy.

FAO. (2017). FAOStat. Retrieve from http://www.fao.org/faostat/en/?#data/QC

FAO-ESA (2016). *Practical Guidance for Using EX-ACT B-VC tool for Banana Value Chain (EX-ACT B-VC)*, http://www.fao.org/fileadmin/templates/ex_act/images/EX-ACT_VC/Guideline_EX-ACT_B-VC__10082016.pdf

Frank, M., Saling, P., Gipmans, M. and Schöneboom, J. (2014). Life Cycle Assessment towards a Sustainable Food Supply – A review of BASF's Strategy. In: *9th International Conference LCA of Food.*

Gornall, J., Betts, R., Burke, E., Clark, R. Camp, J., Willett, K. and Wiltshire, R. (2010). Implications of climate change for agricultural productivity in the early twenty-first century. *Philosophical Transactions of the Royal Society B*, 365(1554), 2973–89.

IFAD. (2009). *Smallholder Agriculture and Food Security in the 21st Century*, (February).

IPCC. (2007). *Climate Change 2007: Impacts, Adaptation and Vulnerability*. Cambridge University Press, Cambridge.

Ingwersen, W. W. (2012). Life cycle assessment of fresh pineapple from Costa Rica. *Journal of Cleaner Production*, 35, 152–63.

Iriarte, A., Almeida, M. G. and Villalobos, P., (2014). Carbon footprint of premium quality export bananas: Case study in Ecuador, the world's largest exporter. *Science of the Total Environment*, 472, 1082–8.

ISO, 14040:2006. *ISO/TS 14040:2006. Environmental Management – Life Cycle Assessment – Principles and Framework*. International Standards Organisation, Geneva.

Lescot, T. (2012). Carbon footprint analysis in banana. In: *Second Conference of the World Banana Forum*, (June), pp. 1–28.

Luske, B. (2010). Comprehensive carbon footprint assessment. Dole bananas. *Soil and More International*, 1–45.

Medeiros DuBois, K., Chen, Z., Kanamaru, H. and Seeberg-Elverfeldt C. (2012). *Incorporating Climate Change Considerations into Agricultural Investments Programmes, A Guidance Documents.* FAO, Rome, 148p.

Mordini, M., Nemecek, D. T. and Gaillard, D. G. (2009). *Carbon & Water Footprint of Oranges and Strawberries A Literature Review.* Swiss Confederation, federal department of Economic affairs (FDEA), 76p. November 2009.

Notarnicola, B., Salomone, L., Renzulli, P., Roma, R. and Cerutti, A. K. (2015). *Life Cycle Assessment in the Agri-food Sector: Case Studies, Methodological Issues and Best Practices.* Springer International Publishing– Technology & Engineering, 390p.

Notarnicola, B., Sala, S., Anton, A., McLaren, S. J., Saouter, E. and Sonesson, U. (2016 July). The role of life cycle assessment in supporting sustainable agri-food systems: A review of the challenges. *Journal of Cleaner Production*, 140, 399–409.

Notarnicola, B., Sala, S., Anton, A. O., Mclaren, S. J., Saouter, E. and Sonesson, U. (2017). The role of life cycle assessment in supporting sustainable agri-food systems: A review of the challenges. *Journal of Cleaner Production*, 140(Pt 2), 399–409.

Ouma, E. and Jagwe, J. (2010). Banana value chains in Central Africa : Constraints and opportunities. *Paper Presented at the Joint 3rd African Association of Agricultural Economists and 48th Agricultural Economists Association of South Africa Conference*, Cape Town, South Africa, 19–23 September 2010.

Roibás, L., Elbehri, A. and Hospido, A. (2015). Evaluating the sustainability of Ecuadorian bananas: Carbon footprint, water usage and wealth distribution along the supply chain. *Sustainable Production and Consumption*, 2, 3–16.

Roibás, L., Elbehri, A. and Hospido, A. (2016). Carbon footprint along the Ecuadorian banana supply chain: Methodological improvements and calculation tool. *Journal of Cleaner Production,* 112, 2441–51.

Rota, A. and Sperandini, S. (2010). *Value Chains, Linking Producers to the Markets.* IFAD, Rome, Italy. www.ifad.org/lrkm/index.htm.

Sikirica, N. (2011). *Water Footprint Assessment, Bananas and Pineapples.* Soil and More International B.V. Dole Food Company, The Netherlands, 41p.

Smith, S., Brassel, F., Cabascango, D., Cepeda, D., Montenegro, F., Ojeda, A. and Rodríguez, E. (2010). Fairtrade bananas : A global assessment of impact final report : April 2010. *Small*, (April).

Smith, P., Bustamante, M., Ahammad, H., Clark, H., Dong, H., Elsiddig, E.A., Haberl, H., Harper, R., House, J., Jafari, M., Masera, O., Mbow, C., Ravindranath, N. H., Rice, C. W., Robledo Abad, C., Romanovskaya, A., Sperling, F. and Tubiello, F., (2014). Agriculture, Forestry and Other Land Use (AFOLU). In: *Climate Change 2014: Mitigation of Climate Change. Contribution of Working Group III to the Fifth Assessment Report of the Intergovernmental Panel on Climate Change* (O. Edenhofer, R. Pichs-Madruga, Y. Sokona, E. Farahani, S. Kadner, K. Seyboth, A. Adler, I. Baum, S. Brunner, P. Eickemeier, B. Kriemann, J. Savolainen, S. Schlömer, C. von Stechow, T. Zwickel and J.C. Minx (Eds)). Cambridge University Press, Cambridge, United Kingdom and New York, NY, USA.

Soussana J. F., (2014). Research priorities for sustainable agri-food systems and life cycle assessment. *Journal of Cleaner Production*, 73, 19–23.

Stoessel, F., Juraske, R., Pfister, S. and Hellweg S. (2012). Life cycle inventory and carbon and water footprint of fruits and vegetables: Application to a Swiss retailer. *Environmental Science Technology*, 46(6), 3253–62.

Svanes, E. and Aronsson, A. K. S. (2013). Carbon footprint of a Cavendish banana supply chain. *International Journal of Life Cycle Assessment*, 18(8), 1450–63.

Thornton, P. and Cramer, L., (2012). Impacts of climate change on the agricultural and aquatic systems and natural resources within the CGIAR's mandate. CCAFS Working Paper No. 23, CCAFS, Copenhagen.

UN Comtrade Database (2017). Retrieved from https://comtrade.un.org/data/

UNEP. (2009). Guidelines for social life cycle assessment of products. Social and socio-economic LCA guidelines complementing environmental LCA and Life Cycle Costing, contributing to the full assessment of goods and services within the context of sustainable development, 104p.

Vermeulen, S. J., Bruce, M. C. and John, S. I. I. (2012). Climate change and food systems. *Annual Review of Environment and Resources*, 37, 195–222.

Chapter 14

Sustainable banana cultivation: from standards to multiple solutions

Sietze Vellema and Kees Jansen, Wageningen University, The Netherlands

1 Introduction
2 Sustainability and the range of standards in banana production
3 The limits of sustainability standards
4 Sustainability and diversity in banana production systems
5 Case study: the variety of banana producers in the Philippines
6 Conclusion
7 Where to look for further information
8 References

1 Introduction

The future and sustainability of banana cultivation has been closely associated with intensive pesticide use in large-scale, monoculture production systems and its impact on human health and the environment. The intensive use of pesticides in export-oriented production has been central to campaigns by civil society. Diseases are serious constraints for sustainable banana cultivation for export as well as in diverse landscapes dominated by smallholders producing banana and plantain for local diets. Of particular concern is the re-emergence and global spread of the Panama disease, a new and virulent strain of the fungus *Fusarium oxysporum* f. sp. *cubense*, the so-called Tropical Race (TR) 4 – which affects different production systems.

The literature on sustainable banana cultivation has been strongly oriented towards the impacts of the large-scale export industry and the expansion of sustainability standards. However, an exclusive focus on global trade may make less visible the large numbers of smallholders producing many different varieties in backyards or small plots under very diverse agroecological conditions. This chapter examines the efficacy of current sustainability initiatives in addressing the challenge of including a broader range of production systems. It looks at the range of sustainability standards in banana production and examines their limits, particularly in the context of the broader role of bananas in diversified production systems. A case study from the Philippines highlights the need to take account of the diverse social and agroecological contexts and broader landscape scales in achieving more sustainable banana production.

http://dx.doi.org/10.19103/AS.2017.0020.24

2 Sustainability and the range of standards in banana production

The research on sustainability efforts in banana largely concentrates on the spread and impacts of voluntary sustainability standards and certification schemes in export-oriented banana cultivation (e.g. Lockie et al., 2015; Melo and Wolf, 2005; Potts et al., 2014). Standards and certification are used to monitor compliance with prescribed and uniform practices, which enables companies to manage supply chains and minimise reputational risks (e.g. concerning product safety). The implementation and use of standardised cultivation and management practices is an important result of the evolving interactions between leading export companies and other social actors addressing the sustainability challenges in the banana export sector. Campaigns by local and global non-governmental organisations (NGOs) flagged the unintended health effects of pest and disease management for workers, local communities and the environment (Barraza et al., 2013; Brisbois, 2016; Brisbois et al., 2017). Studies analyse how standards restructure value chains in international trade (Lockie et al., 2013; Murray and Raynolds, 2000). This section describes how private or voluntary standards have come to play such a prominent role in thinking about the sustainability challenges in banana cultivation, and reveals how external pressures and the choices made by firms influence how the export sector handles sustainability challenges (Jansen, 2004).

2.1 GLOBALG.A.P. and ISO 14001

Currently, leading banana companies apply a range of standards (Table 1). As the name suggests, GLOBALG.A.P. (Global Good Agricultural Practices) has a worldwide coverage. The Food and Agriculture Organization of the United Nations (FAO) reports that in 2013, most of the global certified banana traded by the major multinational banana companies (Chiquita, Dole, Del Monte, Fyffes) were compliant with its requirements (FAO/WBF, 2017). GLOBALG.A.P. is a retail-led standard concentrating on food safety, environmental protection and traceability (Lockie et al., 2015). It was founded in the late 1990s to harmonise different food safety standards and protocols in the food industry. It also includes health and environmental protection standards and is currently the most widely implemented certification scheme in the banana export sector (FAO/WBF, 2017). It has recently been updated to highlight biosecurity measures to prevent the spread of TR4 (https://www.globalgap.org/uk_en/for-producers/globalg.a.p.-add-on/tr4-biosecurity/).

The intergovernmental International Organization for Standardisation (ISO) negotiates the content and management of standards for environmental management with governments and industry and is responsible for developing and enforcing these standards. In the late 1990s, ISO published a series of environmental management standards such as ISO 14001 (BananaLink, 2017). ISO 14001 sets out the requirements for an environmental management system and certifies a process, not a product or environmental impact. Both GLOBALG.A.P and ISO 14001 are generally not visible on the product label itself.

2.2 Other schemes: Rainforest Alliance and Fairtrade

In addition to GLOBALG.A.P. and ISO 14001, three voluntary sustainability standards and certification schemes dominate the export market: Rainforest Alliance, Fairtrade

Table 1 Main voluntary standards in global banana export

Standard	Year of foundation	Scope	Share in global banana production Major production area
GLOBALG.A.P	1997	Food safety and traceability Focus on production processes and risk management systems	Global coverage
ISO 14001	1996	Environmental management systems Focus on process, not product	Global coverage
SA 8000 Social Accountability International	1997	Social performance including child, forced and compulsory labour, freedom of association and right to collective bargaining	Global coverage
Rainforest Alliance/ Sustainable Agriculture Network (SAN) Standard	1987	Sustainable Agriculture Network (SAN) standard, is built around the guiding principles of effective farm planning and management, protection of biodiversity and natural resources, and improved livelihoods	2% Largest share (90%) from Colombia, Costa Rica, Guatemala, Ecuador, Honduras, Panama
Fairtrade International	1988	Fair prices for producers organised in cooperatives. Living wages for farm workers	0.5% Largest share (94%) from Colombia, the Dominican Republic, Ecuador and Peru
Organic standard International Federation of Organic Agriculture Movements (IFOAM)	1972	Private standards for organic products have to comply with public organic standards set at national or regional level	1% Largest share (69%) from the Dominican Republic, Ecuador and Peru (13% from Philippines)

Sources: DeFries et al. (2017), FAO/WBF (2017) and Potts et al. (2014).

and Organic standards (the latter conforming to standards set by the International Federation of Organic Agriculture Movements, IFOAM) (FAO/WBF, 2017). The spread of these standards is documented in a study by the International Institute for Sustainable Development (IISD) and the International Institute for Environment and Development (IIED) published in 2014, which estimates that in 2012, 14% of global banana exports were compliant with these voluntary sustainability standards (Potts et al., 2014). About three-quarters of standard-compliant bananas were cultivated in five countries: Guatemala, Colombia, Costa Rica, Ecuador and the Dominican Republic (Potts et al., 2014). The FAO estimates that Latin America produces more than 90% of global standard-compliant bananas (FAO/WBF, 2017). The IISD/IIED study (Potts et al., 2014) indicates a steady growth of the share of Rainforest Alliance-certified banana in global trade, while the share in production and trade of Organic and Fairtrade standards is quite stable.

A shared feature of these voluntary sustainability standards is that the certificate is also publicly displayed with a label showing the relevant logo on banana sold in export markets. Such visible sustainability initiatives in the banana export sector have largely resulted from public pressure initiated by international environmental and labour organisations targeting the spraying and intensive use of pesticides in plantations (Jansen, 2004; Bellamy et al., 2016). A sequence of global, national and local advocacy campaigns addressed the occupational health of workers or contract farmers as well as the effects on surrounding communities, on the local environment and on product safety (Barraza et al., 2013; Bohme, 2014). These public pressures directed research efforts by the large and multinational banana companies towards developing more cost-effective and environmentally benign pest and disease management practices (Bonicelli et al., 2015). Another consequence was that banana companies began to align with NGOs promoting the use of voluntary standards as an instrument for achieving sustainability, including efforts to achieve a better income for banana workers and smallholder farmers (Frundt, 2009).

Rainforest Alliance has been closely affiliated with the operations managed by Chiquita; all of Chiquita's own banana plantations and a large part of non-Chiquita farms are Rainforest certified (Potts et al., 2014). Chiquita and Rainforest Alliance agreed to collaborate in the early 1990s and to apply the standards developed by the Sustainable Agriculture Network (Jansen, 2004; Were, 2003). Since then, voluntary sustainability standards have become mainstream in the banana export market and Rainforest Alliance is now also used by other leading companies (FAO/WBF, 2017; Potts et al., 2014). The scheme covers the conservation of natural resources and promotes the well-being of workers and local communities. Banana farms that are Rainforest Alliance Certified undergo annual audits to ensure that they comply with social criteria designed to protect workers, families and nearby communities. Research has focused on how effective these schemes have been in changing levels of pesticide use in banana production (Bellamy, 2013; Barraza et al., 2011, 2013; Comte et al., 2018).

Fairtrade certification began to find its way into the banana sector at the end of the 1980s. Fairtrade involvement in the banana sector has a specific history and literature, which looks into the linkages between 'rich' consumers and poor banana producers (Wilson and Jackson, 2016). It grew out of a broader Fairtrade movement that aimed to change environmentally destructive and socially inequitable methods in both production and trade systems, starting by targeting the trade in bananas to European supermarkets from Ghana, the Dominican Republic and Ecuador (Murray and Raynolds, 2000). Fairtrade also seeks to impose restrictions on the use of pesticides and emphasises occupational health.

Fairtrade is embedded in various types of alliances with, for example, trade unionists, environmentalists and churches (Frundt, 2009, 2010), and strives to empower democratically organised producer groups. Fairtrade has also expanded to larger corporate plantations in other countries (Brown, 2013) and is close to private social standards focusing on labour conditions, in particular SA8000, and international agreements negotiated between leading firms and strong trade unions under the auspices of the International Labour Organization (Riisgaard and Hammer, 2011). Fyffes is the biggest supplier of Fairtrade bananas in the European market, under the Turbana label, and is also a leading importer of organic produce (Potts et al., 2014). Fairtrade aligns with the production of organic bananas by smallholder groups (Trauger and Murphy, 2013) as well as with groups or plantations compliant with the Rainforest Alliance standard (Melo and Wolf, 2005).

Major multinational corporations increasingly combine certification to meet sustainability standards with the more managerial types of certification, such as GLOBALG.A.P. and ISO 14001, and also create space to include Organic and Fairtrade bananas. As a result, there is a whole array of standards and management systems involved in sustainability initiatives affecting banana cultivation for export. Sustainable banana cultivation for export is strongly directed by this entangled nature of various voluntary or private standards and certification systems (Aasprong, 2013). Connections exist with public legislation, which sets baseline standards, for example in the areas of occupational health, environmental management or pesticide use (Jansen, 2008; Jansen and Dubois, 2014). Furthermore, the dominance of voluntary standards may alter due to ongoing restructuring of the banana export sector. The spread and influence of voluntary sustainability standards may change because new owners gain influence or other export markets have different demands compared to the European and US markets. The high level of concentration of the international banana trade and the resulting market power of four vertically integrated multinational companies seem to be in decline since 2000 (FAO, 2014). This relates to corporate decisions to no longer own plantations and shift to sourcing bananas from independent banana producers. An FAO report (2014) suggests that major retailers in the United States and Europe use direct container liner services to increasingly purchase from smaller wholesalers or directly from growers. This indicates further fragmentation of the international banana trade and reduced concentration may affect the spread and application of current sustainability standards.

3 The limits of sustainability standards

Private or voluntary standards have come to play a prominent role in thinking about the sustainability challenges in banana cultivation, focusing on how firms' choices and external pressures influence the manner in which the export sector handles sustainability issues (Jansen, 2004). Research has focused on the need to scrutinise claims about the environmental and social effects of standards (DeFries et al., 2017). The expansion of sustainability also motivated critical analyses of the nature and implication of interactions between the Corporate Social Responsibility aspect of companies and civil society (Utting, 2008). The intentions, governance mechanisms and transformative capacities of alliances between leading firms and NGOs applying standards are central to such analysis (Wicki and Van Der Kaaij, 2007; Rueda et al., 2017).

Much of the current literature focuses on export markets and reveals a prevailing interest in the behaviour and practices of a limited number of leading multinational firms operating in specific geographical areas in Latin America. However, in terms of a broader debate about sustainability, this seems too limited a focus. First, global standards are often at odds with local ecological and social requirements for sustainable cultivation (Bain et al., 2013; Lockie et al., 2013). Secondly, global sustainability standards have unintended consequences for the capacity of banana cultivation to supply a variety of local food markets and sustain diversity of production systems in rural areas (cf. Oosterveer et al., 2014). Sustainable banana cultivation must take into account the large number of smallholders producing many different varieties in backyards or small plots under diverse agroecological conditions. As a result, the emphasis on standards limits discussion of a forward-looking sustainability research agenda. The following section discusses a number of broader issues in improving the sustainability of banana cultivation, particularly for smallholders.

4 Sustainability and diversity in banana production systems

4.1 The range of banana and plantain in food security

A wide range of banana and plantain play an important role in food and nutrition security and their production and marketing generate income for resource-poor farmers. They are considered to be a healthy ingredient in daily food provision (Tadele, 2009, 2014) and are important sources of essential nutrients in the human diet. In East Africa, banana is a major staple crop, which is under pressure from pests and diseases (Karamura et al., 2013; Tushemereirwe et al., 2004). In large parts of South-east Asia, banana has many different uses in providing food for local communities (De Langhe et al., 2015) or local markets (Staver, 2009). In the centre of origin of diversity of banana, Indonesia, the richness, availability and diversity of the fruit are visible when travelling across the country or visiting food markets (Sujarwo and Caneva, 2015). In shops and local markets, the diversity of banana and plantain for sale is consistently large (Teo et al., 2005). Research on mixed agroforestry regimes also signifies the importance of banana in ensuring food and nutritional security (Ofori et al., 2014). The issue of food security and the importance of local and regional markets in maintaining diversity in banana and plantain pose a major challenge to sustainability initiatives in banana.

4.2 Banana and plantain in mixed cropping systems

Many different varieties of banana and plantain are grown in mixed cropping systems. Banana and plantain are cultivated as food crops throughout the humid tropics and subtropics in the Americas, Africa and Asia (Heslop-Harrison and Schwarzacher, 2007). They also combine well with cash crops in agroforestry systems, for example coffee in East Africa (Bongers et al., 2015) and plantain as a shade crop in young cocoa farms in West Africa (Norgrove and Hauser, 2014; Ofori et al., 2014). In addition, the cultivation and regular harvesting of banana contributes to a continuous and reliable flow of income for smallholder farmers (Oduol and Aluma, 1990; van Asten et al., 2011). In Tanzania, banana is part of intercropping strategies adopted under conditions of land scarcity and generates year-round revenues (Mwijage et al., 2016). Banana and plantain also fit production systems and livelihood strategies coping with poor soil and climatic conditions (Calberto et al., 2015). The interwoven and dynamic role of banana in mixed cropping systems and the peculiar architecture of agroecosystems affects the emergence, evolution and dispersal of pests and diseases (McDonald and Stukenbrock, 2016), whereby diversity may suppress or temper the development of diseases. This implies research that moves away from a focus on a single crop and accounts for the interactions between different levels in diverse integrated systems or ecological contexts (Chopin et al., 2016). It also entails moving beyond the level of primary production and recognises cross-scale interactions within specific territories (Bamutaze, 2015). Such a perspective includes an interest in land-use diversification and assesses how cropping and landscape approaches balance the competing requirements of intensification and diversification (Castro et al., 2015). The genetic diversity in diverse production environments involving mixed cropping systems will most likely continue to play a role in improving sustainability in both intensive and low-input cultivation of banana and plantain.

4.3 The use of genetic diversity for sustainability solutions

The widely observed role of banana in food security and mixed cropping systems in variable agroecological conditions in the humid tropics is also reflected in its genetic diversity (Bauer, 2012; Francis, 2014). The conservation of this diversity of cultivated varieties is not self-evident and depends on changing production patterns, lifestyles and market demands as well as on pests, diseases and the ability to crop a mix of complementary banana varieties (Kagy et al., 2016). The genetic diversity embedded in mixed cropping systems is seen as a vital source in finding resistance to current pest and disease pressures. So far, however, innovation strongly focused on a limited number of possible solutions for countering pests and diseases in banana cultivation: mainly chemical control, the distribution of clean planting material and, more recently, the breeding of a selection of resistant varieties. Banana and plantain are difficult to improve genetically (Dale et al., 2017): cultivated bananas and plantains are both sterile and the fruit develops without seed, constraining, for example, breeding for disease resistance (Teo et al., 2005). Some authors suggest that this may alter as a result of long-term research using advanced breeding and genetic techniques (Biswas and Yi, 2016; Dale et al., 2017). In addition to breeding and crop improvement efforts, the multiplication and distribution of clean planting material produced with tissue culture techniques and involving farmers' groups and nurseries (Ndubi, 2015; Bandewar et al., 2017) has been part of disease control strategies that support low-input production systems under diverse conditions (Dubois et al., 2013; Niere et al., 2014; Blomme et al., 2013). However, the diversity of varieties for which clean planting material is available is limited. Therefore, it is worthwhile to consider the genetic diversity as a source for providing sustainable solutions for banana cultivation by both smallholders and highly specialised plantations (Côte et al., 2010). However, a means of using the genetic diversity of banana to manage diseases in a sustainable way under variable conditions has not been realised.

5 Case study: the variety of banana producers in the Philippines

The following case study of the Philippines builds on the observed diversity in banana production systems and shifts attention to the variety of banana producers in a single area. It briefly explores the contours of a territorial approach that has the capacity to accommodate multiple solutions for sustainable banana cultivation by these diverse producers. A broader approach is key to achieving resilience in a major production area with multiple actors and interests. In the Philippines, particularly on the southern island of Mindanao, large-scale banana plantations are located adjacent to smallholder and backyard farms growing a range of banana varieties. This situation represents two distinct and isolated spheres for managing pest and diseases such as TR4, namely the large-scale export-oriented plantations of Cavendish bananas and a diverse set of smallholders producing various banana varieties for local food provision. The wide spread of TR4 has been observed in the major production areas for the export market, which affects the viability and future of all types of banana producers and exposes their interdependencies (De La Cruz, 2017). Consequently, it is unlikely that a single solution will suffice to deal with the disease problem confronting these different types of actors and production systems in the Philippines.

The situation in the Philippines shows that collaboration and the search for multiple solutions are not simple. Actors with different interests and positions in banana production are inclined to act independently or to blame each other for the spread of diseases (Montiflor et al., 2015, 2016; De La Cruz and Jansen, 2018). Major banana export companies traditionally manage their own research and development and monitoring systems. However, some large banana companies have cautiously begun to be supportive towards smallholder farmers working in the same area. Some have now decided to collaborate in a public–private partnership to build up a professional laboratory service capacity for the Philippine banana sector, which would benefit both large- and small-scale producers (http://www.fusariumwilt.org/index.php/en/projects/#PromoBanana). Banana companies have teamed up with public research institutes to test and distribute somaclones of more resistant Cavendish varieties. A cooperative of agrarian reform beneficiaries producing Fairtrade banana for export reported to experiment with measures to enhance soil health. As another example, during a multi-stakeholder workshop in the area (Davao City, 2014), different actors recognised newly emerging interdependencies. The private sector has also signalled a willingness to look at local and national government initiatives in surveying the spread of banana diseases and installing quarantine measures, having previously acted quite independently from public organisations. Hence, the spread of TR4 affects social, political and economic relations in this major banana-producing area.

The threat of TR4 to the variety of Philippines banana producers incrementally opens space for research that takes social and agroecological variety as a starting point for finding multiple solutions. During an interaction between research and a range of private and public actors involved in banana production (Davao City, 2016), the challenge of creating socio-ecological resilience was examined (Vellema, 2016). This entails multiple actors working together within particular landscapes as a condition for a coordinated response to disease threats affecting all kinds of actors and production systems. Such a resilience perspective builds on an emerging area of research looking at complex problems with both social and biological dimensions (Folke, 2006; Folke et al., 2007, 2010; McDonald and Stukenbrock, 2016; Côte et al., 2010). This involves a dynamic perspective that is attentive to evolving interactions between human actions, social relations and biological and agroecological processes embedded in specific landscapes (Baker et al., 2013). Such a dynamic and territorial perspective contrasts with the more static character of global sustainability standards, which adapt to the uniformity demanded in export markets (Francis, 2014). The banana company AgroFair operating in export markets also recognises this tension between uniformity and multiplicity (AgroFair, 2015). Addressing this tension implies looking for contradictions, complementarities and fit between global and local actor strategies (Schouten et al., 2016). It also relies on territorial governance skills that can match actions, strategies and collaborations of farmers, governments and companies as a key determinant for sustainability in banana cultivation (Fath et al., 2015; Lamine, 2015; McAllister et al., 2015). This requires assessing whether and how the changing nature of governance and social relations in banana-producing territories creates space for cooperation (Córdoba and Jansen, 2014; Vellema and van Wijk, 2015). The case study of the Philippines demonstrates that a sense of urgency can trigger the first steps towards new forms of collaboration traversing organisational and spatial boundaries. Whether a coordinated response and multiple solutions will be realised depends, among other things, upon the conditions under which the mobilisation of resources for experimenting and testing disease management strategies in the export industry reach smallholder farmers.

6 Conclusion

Reviews of research on sustainable banana cultivation show the existence of two disconnected fields. Sustainability in banana cultivation has been researched either in relation to the intensive use of pesticides and the increased use of standards in the export sector or, to a lesser extent, in the context of the diverse realities of banana cultivation for food security and local markets. The occurrence and spread of pests and diseases is used here as the entry point for exploring options for connecting these two distinct fields. The Philippines case study underscores how the diversity of producers who act independently but are connected in the same territory directs and complicates the search for problem-solving strategies. A key requirement for sustainability in banana cultivation is to connect to differentiated responses by individual actors, in particular export-oriented plantations and small-scale farmers managing mixed cropping systems. A portfolio of different, and perhaps contrasting, solutions may be more conducive for sustainable banana cultivation than a strategy based on a single solution or fixed standards.

7 Where to look for further information

The World Banana Forum (WBF) is a space where the main stakeholders of the global banana supply chain work together to achieve consensus on best practices for sustainable production and trade. http://www.fao.org/world-banana-forum/en/

Bioversity International researches the diversity of banana and its wild relatives, building on smallholder farmers' knowledge to use this resource, to make production systems more productive and resilient to shocks. The Bioversity International Musa Germplasm Transit Centre is home to the world's largest collection of banana germplasm. http://www.bioversityinternational.org/research-portfolio/banana-genetic-resources-and-management-systems/

Banana Link is a not-for-profit cooperative based in Norwich, UK, which works for fair and sustainable banana and pineapple trades. Banana Link raises awareness of the poor living and working conditions faced by plantation workers and small producers in Latin America, Africa and the Caribbean. http://www.bananalink.org.uk/

ProMusa is a platform for sharing news and knowledge on bananas. http://www.promusa.org/

The Committee on Sustainability Assessment advances systematic and science-based measurement tools for understanding, managing and accelerating sustainability. https://thecosa.org/

Wageningen University and Research has an interdisciplinary approach to investigating diseases and their impacts on the future of banana cultivation. http://www.fusariumwilt.org/index.php/en/about/

8 References

Aasprong, H. (2013). Entangled standardizing networks: The case of GLOBALGAP and Fairtrade in St Vincent's banana industry. *International Journal of Sociology of Agriculture & Food* 20: 91–108.

AgroFair (2015). *Sustainability Report 2015*. Barendrecht: AgroFair.

Bain, C., Ransom, E. and Higgins, V. (2013). Private agri-food standards: Contestation, hybridity and the politics of standards. *International Journal of Sociology of Agriculture and Food* 20: 1–10.

Baker, L., Dove, M., Graef, D., et al. (2013). Whose diversity counts? The politics and paradoxes of modern diversity. *Sustainability* 5: 2495.

Bamutaze, Y. (2015). Revisiting socio-ecological resilience and sustainability in the coupled mountain landscapes in Eastern Africa. *Current Opinion in Environmental Sustainability* 14: 257–65.

BananaLink (2017). *Voluntary Standards Initiatives*. Available at: http://www.bananalink.org.uk/voluntary-standards-initiative (accessed 26 June 2017).

Bandewar, S. V. S., Wambugu, F., Richardson, E., et al. (2017). The role of community engagement in the adoption of new agricultural biotechnologies by farmers: The case of the Africa harvest tissue-culture banana in Kenya. *BMC Biotechnology* 17: 28.

Barraza, D., Jansen, K., Van Wendel de Joode, B., et al. (2011). Pesticide use in banana and plantain production and risk perception among local actors in Talamanca, Costa Rica. *Environmental Research* 111: 708–17.

Barraza, D., Jansen, K., van Wendel de Joode, B., et al. (2013). Social movements and risk perception: Unions, churches, pesticides and bananas in Costa Rica. *International Journal of Occupation and Environmental Health* 19: 11–21.

Bauer, V. B. M. (2012). *The East African Banana System: Organization of Planting Material Supply and Produce Marketing*. Weikersheim, Germany: Margraf.

Bellamy, A. S. (2013). Banana production systems: Identification of alternative systems for more sustainable production. *Ambio* 42(3): 334–43.

Bellamy, A. S., Svensson, O., van den Brink, P. J., et al. (2016). What is in a label? Rainforest-Alliance certified banana production versus non-certified conventional banana production. *Global Ecology and Conservation* 7: 39–48.

Biswas, M. K. and Yi, G. (2016). Genes and markers: Application in banana crop improvement. In: Mohandas, S. and Ravishankar, K. V. (Eds), *Banana: Genomics and Transgenic Approaches for Genetic Improvement*. Singapore: Springer Singapore, pp. 35–50.

Blomme, G., Ploetz, R., Jones, D., De Langhe, E., Price, N., Gold, C., Geering, A., Viljoen, A., Karamura, D., Pillay, M., Tinzaara, W., Teycheney, P.-Y., Lepoint, P., Karamura, E. and Buddenhagen, I. (2013). A historical overview of the appearance and spread of Musa pests and pathogens on the African continent: highlighting the importance of clean Musa planting materials and quarantine measures. *Annals of Applied Biology* 162(1): 4–26.

Bohme, S. R. (2014). *Toxic Injustice*. Los Angeles, CA: University of California Press.

Bongers, G., Fleskens, L., Van De Ven, G., et al. (2015). Diversity in smallholder farms growing coffee and their use of recommended coffee management practices in Uganda. *Experimental Agriculture* 51: 594–614.

Bonicelli, B., Cotteux, E. and Douzals, J. P. (2015). Pesticides applications assessment in banana crop conditions: A methodological approach. *Environmental Science and Pollution Research* 22: 17255–9.

Brisbois, B. (2016). Bananas, pesticides and health in southwestern Ecuador: A scalar narrative approach to targeting public health responses. *Social Science & Medicine* 150: 184–91.

Brisbois, B. W., Harris, L. and Spiegel, J. M. (2017). Political ecologies of global health: Pesticide exposure in southwestern Ecuador's banana industry. *Antipode* 50(1): 61–81.

Brown, S. (2013). One hundred years of labor control: Violence, militancy, and the Fairtrade banana commodity chain in Colombia. *Environment and Planning A* 45: 2572–91.

Calberto, G., Staver, C. and Siles, P. (2015). An assessment of global banana production and suitability under climate change scenarios. In: Elbehri, A. (Ed.), *Climate Change and Food Systems: Global Assessments and Implications for Food Security and Trade*. Rome: FAO, pp. 265–91.

Castro, L. M., Calvas, B. and Knoke, T. (2015). Ecuadorian banana farms should consider organic banana with low price risks in their land-use portfolios. *PLoS ONE* 10: e0120384.

Chopin, P., Tirolien, J. and Blazy, J.-M. (2016). Ex-ante sustainability assessment of cleaner banana production systems. *Journal of Cleaner Production* 139: 15–24.

Comte, I., Cattan, P., Charlier, J. B., et al. (2018). Assessing the environmental impact of pesticide use in banana cropping systems. *Acta Horticulturae* 1196: 195–202.

Córdoba, D. and Jansen, K. (2014). Same disease—different research strategies: Bananas and black sigatoka in Brazil and Colombia. *Singapore Journal of Tropical Geography* 35: 345–61.

Côte, F., Tomekpe, K., Staver, C., et al. (2010). Agro-ecological intensification in banana and plantain (Musa spp.): An approach to develop more sustainable cropping systems for both smallholder farmers and large-scale commercial producers. *Acta Horticulturae* 879: 457–63.

Dale, J., Paul, J.-Y., Dugdale, B., et al. (2017). Modifying bananas: From transgenics to organics? *Sustainability* 9: 333.

De la Cruz, J. (2017). *Panama Disease in Banana and Neoliberal Governance: Towards a Political Ecology of Risk*. Wageningen, the Netherlands: Wageningen University.

De la Cruz, J. and Jansen, K. (2018). Panama disease and contract farming in the Philippines: Towards a political ecology of risk. *Journal of Agrarian Change* 18: 249–66.

De Langhe, E., Perrier, X., Donohue, M., et al. (2015). The original banana split: Multi-disciplinary implications of the generation of African and Pacific Plantains in Island Southeast Asia. *Ethnobotany Research and Applications* 14: 299–312.

DeFries, R. S., Fanzo, J., Mondal, P., et al. (2017). Is voluntary certification of tropical agricultural commodities achieving sustainability goals for small-scale producers? A review of the evidence. *Environmental Research Letters* 12: 033001.

Dubois, T., Dusabe, Y., Lule, M., et al. (2013). Tissue culture banana (*Musa spp.*). for smallholder farmers: Lessons learnt from East Africa. *Acta Horticulturae* 986: 51–9.

FAO (2014). The changing role of multinational companies in the global banana trade: Market and policy analyses of raw materials, horticulture and tropical (RAMHOT) products team. Rome: FAO.

FAO and WBF (2017). Voluntary sustainability standards for bananas. Rome: Food and Agriculture Organization of the United Nations and World Banana Forum.

Fath, B., Dean, C. and Katzmair, H. (2015). Navigating the adaptive cycle: An approach to managing the resilience of social systems. *Ecology and Society* 20: 24.

Folke, C. (2006). Resilience: The emergence of a perspective for social–ecological systems analyses. *Global Environmental Change* 16: 253–67.

Folke, C., Carpenter, S. R., Walker, B., et al. (2010). Resilience thinking: Integrating resilience, adaptability and transformability. *Ecology and Society* 15: 20.

Folke, C., Pritchard Jr., L., Berkes, F, et al. (2007). The problem of fit between ecosystems and institutions: Ten years later. *Ecology and Society* 12: 30.

Francis, C. (2014). Book Review: The East African banana system – organization of planting material supply and produce marketing. *International Journal of Agricultural Sustainability* 12: 92–3.

Frundt, H. J. (2009). *Fair Bananas! Farmers, Workers, and Consumers Strive to Change an Industry*. Tucson, AZ: University of Arizona Press.

Frundt, H. J. (2010). Sustaining labor-environmental coalitions: Banana allies in Costa Rica. *Latin American Politics and Society* 52: 99–129.

Heslop-Harrison, J. S. and Schwarzacher, T. (2007). Domestication, genomics and the future for banana. *Annals of Botany* 100: 1073–84.

Jansen, K. (2004). Greening bananas and institutionalizing environmentalism: Self-regulation by fruit corporations. In: Jansen, K. and Vellema, S. (Eds), *Agribusiness and Society: Corporate Responses to Environmentalism, Market Opportunities and Public Regulation*. London: Zed, pp. 145–75.

Jansen, K. (2008). The unspeakable ban: The translation of global pesticide governance into Honduran national regulation. *World Development* 36: 575–89.

Jansen, K. and Dubois, M. (2014). Global pesticide governance by disclosure: Prior informed consent and the Rotterdam convention. In: Gupta, A. and Mason, M. (Eds), *Transparency in Global Environmental Governance: Critical Perspectives*. Cambridge, MA: MIT Press, pp. 107–31.

Kagy, V., Wong, M., Vandenbroucke, H., et al. (2016). Traditional banana diversity in Oceania: An endangered heritage. *PLoS ONE* 11: e0151208.

Karamura, E. B., Jogo, W., Rietveld, A., et al. (2013). Effectiveness of agro-ecological intensification practices in managing pests in smallholder banana systems in East and Central Africa. *Acta Horticulturae* 986: 119–26.

Lamine, C. (2015). Sustainability and resilience in agrifood systems: Reconnecting agriculture, food and the environment. *Sociologia Ruralis* 55: 41–61.

Lockie, S., McNaughton, A., Thompson, L.-J., et al. (2013). Private food standards as responsive regulation: The role of national legislation in the implementation and evolution of GLOBALG.A.P. *International Journal of Sociology of Agriculture & Food* 20: 275–91.

Lockie, S., Travero, J. and Tennent, R. (2015). Private food standards, regulatory gaps and plantation agriculture: Social and environmental (ir)responsibility in the Philippine export banana industry. *Journal of Cleaner Production* 107: 122–9.

McAllister, R. R. J., Robinson, C. J., Maclean, K., et al. (2015). From local to central: A network analysis of who manages plant pest and disease outbreaks across scales. *Ecology and Society* 20: 67.

McDonald, B. A. and Stukenbrock, E. H. (2016). Rapid emergence of pathogens in agro-ecosystems: Global threats to agricultural sustainability and food security. *Philosophical Transactions of the Royal Society B: Biological Sciences* 371: 20160026.

Melo, C. J. and Wolf, S. A. (2005). Empirical assessment of eco-certification: The case of Ecuadorian bananas. *Organization & Environment* 18: 287–317.

Montiflor, M. O., Vellema, S., Concepcion, S. B., et al. (2015). The politics and materiality of coordination: Philippine village leaders responding to the spread of a global plant disease in banana. *BANWA Supplements*.

Montiflor, M. O., Vellema, S., Concepcion, S. B., et al. (2016). Coordination and risk in the Philippine banana industry: Conditions for responding to panama disease. *BANWA Supplements* 5.

Murray, D. L. and Raynolds, L. T. (2000). Alternative trade in bananas: Obstacles and opportunities for progressive social change in the global economy. *Agriculture and Human values* 17: 65–74.

Mwijage, A. N., Gimbi, A. A. and Nyomora, A. A. (2016). Options for sustaining farm productivity in the banana-based farming systems in Bukoba, Tanzania. *International Journal of Agricultural Policy and Research* 4: 79–86.

Ndubi, J. M. (2015). *Getting Partnerships to Work: A Technography of the Selection, Making and Distribution of Improved Planting Material in the Kenyan Central Highlands*. Wageningen, the Netherlands: Wageningen University.

Niere, B., Gold, C. S., Coyne, D., et al. (2014). Performance of tissue-cultured versus sucker-derived East African highland banana (*Musa* AAA-EA) under high and low input systems in Uganda. *Field Crops Research* 156: 313–21.

Norgrove, L. and Hauser, S. (2014). Improving plantain (*Musa* spp. AAB) yields on smallholder farms in West and Central Africa. *Food Security* 6: 501–14.

Oduol, P. A. and Aluma, J. R. W. (1990). The banana (*Musa* spp.) — Coffee robusta: Traditional agroforestry system of Uganda. *Agroforestry Systems* 11: 213–26.

Ofori, D. A., Gyau, A., Dawson, I. K., et al. (2014). Developing more productive African agroforestry systems and improving food and nutritional security through tree domestication. *Current Opinion in Environmental Sustainability* 6: 123–7.

Oosterveer, P., Adjei, B. E., Vellema, S., et al. (2014). Global sustainability standards and food security: Exploring unintended effects of voluntary certification in palm oil. *Global Food Security* 3: 220–6.

Potts, J., Lynch, M., Wilkings, A., et al. (2014). The state of sustainability initiatives review 2014: Standards and the green economy. Winnipeg and London: International Institute for Sustainable Development (IISD) and the International Institute for Environment and Development (IIED).

Riisgaard, L. and Hammer, N. (2011). Prospects for labour in global value chains: Labour standards in the cut flower and banana industries. *British Journal of Industrial Relations* 49: 168–90.

Rueda, X., Garrett, R. D. and Lambin, E. F. (2017). Corporate investments in supply chain sustainability: Selecting instruments in the agri-food industry. *Journal of Cleaner Production* 142(Part 4): 2480–92.

Schouten, G., Vellema, S. and van Wijk, J. (2016). Diffusion of global sustainability standards: The institutional fit of the ASC-Shrimp standard in Indonesia. *Revista de Administração de Empresas* 56: 411–23.

Staver, C. (2009). Increasing the pro-poor impact of banana crop protection strategies: Knowing our clients, involving new players and linking more effectively. *Acta Horticulturae* 828: 361–74.
Sujarwo, W. and Caneva, G. (2015). Ethnobotanical study of cultivated plants in home gardens of traditional villages in Bali (Indonesia). *Human Ecology* 43: 769–78.
Tadele, Z. (2009). Role of orphan crops in enhancing and diversifying food production in Africa. *African Technology Development Forum Journal* 6(3–4): 9–15.
Tadele, Z. (2014). Role of crop research and development in food security of Africa. *International Journal of Plant Biology & Research* 2: 1019.
Teo, C. H., Tan, S. H., Ho, C. L., et al. (2005). Genome constitution and classification using retrotransposon-based markers in the orphan crop banana. *Journal of Plant Biology* 48: 96–105.
Trauger, A. and Murphy, A. (2013). On the moral equivalence of global commodities: Placing the production and consumption of organic bananas. *International Journal of Sociology of Agriculture & Food* 20: 197–217.
Tushemereirwe, W., Kangire, A., Kubiriba, J., et al. (2004). Diseases threatening banana biodiversity in Uganda. *African Crop Science Journal* 12: 19–26.
Utting, P. (2008). The struggle for corporate accountability. *Development and Change* 39: 959–75.
van Asten, P. J. A., Wairegi, L. W. I., Mukasa, D., et al. (2011). Agronomic and economic benefits of coffee–banana intercropping in Uganda's smallholder farming systems. *Agricultural Systems* 104: 326–34.
Vellema, S. (2016). The making of socio-ecological resilience: Requisite variety as a condition for addressing a global banana disease. *Key note at the 8th International Conference on Agribusiness Economics and Management, 'Building Resilience: Agribusiness in Times of Flux'*. Davao City: University of the Philippines Mindanao.
Vellema, S. and van Wijk, J. (2015). Partnerships intervening in global food chains: The emergence of co-creation in standard-setting and certification. *Journal of Cleaner Production* 107: 105–13.
Were, M. (2003). Implementing corporate responsibility — The Chiquita Case. *Journal of Business Ethics* 44: 247–60.
Wicki, S. and Van Der Kaaij, J. (2007). Is it true love between the octopus and the frog? How to avoid the authenticity gap. *Corporate Reputation Review* 10: 312–18.
Wilson, M. and Jackson, P. (2016). Fairtrade bananas in the Caribbean: Towards a moral economy of recognition. *Geoforum* 70: 11–21.

Chapter 15

Organic banana cultivation and sustainability

Frans Wielemaker, Consultant (formerly Director of Research at Dole Fresh Fruit International), Costa Rica

1 Introduction
2 Organic banana production
3 Key requirements: location, soil and use of rotation
4 Soil fertility and fertilization
5 Disease management
6 Managing nematode and insect pests
7 Weed management
8 Discussion
9 Where to look for further information
10 Disclaimer
11 About the author
12 References

1 Introduction

The title of this chapter poses the question: what are the practices that make certified organic banana cultivation and exports possible and what are the crucial agronomic hurdles which limit their expansion? The demand for organic bananas seems unstoppable, but can they ever become mainstream (Fig. 1)? These questions are posed because the substantial amounts of fungicides and fertilizers required in conventional production systems of fresh bananas of the Cavendish variety grown in the wet tropics for export have been criticized by many as environmentally unsustainable. The reasons for the banana industry's focus on Cavendish varieties are varied and involve a desire for standardization in the supply chain, consumer appreciation and their excellent agronomic (high yield) and logistic characteristics (they ship well over long distances).

http://dx.doi.org/10.19103/AS.2017.0020.25

Figure 1 View of the 'by-design' Nueva Esperanza Organic Banana Farm located in Manabí, Calceta, province of Ecuador.

2 Organic banana production

2.1 Demand and markets

The demand for certified organic bananas from the international markets has increased exponentially over the last 25 years, and during 2016, reached 57 million boxes (see Table 1; Wielemaker, 2013; and ongoing market intelligence). This volume represents 8% of total Latin American banana exports, which is huge when compared with any other fresh produce commodity in the world, and which can no longer be considered a niche market. Exact figures through custom statistics are unavailable as no separate codes exist for organic bananas in most exporting and importing countries.

2.2 Sustainability

Sustainability in simple terms means meeting present needs without compromising the future. For banana cultivation, this would imply the need to implement a cultivation system that remains productive without doing long-term damage to the environment. The bottom-line objective for the banana industry as a whole, be it organic or conventional, is to grow bananas while protecting the environment, secure the livelihood and health of farm workers, guarantee a healthy product to the end consumer and secure a fair income to the farming entity. To achieve these goals the banana industry should invest more in research. Cavendish varieties are very susceptible to manifold

Table 1 Export volumes of organic bananas (2016) by country and volume destination by continent

Country	Millions of boxes (18 kg)
Ecuador	20
Dominican Republic	19
Peru	11.5
Colombia	4
Mexico	2.5
Total boxes exported	57
Exports to Europe	33
Export to North America	22
Exports to Asia	2

pests and diseases and have a high fertilizer demand, which makes it a challenge to grow them organically.

2.3 Certification

'Certified Organic' (or 'Certified Biological') status has a legal framework in place in the major markets [European Union (EU), the United States, Canada, Japan and Korea]. The implementation of organic regulations in the United States and the EU has been an important factor for the growth of the organic sector as a whole. In the United States, 'organic' is a legal term defined by the US Department of Agriculture (USDA), backed by Federal Regulation 7 CFR Part 205, describing the 'USDA Organic Standards', which became effective in October 2002. Comprehensive information on what this entails according to US legislation, and how it works in practice can be found at the 'National Organic Program (NOP)' website operated by the Agricultural Marketing Service (AMS) of the USDA (https://www.ams.usda.gov/about-ams/programs-offices/national-organic-program). For the EU, the Council Regulation (EC) No. 834/2007 of June 2007 on 'Organic Production and Labelling of Organic Products' spells out its regulation. This has been published as EU Commission Regulation No. 889 in 2008, and amendments and corrections have been laid out in the EU Commission Regulation No. 354 in 2014. The first and original version, Regulation (EEC) No. 2092, was published back in 1991 (nearly 10 years before the United States). All these regulations can be found on the EU official website on organic agriculture (https://ec.europa.eu/agriculture/organic/index_en). The Japanese Agricultural Standards on organic production became a reality in 2001 and can be found on the website of the Ministry of Agriculture Forestry and Fisheries (http://www.maff.go.jp/e/policies/standard/jas/). The approved logos for these standards are shown in Fig. 2.

These standards have continued to evolve. As an example, petitioning and continuous lobbying has obtained and maintained the listing of ethylene as an approved substance for ripening organic bananas according to the USDA-AMS-NOP rules (Wielemaker and Monge, 2015). Without this compound, organic banana marketing in the United States

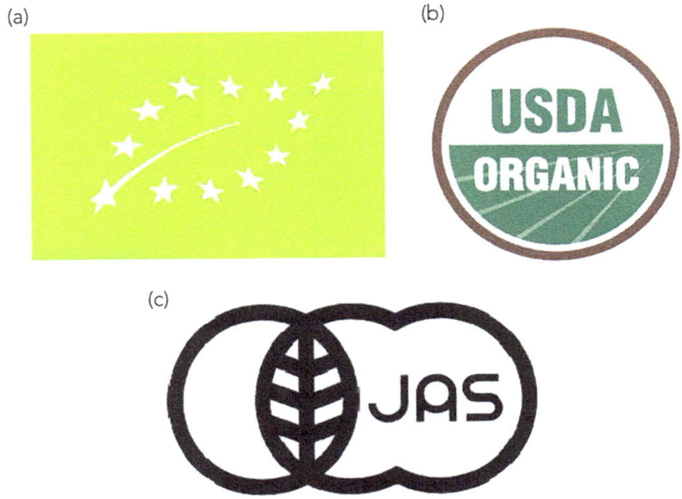

Figure 2 Official seals of the EU (a), the US (b) and Japanese (c) organic (biological) production standards.

would be impossible as there is no suitable alternative for ethylene, which is a simple organic molecule (2HC=CH2) abundant in nature.

2.4 Understanding and interpreting organic standards

A book on the 'complete international history of organic farming' was written a decade ago (Lockeretz, 2007) and provides a description of organic and alternative farming practices from their various roots to contemporary organic standards. Organic farming practices are rooted in maintaining and improving soil health and do not make a 'residue-free' or health claim as many consumers mistakenly think. Organic farming is often seen in a simplistic way, related to what is prohibited and generally focusing on chemicals (fungicides, insecticides, nematicides). There is an implication that organic foods contain fewer chemical residues and are thus safer or healthier, but organic farming makes no claim to that effect, being a production procedure, not a quality or health standard. In fact, pesticide-free production (or non-detectable) would theoretically be easier than the production of organic bananas. Although overall, organic produce does not contain or contains fewer pesticide residues, it is not the subject matter of this chapter to defend or challenge this claim.

For those who wish to look deeper, the US Food and Drug Administration, the US-based Organic Consumers Association (https://www.organicconsumers.org/) and the Organic Trade Association (OTA) (https://www.ota.com/) have data on their websites which substantiate lower residue detections in organic produce when compared with conventional production methods. Ultimately, what is important for many well-informed consumers is banning the use of synthetic pesticides, the prevention of harm to the environment, and safety for farm workers. As consumer awareness increases, so does the demand to know the origins of produce and the method of production (Wielemaker et al., 2009). This has been achieved by using stickers on bananas with a 'Producer Code' by which the farm can be identified virtually.

The International Federation of Organic Agriculture Movements (IFOAM) is a good source for information on the philosophical and practical aspects of organic agriculture. This organization maintains 'The IFOAM NORMS for Organic Production and Processing' (IFOAM, Version 2014) as a reference and basic document for any country wanting to implement organic regulations, as well as a *Training Manual for Organic Agriculture in the Tropics* (Eyhorn et al., 2002). The CODEX Alimentarius has joint organic food standards with the World Health Organization and the UN Food and Agricultural Organization, and these entities also have published information on organic agriculture on their websites, although not specifically on growing organic bananas (http://www.fao.org/docrep/010/a1385e/a1385e00.htm).

In practice, organic banana farms and packing facilities are certified by accredited entities meeting US and EU requirements. These certifying entities inspect the farms and, if necessary, require growers to revise and improve their organic programmes, including highlighting any inputs that are not approved. However, it remains the farmers' responsibility to comply with the rules. Getting a farm certified for organic production requires a substantial amount of work, and documentation and record-keeping are very important. The certifying entity also cross-checks to ensure the volume sold is in accordance with the number of hectares in production and may also collect soil, leaf and fruit samples for residue analysis. Once all requirements are met, the 'Organic' ('Biological') seal can be used on the packaging as a guarantee to consumers. Organic certification typically involves yearly verification and, sometimes, surprise inspections. In conventional agriculture, farmers are supposed to adhere to regulations, but are rarely inspected on what they actually do, and therefore food safety issues often surface too late. Organic certifying entities offer 'combo' inspections combining organic regulation with the food safety certification for 'Global Gap', and many other standards. The implementation of a traceability system for every box of organic bananas is of the utmost importance and it is obligatory to trace any detected residue down to an exact location so it is possible to identify any fraudulent practice or individual.

2.5 Sources and guidance

Understanding and correctly interpreting all the rules and regulations to determine which farm inputs are allowed or prohibited is complex. Fortunately, there are bodies which may be consulted for guidance, such as the Organic Materials Review Institute (OMRI; https://www.omri.org/) and the Washington State Department of Agriculture (WSDA; https://agr.wa.gov/). Both list material inputs which, according to the information received from the suppliers and manufacturers, together with the thorough independent research they perform, comply with the US NOP regulations, for crops, livestock or processing. Unfortunately, no such institute or resource exists to evaluate inputs allowed under European organic law. Annex II of the EU organic legislation is not extensive and has more than one interpretation, which may add to uncertainty and difficulty in coming to a verdict. *The Organic Standard Magazine* (TOS, 2016) is a recommended resource for keeping up to date on developments in global certification. It is an international electronically published journal on developments concerning organic certification, organic accreditation, organic production, processing and marketing trends.

Because growing certified organic bananas for the export market is relatively new, there is only a limited range of published peer-reviewed scientific research, leaving the sector sometimes dependent upon circumstantial evidence and anecdotal stories. Basic experimental designs for simple comparative tests are hardly ever adopted and, just as in conventional research, non-conclusive experiments or those which did not yield significant results are never published.

The proceedings of two dedicated meetings on organic and sustainable banana production have been published by the International Network for the Improvement of Banana and Plantain (INIBAP), now Bioversity International (Rosales et al., 1999; Holderness et al., 2000).

For the novice, a very good introduction to the subject of growing bananas was written by Robinson and Galán Saúco (2010), but only one page is dedicated to organic cultivation and two pages to 'Soil Degradation and the Rise of Organic Nutrition'. It states that 'with organic banana cultivation, achieving long-term sustainability is more important than maximizing yield'. The authors also defined 'leaf diseases as the major problem for organic production besides nematodes and weevils'. Consequently, they indicate 'the search for resistant cultivars [is] a need'. Paull and Duarte (2011) also have recently written a very good review chapter on banana and plantain cultivation, but this lacks any reference to growing bananas organically. Another very good general treatise on bananas is the Consensus Document (#48) on the Biology of Bananas and Plantains (OECD, 2009), which also largely lacks information on organic banana cultivation. The monograph 'Growth Stages of Mono-and Dicotyledonous Plants' by Uwe Meier (2001) is an almost compulsory reference for the growth stages of bananas.

Soto (2016) published 'Bananos, Tecnologías de Producción', in which a chapter is dedicated to the cultivation of organic bananas and another to nutrition, including organic soil amendments, which are the basis for sustainable organic banana production. In general, however, there is only sparse information on growing organic bananas, even at the biennial international ACORBAT (Association for Cooperation in Banana Research in the Caribbean and Tropical America) and CORBANA (Corporación Bananera Nacional de Costa Rica) banana meetings. Hence, there is an urgent need for more research in sustainable and organic banana production methods.

The French Agricultural Research Centre for International Development (CIRAD) has invested in the sustainable production of bananas because the Caribbean islands of Guadeloupe and Martinique, where bananas are grown for the French market, are French territory and thus have to comply with all the EU regulations. These growers are struggling to keep up with environmental regulations and need practical techniques to help develop environmentally more sustainable production systems. This is also applicable to the Australian banana industry where the effects of chemical load run-off towards the Great Barrier Reef are being closely monitored, especially in Northern Queensland.

Taking care of the environment comes with costs, and these are externalized in most conventional production systems. Paul Hawken (2010), in the first 1993 edition of his groundbreaking book *The Ecology of Commerce: A Declaration of Sustainability*, highlighted that 'producing a crop organically has to be costlier because environmental stewardship is part of the cost equation'. Bananas are no exception and growing bananas organically is costlier than conventional production. The following sections discuss the reasons for the expanding production, export and commercialization of organically grown bananas and the challenges of producing them, which puts the increased cost in perspective.

3 Key requirements: location, soil and use of rotation

3.1 Location

Avoidance of Black Sigatoka disease in organic banana production requires a very dry climate (<600 mm precipitation per year). These climate zones are mostly located in arid regions, some near the subtropics. As a result, in Latin America, significant volumes of certified organic bananas for the fresh export market are primarily produced in the Dominican Republic (around Ázua in the South, and Máo, Valverde and Montecristi in the North), Ecuador (Manabí in the North West, and towards the South from Puerta Inca to Santa Rosa), Peru (Northern coastal desert region near Tumbes, Sulláná and Piura regions) and Colombia (bordering the Guajira desert in the North East from Palomino to Riohacha). Organic bananas are also being grown in Colima, Mexico, where rainfall conditions and humidity are not as low and hence vulnerable to Black Sigatoka epidemics (Fig. 3). The challenge under these arid conditions is to find reasonably good soils without elevated salinity levels and which have a guaranteed year-round good quality water supply for irrigation.

3.2 Soil

The first organic banana farms and cooperatives in the Dominican Republic and Ecuador were established on old uncompetitive and half-abandoned conventional banana farms no longer capable of producing decent bananas at a competitive cost price. These

Figure 3 Map with locations where organic bananas are grown today.

unsustainable farms were converted to organic by default and succeeded due to the consumer demand for organic bananas which were in very short supply in the market at that time (early 1990s). Better pricing for these bananas and Fairtrade premiums played an important role in their success. Thus, necessity and market forces played crucial roles in resuscitating these farms and grower organizations.

As organic banana demand increased, new organic farms were started 'by-design' (see Fig. 1). Pasture lands were used as these had no previous history of chemical use and could be organically certified almost immediately. The pasture lands, however, had very compacted soils and had long been 'mined' by cattle without ever having been given anything back (e.g. through fertilization). These areas obviously presented an initial challenge in establishing organic banana farms. Good sub-soiling (up to 0.8–1.0 m deep) and cultivation without turning the top soil were important as bananas do not grow well on compacted soil. These practices guarantee the establishment of good internal drainage and aeration and result in prolific banana root growth.

3.3 Rotations

Due to soil compaction, replanting every 6–10 years has now become a normal practice in Central and South American conventional banana farms. This practice should also be adopted in organic banana farms, but taken a step further by keeping one-seventh of the farm free of bananas for at least a year, putting that area in rotation or a fallow period. A rotation schedule should then become part of the Farm Management Plan, as part of the strategy for maintaining a sustainable system which will keep nematodes and banana weevil borers at manageable levels.

Plant parasitic nematodes which affect the crop are an indication of poor soil health. US and EU organic requirements specifically demand that rotation is part of a farm plan, but certification organizations do not usually spell this out to their clients as they consider bananas to be a perennial crop (as with coffee or cacao). In temperate climates, crop rotation is usually an essential part of a nematode management plan and several mostly annual crops (especially potatoes) are grown alternately, precisely for that reason and often mandated by official regulations.

To rejuvenate the soil, farms should set aside part of a plantation (e.g. 14%) each year and either plant a rotation crop or incorporate a fallow period. This should be considered in every farm plan, organic or conventional. Understandably, most banana farmers prefer to plant bananas 'fence to fence', using the whole growing area each year which gives a better return, albeit only in the short term. However, setting part of the farm aside for 8–12 months will suppress nematodes and pests in a sustainable way. This yearly rejuvenation can be taken as the opportunity to grow a large amount of biomass which may be incorporated or composted. It also offers the opportunity to alleviate soil compaction (e.g. by deep ripping and disking), facilitate the application of organic amendments, change banana variety, redesign and improve drainage systems and start anew with optimum plant distribution and densities.

When a section of a farm is rejuvenated in this way, it provides the opportunity to begin to convert to organic production. Transition according to organic regulations normally takes 3 years, or 2 years plus the duration of the plant crop. Due to the complex infrastructure of a banana plantation, the combination of rotation with land preparation is an expensive operation as it may require the partial removal of irrigation equipment and cableways to give tractor access.

Rotation with short-cycled crops such as forage sorghum or wild brassica, as a fallow period for at least 6 months, will revitalize an area and reduce the nematode population. *Brassica* cover crops, once fully grown and disked into the soil, release isothiocyanide and disulphide containing metabolites which reduce nematode populations. Simply fallowing with natural voluntary weed growth will not reduce nematodes significantly as many weeds are alternative hosts.

Fast-growing rotation crops which build a large volume of biomass, referred to as 'green manure', include the velvet bean (*Mucuna* spp.) and Kudzú (*Pueraria* spp.). These can be planted after soil preparation as they grow quickly and smother weeds. Subsequently, these can be slashed just before planting the banana plants 'through' the mulch which keeps weeds under control for a considerable time. The seeds of these green manures are available commercially and locally where traditional agricultural practices are still valued and are part of the knowledge system. Another fast-growing plant which produces substantial biomass in a short time is *Ricinus communis*, as its 'cake' – after castor oil extraction – is a good nematode suppressant.

4 Soil fertility and fertilization

4.1 Soil types and nutrition

Recommendations on organic amendments, fertilizers, rates and frequencies should be based on soil and leaf analysis with samples taken at least twice a year in addition to observations by the experienced farmer. It is of the utmost importance that basic plant nutrition is understood, monitored and adjusted to the needs of growing banana crops in an aging plantation. A good soil map helps to tailor the application of organic fertilizers and amendments. Aerial monitoring by drones or aircraft can also provide a good overview of the leaf colour of the various areas of a farm.

The arid regions where organic bananas are mainly grown have predominantly alkaline soils (pH > 7.0), which inherently results in serious micro-element deficiencies, particularly zinc and boron, which may be supplemented through foliar applications. These soils are often quite low in organic matter content (<0.5% or even less) which does not meet the basic requirements for a healthy organic soil capable of delivering good root growth and nutrient retention. These soils range from clay loams and sandy loams to sandy soils and are mostly heavily weathered. This is far from ideal for good banana production. It is therefore essential to practice the organic principle of nurturing the soil and improving fertility. At the very least, nutrients exported when bananas are harvested need to be replaced, taking into account the relative (in)efficiency of organic fertilizer conversion.

Production of organic bananas on a per hectare per year basis is often low (between 1500 and 2850 boxes of 18 kg/ha/year) in comparison to conventional bananas (between 2500 and 3200 boxes of 18 kg/ha/year). Conventional bananas are grown on rich alluvial soils in high rainfall, humid tropical lowland regions with high day and night temperatures and employ large amounts of synthetic fertilizers. Organic bananas grown in the subtropics endure extremes of temperature which affect good banana growth, high UV radiation which scorches unprotected leaves and fruit, and often high velocity winds which shred the leaves. All these factors negatively influence yields as a trade-off for the arid conditions necessary to avoid Black Sigatoka.

4.2 Nitrogen

In conventional banana plantations, synthetic urea (46% N), ammonium sulphate (21% N), calcium nitrate (15.5% N), di-ammonium phosphate (18% N), mono-ammonium phosphate (11% N) and chemical mixtures or physical blends with phosphorus (P) and potassium (K) are frequently used to supply nitrogen (N) to the crop. In organic cultivation, it is much more difficult and costly to supply the high N requirements of bananas.

The sources of N in organic banana cultivation are basically composted vegetative matter and manures (0.5–2.0% N at the most), and various derivatives of the rendering industry. Animal by-products are converted to amino acids derived from hydrolysed feather meal (13.5% N) and leather meal (12.0% N) protein. As N is not accumulated in the banana plant and the requirement in an established plantation is continuous, it must be supplied frequently or as a slow-release application, keeping in mind that N, independent of its synthetic or organic origin, may volatize or leach from the soil if not managed well.

4.3 Phosphorus

To meet P requirements in conventional banana production, synthetically produced fertilizers are employed such as mono-ammonium phosphate (48% P_2O_5) and di-ammonium phosphate (46% P_2O_5), and chemical mixtures or physical blends with N and K. Although bananas do not need much P, this nutrient is only available in very low quantities in high pH soils. It therefore needs to be monitored closely and corrected, particularly as only rock phosphate is allowed in organic agriculture. This mined material only becomes available for the plant under acidic conditions, which only occur around the roots in alkaline soils.

4.4 Potassium

Potassium is a very important nutrient for banana growth and the formation of fruit which is rich in K. The main K sources used in conventional banana production are: potassium chloride (60% K_2O), potassium sulphate (48% K_2O), Sul-Po-Mag (22% K_2O) and chemical mixtures or physical blends with N and P.

Potassium levels in the soil should be monitored because excess amounts may leach, but the amount available to the plant should be controlled so that it is replenished before the plant becomes deficient. The organic farmer is restricted to what is permitted for improving nutrient deficiencies. Mined Sul-Po-Mag (22% K_2O) is permitted under organic standards.

Potassium sulphate (48% K_2O) is also permitted in organic agriculture, but only in exceptional cases. It is the preferred source where sodium levels are high in order to flush out the sodium as sodium sulphate with irrigation water and precipitation. Mined potassium sulphate can, however, only be used in organic banana production when the specific need is proven. This is the case when mined Sul-Po-Mag cannot be used due to an imbalance in the soil between K and Mg and where the additional Mg would make the imbalance worse. This imbalance occurs in many subtropical arid soils with a high pH. Only then will the certification entity approve the use of potassium sulphate, which is soluble in water and can be applied through the irrigation system. Sul-Po-Mag is not soluble and is thus applied by hand.

4.5 Organic and non-organic fertilizers

Nitrogen fertilizers used in conventional and organic agriculture become available to the plant only after being processed through digestion by microorganisms, which then make them available to the plant in the form of NO_3^- and NH_4^+. However, organic fertilizers need to be broken down first as they contain proteins, amino acids and minerals bound to complex organic materials. This means that organic fertilizers have a delayed release compared to conventional fertilizers used in conventional agriculture which do not need to be broken down. However, when organic N fertilizers are applied at consistent intervals, the soil will be supplied with a steady flow of N. This is particularly the case when the soil is rich in organic matter which will serve both as a sponge and buffer, while conventionally fertilized bananas may experience considerable losses of nutrients due to nitrification and lixiviation. The actual liberation of N from an organic fertilizer or amendment depends on the C/N ratio. In the long run, compost is beneficial, but it can be problematic in the shorter term. This is because compost with a high C/N ratio may fix the N in the soil, resulting in a temporary 'starvation' and yellowing of the banana crop. Quick fixes as in conventional farming with synthetic fertilizers do not exist in organic banana cultivation which requires good management and programming.

4.6 Fertilization and irrigation

Although fertigation – a combination of irrigation and fertilizer application – is not possible with most organic fertilizers, some nutrients can be applied through the irrigation system, depending on the system and nozzle aperture. Irrigation systems in organic farms need to wet organic fertilizers, amendments and compost applied to the top of the soil as the nutrients will not otherwise move towards the banana roots. Since it does not sufficiently wet these materials, drip irrigation is not an option for organic banana farms, thus generating a difficulty as water is usually limited in the arid areas suitable for organic banana production. For this reason, 'double tram-line planting' with micro-spray irrigation just above the soil level, (which ensures wetting the organic fertilizers and compost but not the inter-row), is the preferred farm layout (Fig. 4). This concentrates most water in the banana rows and facilitates mechanical weed control in the inter-row.

4.7 Compost production and usage

Soils in the majority of conventional banana plantations have been degraded and there is widespread soil erosion. It is necessary to protect banana soils from erosion by using cover crops and returning the organic matter content to adequate levels. In an organic farm stewardship of soil fertility is the foundation for growing a healthy crop and stabilizing yields. Compost and other organic amendments improve the physical, chemical and biological properties of soils and ultimately increase banana yields. They should always be part of the organic farm fertility program. All the organic waste from a banana farm packing plant should be composted for on-farm use and other ingredients should be obtained from the region. Bartering reject bananas in exchange for manure from cattle farming operations is often an option.

Identifying, exploring and deploying waste streams is crucial for organic banana production. As an example, the water-hyacinth (*Eichornia grasipes*), which is a nuisance in rivers and lakes throughout the world, is an excellent ingredient for composting, provided

Figure 4 Double tram-line planting with micro-spray irrigation just above the soil level, which guarantees wetting organic fertilizers and compost but not the inter-row.

it can be harvested easily and cheaply from a nearby location. As more banana farms convert to organic production, the demand for waste streams for compost will increase and potentially become more expensive. The same is true for organic fertilizers from the rendering industry.

4.7.1 Benefits of organic amendments

True organic growers will recognize the importance and long-term benefits of organic amendments because they know these will supply significant quantities of stable organic matter and a variety of macro- and micronutrients over time. They will improve the soil structure and porosity, reducing compaction and therefore improving aeration, permeability and increasing moisture infiltration and reducing run-off. They will also improve the absorption and moisture holding capacity of light soils, reducing water loss and nutrient leaching. Compost stimulates root proliferation, which will expand into a bigger soil mass and thereby provide access to more nutrients. Amendments will also increase the numbers of earthworms and improve soil microbial activity, reducing nematode populations and the impact of soil-borne diseases. Lastly, it will stabilize soil pH, improving nutrient uptake and improving the soil's ability to hold nutrients through improvement of the cation exchange capacity (CEC).

4.7.2 Making compost

It should be kept in mind that making compost requires equipment, labour and management. The composting site, storage of raw materials and storage of finished product can occupy a considerable area of land and building space. Excessive odours and flies may be a problem, especially in the early stages of learning about best practice. Foul odours are generated when the process is not managed correctly, particularly when using fresh manures. Ideally, there should be zero detectable nitrates in the finished compost. In organic production, fresh manures are not allowed to be applied 90 days prior to the projected harvest date due to food safety regulations. There are numerous books on composting, some being must-reads, basics or classics, like *The Rodale Book of Composting* edited by Martin and Gerchuny (1992), published by Rodale Press. The origin of this classic text goes back to Mr J.L. Rodale, one of the founders of the organic movement in North America.

Compost can be home-made or bought from third-party suppliers, but obtaining a large enough supply can be a challenge. The origins and quality of the ingredients should be thoroughly checked. It is advisable to use reliable suppliers who ideally should have their compost production facilities certified. If a farmer has spare land, biomass for compost production should be grown. In the arid tropics where most organic bananas are grown, composting facilities are often out in the open. Nevertheless, it is preferable to make compost under a fixed or temporary roof during the few months of rain or to cover open-air windrows (lines of cut crop residue) with tarpaulins to prevent excess moisture.

There are various types and sizes of compost turning equipment such as from-the-side tractor-pulled machines and self-powered over-the-row windrow turners. Frequent turning, especially in the first few weeks, is important to maintain proper aeration in the windrow and thus the aerobic conditions that will minimize bad odours. Anaerobic conditions (oxygen deficient) should be prevented at all times. If a roof cannot be afforded, a special UV-protected fabric cover is recommended to keep excess moisture out. At all times, it is important to keep the C/N ratio above 25 in the final finished product. Nitrogen in compost is mostly bound as organic protein and will over time convert to nitrate nitrogen. There should be no nitrite in the finished compost, as this is toxic to soil life.

All compost leachates (or compost teas) from outdoor concrete slabs should be collected and used as an inoculation sprayed onto newly formed windrows in order to facilitate the start of the composting process. It can also be injected through the irrigation system of the farm in newly planted areas. A finely ground source of rock phosphate may also be spread onto new windrows to be thoroughly mixed with the compost, as the microbial activity will make the nutrients available to plant roots.

Aerobic composting, through frequent turning, does not contribute to carbon dioxide emissions and prevents the production of methane and nitrous oxide, which all contribute to global warming. Aerobic composting is considered part of the natural carbon cycle and sequesters carbon with the compost and humus, becoming a long-term storage sink which locks up carbon as the compost is spread in the field. Carbon is essential for microbial activity and thus for fertility. This will be discussed in more detail below.

4.8 Maintaining root-system health

Apart from providing a physical support to anchor the banana plant, healthy roots are the basis for good plant growth and sustainable production. The root system of tall followers (1 m, or F-10 stage plants) provides the trunk with a multitude of connected live roots which stabilizes the plant and improves nutrient uptake, even after they are pruned just above the growing point (called 'molars' – see Fig. 5).

Figure 5 Extensive and healthy new root proliferation connected to pruned followers (called 'molars').

The banana plant is nurtured through mass vascular flow when water with nutrients in the soil solution is vacuumed from the roots to the foliage by transpiration ('the pull'). Some of the carbon captured by the leaves through photosynthesis is then transported to the roots and exudate. This subsequently feeds and stimulates the microbes around the roots. To stimulate microbial activity, carbon can also be added to the soil, for example through compost, but the most efficient way is by using water-soluble humic acid applications. Humic acids are derived from rare natural deposits and are basically concentrated organic matter.

There are various organically approved humic-acid-based soil amendments. They contain natural organic acids (humic acids, humic acid precursors, fulvic acids and humins). The carbon which these humic acid products contain activates soil microbes, which then liberate various major, minor and trace elements to the aqueous rhizosphere. They also improve soil structure and water movement, and are especially effective on sandy soils, as well as increasing the CEC. Pure humic acids contain 9% P (as P_2O_5) and 14% K (as K_2O), but zero N.

Some organic fertilizers are made from ocean-caught fish. These fertilizers contain amino acids and minor elements. These usually contain 5% N, 1% P_2O_5 and 1% K_2O. However, these are expensive as they are usually based on saltwater fish and can contain up to 4% chlorine (and sodium as well as heavy metals), which requires chemical analyses and certification before application.

Other organic fertilizers are based on seaweeds, which are harvested in northern hemisphere cold waters and made into a powder which contains 17% soluble K. It is important these are certified as the seaweed should have been produced and harvested sustainably. Kelp-based fertilizers are made from large seaweeds belonging to brown algae (Phaeophyceae) and are harvested in shallow cold ocean waters from underwater kelp forests. Such fertilizers contain up to 60 mineral elements. Brown kelp, for example, contains a high concentration of iodine and various plant hormones as well as amino acids which improve root growth.

Special off-the-shelf or home-made organic N-containing products which are often in liquid form, such as seaweed extracts and vermicompost tea, can be applied as foliar sprays, sometimes together with micro and trace elements. Vermicompost (or worm castings from cultivated worms, usually California Reds) is probably the best organic amendment and can be home-made, but this requires substantial effort.

Composted chicken manure from egg-laying chickens can be used, but the electrical conductivity is often high, which is a problem on soils with a high salinity. Naturally occurring decomposed bat or bird manure (guano) is available from various sources. It is available in large quantities from Chilean and Peruvian coastal islands where it is mined seasonally and sustainably to avoid disturbing bird colonies. Guano contains substantial amounts of sodium and it is therefore not advisable to use it on soils with salinity problems. The USDA NOP rules allow up to 20% of the annual N requirement of a crop to be supplied as sodium nitrate (or Chilean nitrate with 16% N), a mined mineral from the Chilean desert, which also contains a lot of sodium, but the European Union organic standards prohibits it, thereby complicating export of organic bananas to the EU.

All hand-applied organic fertilizers or any organic amendment such as compost should be applied in a complete solid circle over the entire 'drip-line' area in the first crop (plant crop). Once the next-generation production unit is selected, it should be applied in an arc, three quarters around each production follower with at least a 1.0 m radius to prevent roots from 'burning' due to nutrient excess close to the mat.

5 Disease management

5.1 Black Sigatoka

The main problem of Black Sigatoka, apart from the fact that it necrotizes the foliage and reduces yields, is that it also induces premature 'ripes and turns' of bananas in transit to international markets (see Fig. 6). Furthermore, reduction of the photosynthetic area negatively affects the reserves of the corm (bulb) of the plant, delays ratooning (cycling to the next generation) and reduces bunch size (number of hands and filling of the fingers), and thus overall stem weight. The manicuring of infected leaves (or parts) to reduce the inoculum load also opens the banana canopy, which reduces shade at the soil surface and causes abundant competitive weed development. In combination, these factors rapidly place a banana farm in serious economic difficulties due to diminished yields and high costs.

5.1.1 Control in conventional banana production

To control Black Sigatoka in conventional banana plantations in the wet tropics, fungicide applications are needed almost weekly. This is clearly not sustainable and needs to change. The number of applications depends on climatic conditions in terms of rainfall, humidity and minimum night temperatures. When the evaporation does not accumulate to >45 mm per week and nightly temperatures are >20°C, Black Sigatoka will gain a foothold in Cavendish plantations, and significantly hampers organic banana production.

In the wet tropics, every new emerging leaf needs to be protected with a fungicide such as chlorothalonil or mancozeb to control Black Sigatoka as the myriad of systemic fungicides have lost their efficacy due to the selection of strains with reduced sensitivity. A comprehensive review of Sigatoka disease control procedures in conventional bananas from its beginnings when this disease arrived in the New World has been written by

Figure 6 Cavendish plant with severe disease symptoms caused by Black Sigatoka in the wet tropics.

Guzman (2006). Orozco (2013) edited a comprehensive monograph on Black Sigatoka disease which also includes the various monitoring and control practices needed to keep the disease in check. Diaz-Trujillo et al. (2017) discuss recent research on the development of fungicide resistance in Black Sigatoka. In conclusion, due to the extreme susceptibility of Cavendish bananas to Black Sigatoka, all attempts to produce them organically for the fresh banana export market have failed across the wet tropical regions.

5.1.2 Control in organic banana production

The philosophy of organic agriculture is that it achieves sustainability 'by use of cultivation techniques that encourage soil conservation and enhance soil fertility, resulting in healthy plants that withstand pests and diseases, while growing a nutritious and healthy crop'. Positive results can generally be seen in the field when organic practices are employed, but these do not work for controlling an air-borne disease like Black Sigatoka. It is erroneous to believe that enhanced soil fertility and soil health obtained through organic practices can cause banana plants to withstand this very aggressive disease which affects the foliage. Pests and diseases of the root system are a totally different story as microbial life can be enhanced by organic matter applications. This stimulates microbial activity which helps in managing nematode populations and soil-borne diseases.

Despite all claims to the contrary, no effective organic fungicide exists for the control of the Black Sigatoka disease in the wet tropics. All products on the market today which are 'justified' by anecdotal stories or circumstantial evidence suffer from poor experimental design, thus making sound statistical analyses and conclusions on their efficacy impossible. Antagonistic phyllosphere microbes isolated from older banana leaves do prevent *P. fijiensis* spore germination under laboratory conditions, but lack efficacy under field conditions. This might be due to lack of appropriate food substrates on the young emerging 'cigar' leaf before new splash and air-borne inoculum, which causes new infections, arrives.

Copper-based fungicides, of which some are allowed in organic agriculture, have limited efficacy for Black Sigatoka. In fact, they are phytotoxic to young banana leaves, depending on formulation, concentration and droplet size. In addition, they cannot be applied frequently as organic rules prohibit continuous copper applications to prevent accumulating excessive copper in the soil which is poisonous to soil life and affects root development. Aerial applications of mineral oil are permitted in organically produced bananas and will slightly delay expression of the disease, but do not control Black Sigatoka, and certainly not under weather conditions conducive to the disease. Similarly, despite claims, vegetable-based oils (such as palm oil and others) have insufficient efficacy for adequate disease management.

So far, the only practice which mitigates the effects of the disease somewhat is the meticulous, twice-weekly removal of all affected ('burnt') necrotic leaf areas from the banana plant. Once this practice (see Fig. 7) is implemented, together with de-handing (removal of four, five and up to six distal hands) to bring the harvesting date forward, the incidence of premature ripening can be reduced. However, this practice is extremely expensive in labour costs and reduced production and cannot therefore be regarded as a sustainable method in the long term.

In general, it is important to note that Black Sigatoka is highly aggressive as the spores readily infect the new unrolling youngest leaf (or 'cigar' leaf) of every banana plant. In the

Figure 7 Serious Black Sigatoka disease of bananas in the wet tropics with leaves neatly manicured to reduce the likelihood of premature ripening.

wet tropics, there is never a significant dry spell that could reduce the rate of the epidemic; therefore, the only way to organically produce export Cavendish bananas is to avoid Black Sigatoka by planting in very dry environments (see Fig. 8). Overhead micro-sprays will still facilitate Black Sigatoka disease development (Wielemaker and Lobo, 1990).

5.2 Crown rot and mould

Another major challenge for organic banana growers is the occurrence of the post-harvest diseases such as crown rot and crown mould. This complex of diseases is caused by an array of fungi. The most common and prevalent pathogens are species from the genera *Fusarium, Colletotrichum, Gloeosporium* and *Chalara* (previously known as *Botryodiplodia*). The latter is especially aggressive (see Fig. 9). Sanitation at the packing plant is important to lower inoculum pressure and avoid contamination. As the crown tissue of a cut hand is, in essence, a large wound, it is hard to deliver a banana cluster to the end consumer with a healthy, fresh and clean crown.

5.2.1 Treatment methods

Conventional bananas are treated with fungicides such as thiabendazole, imazalil or azoxystrobin, which are either applied alone or in combination. Mechanical fogging chambers with fixed nozzles are frequently part of the packing line, treating each tray of banana clusters for a few seconds by using dipping tanks, cascades, backpack sprayers

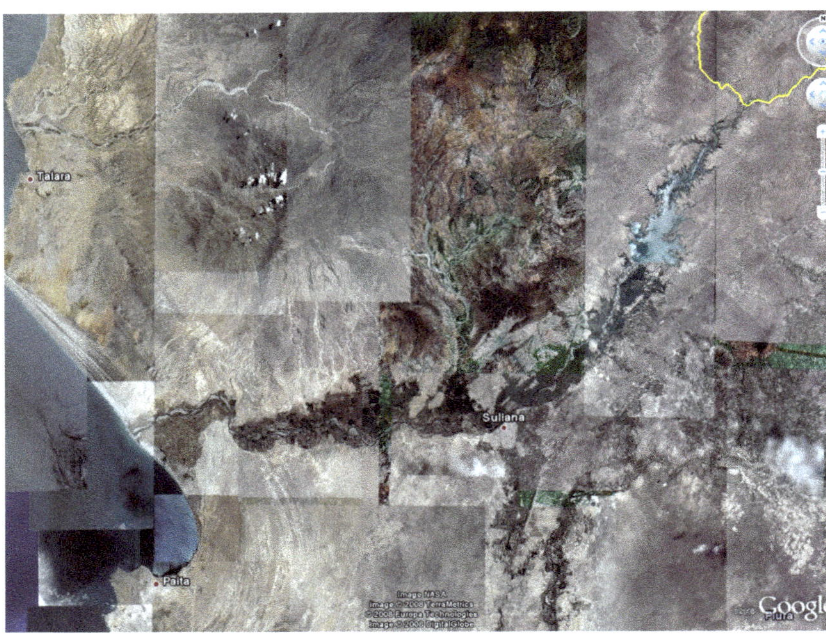

Figure 8 Google Maps image showing where organic bananas are cultivated in the desert on the Northern coast of Peru where water is provided by the Poechos Dam.

or brush applications. Organic banana crowns are treated with natural products such as citrus seed extract.

A more recent and hands-on method which effectively preserves the crowns of organic bananas is the application of a paraffin-based polyolefin wrap (Parafilm or ParaSeal) to each individual crown (see Fig. 10). This wrap deprives the rot and mould spores of oxygen, creating a miniature modified atmosphere (MA) environment around each crown. These materials are not cheap, and the operation is laborious, but it saves the higher cost of alternatives such as vacuum bags and the labour cost involved in opening bags at the ripening facility. It therefore enables organic bananas to be shipped in standard reefer containers whilst still allowing normal ripening. The paraffin polyolefin wrap protection stays on the crown up to the retailer stage, while controlled/MA treatments are removed several days earlier at the start of the ripening procedure. The covered wound prevents latex oozing from the crown and maintains the turgor of the tissue, giving the crown a fresh look.

One organic solution for controlling post-harvest rots and moulds during transit to export markets without the use of fungicides can be achieved by packing bananas in a tightly closed plastic bag within the carton box. This long-standing practice is called vacuum bag packing and is also used in the conventional banana trade because it prolongs the green-life of the fruit. This packing technique removes oxygen, thus increasing the level of carbon dioxide, creating an MA which inhibits pathogen growth. The change in the proportion of these gases is caused by the respiration of the bananas. As the bags are tightly closed, the bananas use the oxygen which remains inside the bag. The oxygen level inside the bag therefore drops slowly but steadily. This lowers the respiration rate to the metabolic minimum required, thus extending the green-life of the bananas substantially.

Figure 9 *Chalara* sp. (previously known as *Botryodiplodia* sp.) is a very aggressive post-harvest disease.

Figure 10 Crowns individually wrapped in paraffin-based plastic polyolefin.

This practice of using vacuum bag packing is recommended for transporting organic bananas and conventional bananas as it does not involve the use of any chemicals. If the bag is not completely airtight, the benefits will not be not achieved. The oxygen level should not fall below 2% (which prevents the fruit from ripening normally) and causes 'black peel' symptoms.

Newer shipping containers with tighter closing doors and lower air leakage rates can put a banana load at risk and provoke 'black peel' due to lack of oxygen, particularly if the bananas are already packed in vacuum bags. As noted, the bags need to be manually cut open before entering the ripening room. Special micro-perforated polyethylene bags have been developed with variable O_2 and CO_2 permeability through micro-perforations in the plastic. This is known as modified atmosphere packaging (MAP). Several branded technologies exist in the market. In contrast with vacuum bags, fruit can be ripened without having to cut the MAP bags open, which saves labour and reduces costs.

Instead of reaching the equilibrium of 2% oxygen by use of vacuum or micro-perforated plastic bags, there are now 40 ft shipping containers that do exactly the same at the macro scale. All these units work on the principle of reducing O_2 or generating CO_2. Where the gaseous atmosphere is changed by adding more N than that already exists in the ambient air to replace the oxygen in the shipping container, it is referred to as controlled atmosphere (CA). There are detailed procedures for establishment of MA or CA to reach optimum N_2, O_2 and CO_2 levels, in reefer containers, in separate hatches or entire cargo holds.

MA, MAP and CA techniques are allowed under organic regulation and represent perfect solutions as they involve changing only the gaseous atmosphere around the fruit and preserve freshness without applying chemical post-harvest fungicides. They involve more costs in labour, materials and equipment and are not yet generally used not even for conventional bananas to the major markets of North America and Europe. Distant markets, however, are usually supplied using CA.

6 Managing nematode and insect pests

6.1 Nematodes and weevil borers

In addition to Black Sigatoka, there are other challenges in growing bananas sustainably and organically, including nematodes and weevil borers. *Radopholus similis* is the most prevalent nematode affecting bananas in Central and South America, although other nematodes also contribute to root infestations. These infestations result in loss of water and nutrient uptake and consequent substantial losses such as lower bunch weight caused by fewer hands and shorter fingers, loss in ratooning ratio (bunches per mat per year), and immediate production unit losses due to fallen plants ('tip-overs').

Conventional banana growers control these nematodes and the banana weevil borer (*Cosmopolites sordidus*) at the same time, using synthetic nematocides. The organophosphates fenamiphos, ethoprop, terbufos or cadusafos are rotated with the carbamates, carbofuran or oxamyl to avoid the selection of populations with reduced sensitivity. If these chemicals were not applied, a vast number of conventional banana farms would not be financially viable. The use of nematocides is the main reason that it has been possible to grow Cavendish monocultures for decades without crop rotation or a fallow period. But if conventional banana growing is to become more environmentally friendly, the use of synthetic nematocides will need to be banned or drastically reduced.

The organic approach to maintaining nematodes at acceptable levels is based on the principle that healthy soils will result in healthy roots. Annual applications of 20 tons of good mature compost per hectare have been shown to keep *Radopholus* populations under control. Biological nematocides exist and are based on the naturally occurring nematofagous fungal genus *Paecilomyces*. In addition, root-health-improving microorganisms based on *Streptomyces*, *Pseudomonas*, *Bacillus* and *Trichoderma* are also frequently applied.

Populations of weevil borers in organic banana plantations can be reduced by using 'sandwich' traps made of fresh pseudostems. Adding a pheromone to the trap, of which several commercial versions exist, greatly increases the catch rates, keeping populations at manageable levels over time. Applications around the base of the mat of commercially available biological products based on *Beauveria* and *Metarhizium* fungi that infect the weevils also help to reduce populations. Both fungi can be 'home-grown' in relatively simple facilities.

6.2 Insects

There are numerous insect pests which affect bananas. Those affecting the bunch are the most critical and the most difficult to control. The bunch pests of primary concern are thrips, scales, mealy bugs and beetles, although there are several other insects which also cause damage. In conventional banana cultivation, plastic bags (or strips or scarfs) are impregnated with insecticides (active ingredients such as chlorpyrifos, buprofezin, bifenthrin and imidacloprid) and placed over the growing bunch for about 3 months. The impregnated bag concept, which only protects the bunch, is a good example, in the context of conventional banana production, of integrated pest management as it avoids blanket spraying, but obviously cannot be used in organic banana production. Plastic bags in organic banana cultivation may contain non-synthetic botanical insecticides. They protect the banana bunches from bird and bat scratches, but also stimulate the rate of fruit development which results in earlier harvest and a shorter cycle.

6.2.1 Red rust thrips

The red rust thrip (*Chaetanaphothrips* spp.) is probably the most difficult pest threat in organic banana production. The cosmetic damage done by this insect rapidly renders the fruit unmarketable due to consumer rejection (see Fig. 11). The threshold in terms of number of insects per bunch is therefore extremely low. Frequent spray cycles are required for the young emerging banana bunches to control these thrips in organic banana farms. The most effective botanical insecticides are pyrethrin, derived from the *Pyrethrum* flower, and spinosad, the active ingredient of which is a fermentation product of the soil bacterium *Saccharopolyspora spinosa*. Once infestations become severe, spraying the entire pseudostems, with or without fruit, is required. All these procedures will also reduce the infestation of the flower thrips (*Frankliniella* spp.), which can also cause substantial damage to the fruit in organic banana farms.

According to the US EPA and thus NOP organic regulations, organic farmers can no longer use a number of natural insecticides. These include nicotine and rotenone, since manufacturers have not considered it cost-effective to provide the required toxicological data package to support re-registration of these botanicals. The botanical insecticides Sabadilla and pyrethrine may soon also be banned due to the lack of supporting toxicological data. Such data is extremely costly to generate and not justifiable since, as natural substances, these products cannot be patented as they are naturals and no return on investment can be obtained.

6.2.2 Banana fruit scarring beetle

The banana fruit scarring beetle Colaspis (*Colaspis ostmarki)* is another challenging insect for organic banana growers. Colaspis adults can fly and both sexes harm the fruit. At

Figure 11 Severe damage caused by red rust thrips, which causes cosmetic damage and makes the banana unmarketable.

present, the only treatment is prevention by using a 'non-woven' fabric bag, made of polypropylene. The bag needs to be placed very early over the young emerging banana flower bell and should be closed tightly at both ends. To protect the fruit in an organic plantation, bagging cycles should be made two to three times per week.

6.2.3 Mealy bugs and scale insects

Mealy bugs and scale insects are difficult to keep under control. There are various mealy bugs belonging to the genera *Pseudococcus, Dysmicoccus, Planococcus* and *Ferrisia*. Adult mealy bugs leave a sticky secretion – called 'sooty mold' – which facilitates the development of black moulds and drips onto lower hands, thereby exacerbating the problem.

The coconut scale *Aspidiotus destructor* and the orchid scale *Diaspis boisduvalii* are common scale insects in Latin America. Scales generally move slowly but will eventually reach the fruit if not controlled in a timely manner. Mealy bugs and scale insects do not necessarily cause a lot of damage. However, they are classified as quarantine organisms considered 'actionable bugs' in many ports of arrival, resulting in containers being put on hold for several days until the scales are identified by a qualified entomologist. Pending identification, the whole container either needs to be fumigated or returned to the country of origin. For organic bananas, fumigation with a synthetic fumigant is not allowed and, as an effective natural fumigant does not exist, the containers would be refused and need to be returned or destroyed.

The presence of small numbers of mealy bugs and scales is very hard to detect as they hide between the fingers and necks of the bananas and even in the remnants of the flower scars on the banana tip, thus requiring thorough inspection (Fig. 12). In addition to 'actionable'

Figure 12 Organic bananas destined for Japan and Korea need thorough inspection before sending them to the markets as a single insect may cause quarantine problems at the port of entrance.

bugs on the harvested fruit, 'hitchhikers' which may crawl or fly into the container when the doors are open during loading and may cause similar quarantine problems. Screening nets are therefore installed to prevent their incursion and transportation. These insects present a considerable challenge for organic banana growers as one single bug detected at the port of entry will result in very high economic losses.

7 Weed management

Keeping weeds at bay presents a considerable challenge in organic banana growing as weeds compete with the bananas for nutrients. The basic objective of weed control is to ensure the competition for nutrients favours the banana crop. The proliferation of climbing weeds must be prevented to avoid the need for excessive clearing of the pseudostem and bunch.

It is important to keep the ground surface of a banana farm covered, ideally with a cover crop that will additionally help to prevent erosion. Conventionally grown bananas will receive at least four annual herbicide cycles using the following herbicides in rotation: glyphosate, glufosinate-ammonium, paraquat, diquat, a mixture of paraquat with diquat, diuron, glyphosate trimesium or indaziflan. Ametryne is exclusively used around banana plantations as it is phytotoxic to bananas. Applying a pre-emergent pre-plant seal of oxyfluorfen herbicides helps conventional banana farmers through at least the first month after planting on bare soil as it prevents weed seeds from germinating.

None of these synthetic herbicides are permitted in organic banana production. In a well-managed conventional banana plantation free of the defoliation caused by Black Sigatoka, weed control cycles can be reduced as shading suppresses weeds. Only patches with the most obnoxious grass weeds then need to be kept under control.

7.1 Cutting and mowing

Apart from concentrated vinegar applications, there are no effective non-chemical synthetic weed control methods available to the organic banana grower. Weed control therefore requires the application of cultural methods. Only physically cutting or mowing weeds is permitted in organic banana farming which constitutes a significant cost in labour. Grass weeds are more problematic than broad leaf weeds, as they compete more aggressively with the relatively superficial root system of a banana plant.

Invasive grasses such as Elephant grass and Nutsedge grass should be controlled or even pulled up by hand and taken out of the plantation before they set seed. Weeds may harbour pests or snakes and should therefore also be kept low for both pest management and safety reasons. They may also host viruses such as the Cucumber Mosaic Virus which can be transmitted to the bananas by insects, causing yield reduction and giving rise to blotched ripening, increasing the likely need to remove virus-infected banana plants.

Weeds in organic banana plantations are mainly controlled by machete and power trimmers which use a spinning nylon string. Care should be taken not to damage newly planted banana plants or followers. Control cycles should be sufficiently frequent to prevent any weeds from seed setting and dispersal. Once a newly planted organic banana plantation closes its canopy, the ground surface is fairly well shaded and weed growth becomes less abundant.

7.2 Ground cover: crop residues, cover plants and other techniques

As noted, controlling weeds is challenging before the closed canopy stage is reached. Once the harvest starts, all crop residues should be well distributed over the plantation floor to smother weeds as much as possible. These residues should not be moved away from the base of the mats or put onto rows. Weeds should be kept as low as possible in a circle of at least 1 m in diameter around each plant. Cleaning the base of the mat from leaf debris should only be allowed just prior to an application of an organic fertilizer.

Growing a ground cover using selected plants which are adapted to grow under the closed banana canopy is a good practice for controlling weeds, preventing soil erosion and promoting healthy soil conditions for long-term production. The most effective way is to manage and encourage the growth of suitable native broadleaf weeds as a mixed stand cover crop. Ideally the cover crop should be adaptive to seasonable changing conditions such as shade and full sunlight exposure, cold and heat stress, water stress tolerance and possible recovery after flooding.

The selection, multiplication, introduction and establishment of a cover crop in a plantation is neither easy nor cheap. It should tolerate or outgrow grass/weed competition and grow under and through banana crop residues. It should also have a shallow root system which does not directly compete with the banana crop. It must not climb into the banana pseudostem or followers, harbour diseases, attract pests, become a host plant or impede workers from tending banana plants. A cover crop must prevent soil erosion, suppress weeds, improve physical soil characteristics and stabilize soil temperature. It must also alleviate soil compaction and improve the infiltration rate of irrigation and rain water carrying nutrients when applied to the roots.

Wielemaker et al. (1997) showed that the cover crops 'Pinto's Peanut' (*Arachis pintoi*) and 'Mouse Ear' (*Geophila repens* or *G. macropoda* and *Oreja Ratón*) do not compete with the banana crop in terms of production in a conventional banana farm. This replicated trial, conducted over six and a half years, compared these two cover crops with the standard operating procedure using the herbicides glyphosate and paraquat, and with mechanical weed control. Mouse Ear proved to be the best of the two cover crops as it maintains a low growth habit and is not a host to banana nematodes. Indigenous Mouse Ear offers a very good cover crop for banana plantations as banana leaf debris did not suffocate the established ground cover which grew through it without any difficulty. This was a new development since any plants growing in a plantation other than bananas had previously been considered to be weeds (Fig. 13).

Other methods of controlling weeds have been tried such as solarization with transparent plastic tarpaulins which 'cook' the epidermis of the weeds. When the cover is removed, sunlight will then complete the withering of the foliage. Weed control with a small tractor-drawn tarpaulin under which steam is inserted has been successful in row plantings. A somewhat similar result is achieved with gas-fed flame-throwers and to a certain extent, also with spraying vegetable oils. However, weeds treated by the methods described above tend to come back quite quickly as the root systems are not affected.

Figure 13 The indigenous 'Mouse Ear' (*Oreja Ratón*) cover crop adapts itself very well to the banana plantation floor.

8 Discussion

The banana industry overall needs to undergo radical change if it is to manage Black Sigatoka in an environmentally friendly way. It will be necessary to invest in the sustainability of conventional banana cultivation in the wet tropics and convert large areas to organic production as soon as is feasible. An alternative solution is to develop varieties with durable resistance, through either classical breeding or induced mutagenesis enabling the selection of soma-clones resulting from *in vitro* plant production. But unfortunately, even after many decades, a Cavendish clone with resistance to Black Sigatoka has never been successfully developed. The few clones with reduced susceptibility had very low brix values (a measure for sweetness) and were therefore not commercially viable. The only pure mutation trait ever found in Cavendish was dwarfism. Such individuals were found in Giant Cavendish plantations and became the Dwarf Cavendish (or Ecuador Dwarf) variety which changed the industry as these are more resistant to high wind speeds and don't topple over. This actually occurred before tissue culture became the mainstream method of producing planting material.

The chance of finding a Black-Sigatoka-resistant plant today with *in vitro* multiplication should in theory be much higher due to somaclonal variance. Based on sheer selection of millions of *in-vitro*-produced Giant Cavendish plants, the Taiwan Banana Research Institute found a few lines with reduced susceptibility to the soil-borne fungus *Fusarium oxysporum* f.sp. *cubense* Tropical Race 4. These are seen as a 'stop-gap' solution to the TR4 problems in Asia and Africa and are called Giant Cavendish Tissue Cultured Variants (GCTCV), such as GCTCV-119 and, GCTCV-218 and GCTCV-219. Unfortunately, somaclonal variation approaches have never resulted in any commercially interesting plants with reduced susceptibility to Black Sigatoka.

The Honduran Foundation of Agricultural Investigation has recently announced a renewed classical breeding program (Fig. 14). Although this renewed effort to improve Cavendish is focused on the TR4 threat, the male parent(s) with the presumed TR4 resistance trait(s) may also provide (partial) resistance to Black Sigatoka. Depending on the level of resistance obtained, it may contribute to growing conventional bananas with reduced fungicide inputs, or even to converting large areas to organic banana cultivation in the wet tropics. This could become a major game changer. Such a new variety will, however, always be a near copy and never a true-to-type copy of Cavendish. Classical breeding programmes are also conducted by CIRAD in Guadeloupe, and by the Brazilian Agricultural Research Corporation in Cruz das Almas. This represents a modest set of breeding programmes for such an important crop, and any new initiative contributing to more sustainable banana production should be welcomed.

Because genetic engineering is prohibited in organic agriculture, it is unfortunately not an option to adopt an engineered Black-Sigatoka-resistant banana variety (Manning 2004; Ronald and Adamchak, 2008). It will, however, become an option in conventional banana production in the near future. The result would be a 100% Cavendish-true-to-type banana variety. This would be a major change for the conventional banana industry as it enables banana production in the wet tropics with limited fungicide inputs for Black Sigatoka control. New technologies such as cisgenesis (using genes from wild inedible but resistant bananas) may make the genetic modification technologies more acceptable to the general public (Dale et al., 2017). In this respect, the recent CRISPR Cas-9 technology is much more precise because it edits only particular gene(s), and this technology could be used to engineer Cavendish bananas resistant to Black Sigatoka without inserting foreign DNA. It is therefore a promising technology for improving disease resistance, not only to Black Sigatoka, but also to Fusarium wilt (TR4) and Banana Bunchy Top Disease (BBTV). Wageningen University and Research in the Netherlands, and the Queensland University of Technology in Australia are working on engineering technologies towards these goals.

Figure 14 Classical breeding showing pollination of Cavendish with Cavendish to create a tetraploid, later to be crossed with a disease-resistant diploid.

Other potential developments are in nutrition. Conventional banana nutrition production uses slow-release polymer-enclosed and sulphur-coated urea (or ammonium sulphate). These controlled release formulations (CRF) are designed to mimic the slow release achieved with organic fertilizers, supplying the same NO_3^- and/or NH_4 as nutrients to the roots. This approach offers a potential solution to alleviating the shortage and high cost of organic fertilizers and amendments from sometimes limited sources (such as guano and kelp). However, this approach is opposed by some organic experts. This is because urea is synthetically derived by a carbon emitting industry from the atmosphere which contains 78% N. A key challenge is to produce CRF fertilizers in a more sustainable way. There is a need for new thinking if organic banana production is ever to replace conventional production once the Black Sigatoka issue is solved.

9 Where to look for further information

Soto (2016), *Bananos, Tecnologías de Producción* contains substantial information on organic soil amendments. A seminal work titled *Agricultura Orgánica* by the German scientist Albrecht Benzing (2001) is probably one of the best books on organic agriculture written in Spanish and is applicable to the Andean region of Latin America. Gary Zimmer (2000), *The Biological Farmer* is geared towards farming in North America, and gives considerable insight into caring for soil and making it productive.

In *Dirt, The Erosion of Civilizations*, Montgomery (2007) expertly describes how little care humanity has taken of soils, from ancient civilizations to modern times. Once more, the focus on soil and soil health in organic agriculture is the basis of this farming practice. Montgomery and Biklé (2016) in a new book, *The Hidden Half of Nature: The Microbial Roots of Life and Health*, explain how microbes work in the rhizosphere, again emphasizing soil as the substrate of life and plant growth. The Soil Science Society of America has a very good website promoting the importance of education on soils and soil life and the requirement of a 'rich' soil for agricultural production. To enhance soil life, microbial inoculants can be bought off the shelf or 'made in-house' by trapping efficient microbes in nearby forests.

Nutrients present in the soil are released by the increased microbial activity. Vesicular arbuscular mycorrhizal fungal microorganisms solubilize phosphate and *Azotobacter* bacteria fix N. The banana root microbiome needs much more study to increase our understanding. Draye (2010) in a chapter called, 'Banana Roots, Architecture and Genetics', made a very good contribution to this field in the book *Plant Roots, The Hidden Half*. A symposium called, *Banana Root System: towards a better understanding of its productive management* also provides a good start to addressing this subject (Turner and Rosales, 2004).

The best book, perhaps the only one exclusively dedicated to banana nutrition, is published in Spanish by the Colombian association of banana producers (AUGURA), and is written by Cenibanano Research Centre scientists Dr Javier Sanchez and Dr John Jairo Mira (Sanchez and Mira, 2013). This gives the most comprehensive overall insight into banana nutrition. Other good sources are *Nutrient Management Plan for Organic Systems, an Implementation Guide* by Dufour et al. (2014) and the USDA/National Resources Conservation Service (NRCS, 2016) publication, which has information on their website on organic agriculture, similar to the Soil Science Society of America (SSSA, 2016).

10 Disclaimer

The author has tried to describe as accurately as possible the practices and challenges of the conventional banana industry while discussing the alternatives and existing limitations in growing bananas organically. All photographs were taken by the author himself and should be considered copyrighted without exception.

11 About the author

The author graduated from the Tropical Agricultural College in Deventer in the Netherlands. He has worked in research and development positions in the Latin American banana industry since 1978, of which 33 years were with Dole Fresh Fruit International, mainly in the field of pest and disease control, especially the many aspects of Black Sigatoka control. Motivated by the desire to change the industry, he pioneered Dole´s Organic program in 1995 and, with his team, made Dole today´s leader in organic banana (and organic pineapple) production and marketing. He finally held the position of Director of Research at Dole's headquarters in San José, Costa Rica. Currently, he is active as an independent consultant (AGROASESORIA SA) working with clients throughout Latin America. He is co-founder and director of an organic banana export company (through CONDOR SAC), as well as an organic materials input company (LIVING GROUND SAC) in Peru, marketing key inputs to organic banana farmers throughout Latin America and marketing certified organic bananas in the United States and EU (through TRANSASTRA SA).

12 References

Benzing, A. (2001). *Agricultura Orgánica: Fundamentos para la Región Andina*. Neckar-Verlag, Germany, 682pp.

Dale, D., Paul, J.-Y., Dugdale, B. and Harding, R. (2017). Modifying bananas: From transgenics to organics? *Sustainability*, 9(3): 333–46.

Diaz-Trujillo, C., Chong, P., Cordovez Da Cunha, V., Guzman, M., De Wit, P. J. G. M., Stergiopoulos, I., Meijer, H. J. G., Arango Isaza, R. E., Scalliet, G., Sierotzki, H., Peralta, E. L. and Kema, G. H. J. (2017). A new resistance mechanism to azole fungicides in the fungal banana Black Sigatoka pathogen *Pseudocercospora fijiensis* is driven by increased expression of Pfcyp51 through multiple promotor repeats. *Molecular Phytopathology*. doi:10.1111/mpp.12637.

Draye, X. (2010). Chapter 16: Banana roots: Architecture and genetics. In: Waisel, Y., Eshel, A. and Kafkafi, U. (Eds), *Plant Roots: The Hidden Half*. CRC Press, Marcel Dekker, NY, pp. 261–77.

Dufour, R., Brown, S. and Troxell, D. (2014). *Nutrient Management Plan for Organic Systems: An Implementation Guide*. USDA, NCAT and Oregon Tilth. www.ncat.org

Eyhorn, F., Heeb, M. and Weidmann, G., eds. (2002). Training manual for organic agriculture in the tropics. http://www.s-ge.com/global/export/en/filefield-private/files/182754/field_blog_public_files/69934

Guzman, M. (2006). Current situation and future perspectives of the Black Sigatoka management in Latin America. *2006 ACORBAT Brazil, Proceedings*, pp. 83–91.

Hawken, P. (2010). *The Ecology of Commerce: A Declaration of Sustainability*. Harper Business, New York, NY.

Holderness, M., Sharrock, S., Frison, E. and Kairo, M., eds. (2000). Organic banana 2000: Towards an organic banana initiative in the Caribbean. *Proceedings of Meeting held in the Dominican Republic*, November 1999, Published by INIBAP.

Howard, A. (1943). *An Agricultural Testament*. Oxford University Press, New York, NY.

Lockeretz, W. (2007). *Organic Farming: An International History*. CABI, Wallingford, UK.

Manning, R. (2004), Super organics: The future of food. *WIRED Magazine*.

Martin, D. L. and Gerchuny, G., eds. (1992). *The Rodale Book of Composting*. Rodale Press, Emmaus, PA.

Meier, U., ed. (2001). Monograph 'Growth stages of mono-and dicotyledonous plants'. Available at: http://www.jki.bund.de from the BBCH (Biologische Bundesanstalt, Bundessortenamt and Chemical industry). Also available in Spanish.

Montgomery, D. R. (2007). *Dirt: The Erosion of Civilizations*. University of California Press, Berkeley, CA, 285pp.

Montgomery, D. and Bikle, A. (2016). *The Hidden Half of Nature: The Microbial Roots of Life and Health*. W.W. Norton & Company Publisher, New York, NY.

National Research Council (2010). *Toward Sustainable Agricultural Systems in the 21st Century*. The National Academy of Sciences and National Academy Press, Washington DC.

NOP (2016). National Organic Program's official website. https://www.ams.usda.gov/rules-regulations/organic/nosb

NRCS (2016). National Resources Conservation Service. http://www.nrcs.usda.gov/

OECD (2009). Consensus Document (#48) on the Biology of Bananas and Plantains. http://www.oecd.org/science/biotrack/46815938.pdf

Orozco, M., ed. (2013). *La Sigatoka Negra y su Manejo Integrado en Banano*. INIFAP, Colima, Mexico.

Paull, E. and Duarte, O. (2011). Chapter 8: Bananas and Plantains. In: *Tropical Fruits*, volume 1, 2nd Edition. CABI, Wallingford, UK, pp. 185–220.

Robinson, J. C. and Galán Saúco, V. (2010). *Bananas and Plantains*. CABI, Wallingford, Oxford.

Ronald, P. C. and Adamchak, R. W. (2008). *Tomorrow's Table: Organic Farming, Genetics, and the Future of Food*. Oxford University Press, Oxford, UK.

Rosales, F. E., Tripon, S. C. and Cerna, J., eds. (1999). *Organic/Environmentally Friendly Banana Production*. INIBAP and CRDI, Montpellier, France. ISBN: 2-910810-99-2.

Sánchez, J. D. and Mira, J. J. (2013). *Principios para la Nutrición del Cultivo de Banano. Centro de Investigación Cenibanano*. Published in Spanish by Augura, Colombia. www.augura.com.co

Soto, M. (2016). *Bananos, Tecnologías de Producción*. Editorial Tecnológica de Costa Rica, Cartago, Costa Rica.

SSSA (2016). Soil Science Society of America. https://soilsmatter.wordpress.com/

TOS (2016). *The Organic Standard Magazine*. Organic Assignments C.B., Vigo, Spain. http://organicstandard.com

Turner, D. W. and Rosales, F. (2004). Banana Root System: Towards a better understanding of its productive management. *Symposium Proceedings of Meeting* held November 2013 in San Jose, Costa Rica. Published by INIBAP.

Wielemaker, F. (2013). Retos de la investigación en Banano Orgánico y Avances Relevantes en Control de Enfermedades y el Manejo de la Nutrición. Presentación en la 3ra Reunión Técnica Bananera de Colombia, Universidad Magdalena, Santa Marta, 6 and 7 Junio 2013. CENIBANANO.

Wielemaker, F. and Lobo, J. R. (1990). Practical notes on Black Sigatoka control. In: Fullerton, R. A. and R. H. Stover (eds), *Sigatoka Leafspot Diseases of Bananas*. INIBAP, pp. 107–14.

Wielemaker, F. and Monge, L. (2015). Petitioning for continued listing of Ethylene on the 2017 Sunset list comment by the authors submitted at the 2015 Fall NOSB Meeting in Stowe, Vermont, Tracking Number:1jz-8l8n-7h6a. See: http://www.regulations.gov/document?D=AMS-NOP-15-0037-0028.

Wielemaker, F., et al. (1997). Banana production with 'Pinto's Peanut' (*Arachis pintoi* cv *Amarillo*) and 'Oreja De Raton' (*Geophila repens*) as cover crops. *1er Taller International Sobre Control Biologico y Manejo Integrado de Plagas en el Cultivo de Banano*, 16–19 November 1997, Costa Rica. Fundacion AMBIO.

Wielemaker, F., Monge, L. and Saenz, L. (2009). The Traceable and Authentic Organic Banana from Dole: Building a link between the Consumer and the Organic Farm using the Dole Producer Code (DPC). Presentation at the 2009 Annual Trade Show and Conference in Chicago of the Organic Trade Association (OTA).

Zimmer, G. F. (2000). *The Biological Farmer*. Acres USA Publishers.

Index

1-aminocyclopropane-1-carboxylic acid 258
6-Benzylaminopurine (BAP) 131
5S 173
18L knapsack pump diaphragm sprayer 248

AAA varieties 13, 35, 46, 52, 65, 99, 102, 105, 133
ABB bananas and cultivars 6, 9, 10, 13, 14, 43, 46, 65–66, 105, 153
Acclimatisation stage 135
ACORBAT. *see* Association for Cooperation in Banana Research in the Caribbean and Tropical America (ACORBAT)
Active photosynthetic radiation (PAR) 209–210
Aerobic composting 329
African plantains 13–14
'Agbagba' False Horn plantain 103
Agricultural Marketing Service (AMS) 319
Agrobacterium tumefaciens 52
Agroforestry systems 36
Aguirre, Fernando 21
Alliances and networks 68
Ammonium sulphate 326
AMS. *see* Agricultural Marketing Service (AMS)
Antagonistic phyllosphere microbes 332
Antibody-based diagnostic 46
Arbenz, Jacobo 26–27
Archaeobotanical evidence of banana cultivation 11–12
Ascorbic acid 131
Association for Cooperation in Banana Research in the Caribbean and Tropical America (ACORBAT) 322
Axial sucker 205, 207
Ayr Research Facility 152, 153, 155
Azoxystrobin 334

BA. *see* Benzyladenine (BA)
Babuvirus 74
Backyard garden 36
Bacterial contamination 130
Bacterial wilt resistance 52
Bacteriophage Phi29 DNA polymerase 46
Banana backyard systems 36
Banana bunch top disease (BBTD) 74
Banana bunchy top virus (BBTV) 73, 121, 135, 343
Banana cultivars 183–184
Banana cultivation
  in Africa 33–57
    cultivar composition and distribution 34–35
    importance and challenges 33–34
    improving banana varieties 50–53
    marketing as constraint 54–55
    moisture stress 41
    nutrient deficiencies 40–41
    pests and diseases 41–49
    production systems 35–39
    in Uganda 39–40
  in subtropics 195–219
    access to market 218–219
    cultivars 216–217
    description 195
    greenhouse 207–216
    influence of climate 196–199
    plant protection 218
    problems 200
    removing healthy leaves 207
    solving problems 200–207
  see also Organic banana cultivation; Sustainable banana cultivation
Banana industry
  Cavendish era 27–28
  Cavendish in crisis 28–29
  description 21–22
  hunt for new varieties 29–30
  monoculture
    development 23–24
    dynamics 24–25
    varieties and Cavendish 22–23
  post-war developments 25–27
Banana intercrop system 36–37
Banana monocrop systems 37
Banana Research Program 53
Banana streak virus (BSV) 46, 73, 124
Banana suckers 117, 118–119
Banana value chain
  challenges 67–79
    pre-production 67–70
    processing 75–79
    production 70–75
  description 65–66
  in Ecuador 296–297
  enabling environment 79–80
  Peruvian 294–296
  and sustainability 282–283, 289–290
Banana weevil (BW) borer 42–43, 72–73, 121, 179–180, 337
Banana Xanthomonas wilt (BXW) 47
'Bananos, Tecnologias de Produccion' 322
Banavac plastic bag 262
BAP. *see* 6-Benzylaminopurine (BAP)
Barcode scanning system 128
Bateson–Dobzhansky–Muller model 10
BBTD. *see* Banana bunch top disease (BBTD)
BBTV. *see* Banana bunchy top virus (BBTV)
Beauveria bassaiana 45
Benzyladenine (BA) 131
Benzylaminopurine 123
BG Door International BV 271
'Big Mike.' *see* Gros Michel
Biodiversity integration 182
Biology of Bananas and Plantains 322

Biomass 175–176
Bioreactors 132
Biosecurity 146
Biotropic GmbH 291
Bioversity International 40, 147, 322
Black, Eli 28
Black leaf streak disease. *see* Black Sigatoka
Black Sigatoka 45, 73, 74, 144, 218, 261, 331–334, 342, 343
Bottom-up kaizen methods 179
Brazilian Agricultural Research Corporation 343
BSV. *see* Banana streak virus (BSV)
Bunch
  age 237
  covers 207
Bunch formation
  aerial true stem growth 108–110
  description 93–94
  floral phase 97–99
    overview 97
    timing of events 97–99
  formation of hands and fruit 102–108
    genotypes 103–106
    seasonal and edaphic conditions 106–108
  gross morphology and 94–95
  inflorescence 99–102
    fruit-forming and non-fruit-forming 102
    thyrse with cincinni 100–102
    vegetative apex and external indications 99–100
  phases of development 95–97
    methodology 96–97
    phases and 'clocks' 95–96
  photosynthate availability and 110–113
    discussion 112–113
    materials and methods 110–111
    overview 110
    results 111
Business process re-engineering 173
BW. *see* Banana weevil (BW) borer
BXW. *see* Banana Xanthomonas wilt (BXW)

CA. *see* Controlled atmosphere (CA)
Cableway infrastructure network 240–241
CAC. *see* Codex Alimentarius Commission (CAC)
Caffery, Jefferson 24
Calcium nitrate 326
Calibration and maintenance 277
Callimusa 3–4
Carbamate insecticides 44
Carbofuran 44
Carbon footprint (CFP) 182, 188, 281, 285–286
Carbon Trust 284
Carson, Rachel 159–160
Cation exchange capacity (CEC) 328
Cavendish banana and cultivar 15, 22, 27, 29, 97, 99, 106, 124, 133, 182, 217–218, 310, 334, 342
Cavendish cultivar 15, 217–218
CCLAC. *see* Coordinating Committee for Latin America and the Caribbean (CCLAC)
CEC. *see* Cation exchange capacity (CEC)
Celloclim® 211
Cell suspension 126
Certification, principles of 291
Certified Organic status 319
CFP. *see* Carbon footprint (CFP)
Chaetanaphothrips spp. 338
Chiquita 21, 23–28, 29, 144, 155, 237, 306
Chiquita 2000 271
Chiquita International Services Group 271
Christian Science Monitor 26
Cincinnus 100–102
CIRAD 186. *see* French Agricultural Research Centre for International Development (CIRAD)
Circular conveyors 248
Climatic changes 199
Climatic hazards 200
'Clowd' diseases 49
Codex Alimentarius Commission (CAC) 159, 160, 252, 321
Coefficients of determination 111
Cohn, Irving 25
Colaspis ostmarki 338
Commelina caroliniana 101
Continuous improvement 173, 179
Controlled atmosphere (CA) 336
Controlled release formulations (CRF) 344
Conventional banana production 331–332
Conventional breeding 50–52
Conventional planting material 119, 122–124
Cooking bananas 65, 75, 79
Coordinating Committee for Latin America and the Caribbean (CCLAC) 252
Copper-based fungicides 333
CORBANA. *see* Corporacion Bananera Nacional de Costa Rica (CORBANA)
Corporacion Bananera Nacional de Costa Rica (CORBANA) 322
Corporate Social Responsibility 307
Cosmopolites sordidus 42, 72, 149, 218, 337
Cost–benefit analysis 298
Cost competitiveness 78
Council Regulation (EC) No. 834/2007 319
Cover plants 341
CRF. *see* Controlled release formulations (CRF)
CRISPR Cas-9 technology 343
Crop
  growth simulation models 228
  residues 341
  response 228
Crown rot and mould 334–337
Curve-fitting 111
Cytokinins 123, 131–132

Dale, James 29
DArT markers 6

Integrated crop management (ICM) 172, 174, 182, 185
Integrated pest management (IPM) 73
Interko/TH system 268
International Federation of Organic Agriculture Movements (IFOAM) 305, 321
International Institute for Environment and Development (IIED) 305
International Institute for Sustainable Development (IISD) 305
International Institute of Tropical Agriculture (IITA) 40, 50
International Labour Organization 165, 306
International Network for the Improvement of Banana and Plantain (INIBAP) 322
International Organization for Standardisation (ISO) 304
International Plant Protection 160
International Trade Center (ITC) 160, 165, 166–167
International Transit Centre 147
In vitro banana propagation 132
IPM. *see* Integrated pest management (IPM)
Irrigation system 41, 200
ISO. *see* International Organization for Standardisation (ISO)
ISO 14001 290, 304, 306
ISO 14040 284, 289
ISO 14044 284
ISO 14067 289
ITC. *see* International Trade Center (ITC)
ITC Standards Map 169

Japanese Agricultural Standards 319

'Kaizen' business improvement 173
Kellogg, Frank Billings 24
Kokkeel, Willem 263

Lady Finger cultivar 73, 217–218
LCA. *see* Life cycle assessment (LCA)
L-cysteine 131
Leaf
  and dust abrasion 198
  pruning 175
  tearing 198
'Lean' organization 173
Life cycle assessment (LCA)
  banana value chain and sustainability 282–283, 289–290
  climate change and value chains impact 283–284
  environmental impact and carbon footprint 286–287
  environmental issues awareness in agriculture 281–282
  integrated socio-economic carbon footprint banana value chain tool 292–297
    case studies 294–297
    FAO EX-ACT 293–294
  labelling and certification 290–292
  strategic research and development orientations 298
  systematic carbon footprint appraisal 298
Linear conveyors 248

MA. *see* Miniature modified atmosphere (MA)
Makerere University 40
MAP. *see* Modified atmosphere packaging (MAP)
Maroochy Research Facility 152, 153
Marquez, Gabriel Garcia 24
Mass spectrometry 184
Maximum Residue Limits 164
'Mbang Okon' French Horn plantain 103
Mbwazirume Highland banana cultivar 45
McDonald 25
Mealy bugs 176, 185, 337, 339–340
Meloidogyne spp 42
Meristematic cells 99
Meristem culture 129–130
Metabolomics information 156
MGIS. *see Musa* Germplasm Information System (MGIS)
Micropropagation 124, 125–126, 133
Miniature modified atmosphere (MA) 335, 336
Ministry of Agriculture Forestry and Fisheries 319
Miss Chiquita 25
Model T 25
Modified atmosphere packaging (MAP) 336
Moisture stress management 41
Moko/Bugtok disease 74–75
Molecular breeding 52
Molecular fingerprinting 156
Mono-ammonium phosphate 326
Monocropping systems 36
Mouse Ear 341
Mulching 41
Musa acuminata 3–5, 6, 9–11, 15, 94–95, 101, 118, 196
Musa balbisiana 3–4, 6, 9–11, 15, 47, 94, 101, 118, 124
Musa cropping systems 37–39
Musa Germplasm Information System (MGIS) 147, 151, 156
MusaNet Technical Guidelines for the Safe Movement of Musa Germplasm 124
Musa paradisiaca 3
Musa rubra Kurz 4
Musa sapientium 3
Musa velutina 101
Mycosphaerella fijiensis 74

NAADS. *see* National Agricultural Advisory Services (NAADS) Act
Nakasongola District Farmers Association 79

'Nakyetengu' 52
NARO. *see* National Agricultural Research Organization (NARO)
National Agricultural Advisory Services (NAADS) Act 80
National Agricultural Research Organization (NARO) 40, 43, 50, 52, 53, 55
National Land Policy 79
National Organic Program (NOP) 319, 321, 331, 338
National Research Council 166
Nematicides 179
Nematodes 42, 72–73, 121, 123, 144, 186–187, 337
Newsweek 26
New York Times 23, 25, 26
NGOs. *see* Nongovernmental organisations (NGOs)
Nitrogen 40, 71
Non-banana-growing areas 53
Nongovernmental organisations (NGOs) 304, 307
NOP. *see* National Organic Program (NOP)
Northern Uganda, non-banana-growing areas of 53
Nucleic acid sequencing 184
NUE. *see* Nutrient use efficiency (NUE)
Nutrient
  cycling 71
  replenishment 40–41
Nutrient use efficiency (NUE) 226, 229, 230
Nutrition and soil management
  assessing nutrient requirements 224–229
    accounting for relationships 225
    balances 225–227
    flows to plants 227
    foliar analysis and crop NUE 228
    role of models 228–229
    status 224–225
    tillage and drainage 228
  description 223–224
  fertilizer use recommendations 229–230
  precision application of nutrients 230–231

'Obino L'Ewai' French Horn plantain 104
Off-types 132–134
One Hundred Years of Solitude 24
Organic banana cultivation
  certification 319–320
  compost production and usage 327–328
    making compost 328–329
    organic amendments benefits 328
    root-system health 329–331
  control in 332–334
  demand and markets 318
  description 317
  disease management 331–337
    Black Sigatoka 331–334
    Crown rot and mould 334–337
  fertilization and irrigation 327
  insects 337–340
    banana fruit scarring beetle 338–339
    mealy bugs and scale 339–340
    red rust thrips 338
  location 323
  nematodes and weevil borers 337
  nitrogen 326
  organic and non-organic fertilizers 327
  organic standards 320–321
  phosphorus 326
  potassium 326
  rotations 324–325
  soil 323–324
  soil types and nutrition 325
  sources and guidance 321–322
  sustainability 318–319
  weed management 340–341
    cutting and mowing 340–341
    ground cover 341
  see also Banana cultivation; Sustainable banana cultivation
Organic certification 160, 166
Organic Consumers Association 320
Organic farming 320
Organic fertilizers 176, 330–331, 341
Organic Materials Review Institute 321
Organic Production and Labelling of Organic Products 319
Organic Standards Magazine 321
Organic Trade Association (OTA) 320
Organochlorides 44
Organophosphates 44
Origin of banana 4–9
Oritos 250
OTA. *see* Organic Trade Association (OTA)

Pacific plantains 12
Packaging system 249–252
Packing house operations 241–245
Panama disease. *see* Fusarium wilt
Panama-disease-resistant commodity variety 29
PAR. *see* Active photosynthetic radiation (PAR)
Paraffin polyolefin wrap protection 335
Pararetroviruses 124
PAS. *see* Publicly Available Specification (PAS)
PCR. *see* Polymerase chain reaction (PCR)
Pentalonia nigronervosa 74
Peruvian banana value chain 294–296
Pests
  of bananas 41–43
  management 43–45
    biological control 44–45
    chemical control 44

clean planting material 43
crop rotation 44
post-harvest residues destruction 43–44
weevil trapping 44
Phenolic production 131
Phenotyping 156
Photoperiod response 96
Photosynthate availability 112
Phytoliths 11–12
Phytosanitary barriers 219
Pinto's Peanut 341
Pisang Awak A genome 14–15
Pisang Mas. *see* Oritos
'Pisang Siem' 15
Plaesiusjavanus Erichs 44
Plantain – cacao systems 36
Plant and rhizosphere microbiome 182, 186
Plant diversity 185
Planting date for bananas 202–203
Planting density and spatial arrangement 204–205
Plant propagation methods
bits and suckers 119–122
advantages 120
disadvantages 121–122
description 117–118
evolution 118
liquid culture systems 132
morphology and vegetative propagation 118–119
off-types 132–136
elongation and rooting 134
nursery production 134–135
screening and roguing 136
temperature 135–136
propagation using conventional planting material 122–124
tissue culture 124–132
advantages and disadvantages 125
after culture initiation 131
case study 126–127
indexing banana for bacteria 130–131
material selection for culture initiation 127–130
micropropagation 125–126
multiplication 131–132
Polyethylene (PE) covers 209, 210–211, 213
Polymerase chain reaction (PCR) 46
Postharvest treatment 248–249
Potassium 40, 71–72
chloride 326
sulphate 326
'Poyo' Cavendish 103
Pratylenchus goodeyi 42, 72
Pratylenchus spp. 149
Precision breeding. *see* Molecular breeding
Pre-production challenges
access to quality planting material 68
labour for land preparation 67–68
land tenure and fragmentation 69–70
social constraints 68–69
Primiphosethyl 44
Processing challenges
marketing challenges 77–79
post-harvest losses 75–76
value adding 76–77
Production challenges
access to fertilisers and pesticides 70–71
bacterial, fungal and viral diseases 73–75
climate variability and change 75
crop management practices 72
labour for plantation management 70
pests 72–73
soil fertility and water/moisture stress 71–72
Product sensors 275
ProMusa website 123
Proto-cultivars 5
Provitamin A enhancement 52
Pseudostem trap 44
Publicly Available Specification (PAS) 284
Pueraria phaseoloides 175

QBAN. *see* Quality Banana Approved Nursery (QBAN)
Quadratic equations 111
Quality Banana Approved Nursery (QBAN) 124
Quality planting material 68, 79
Queensland Department of Agriculture and Fisheries 152
Queensland University of Technology 343

Radopholus similis 42, 72, 149, 337
Rainfall 75
Rainforest Alliance 163, 164, 165, 166–167, 170–171, 182, 209, 304–307, 306
Rainforest/Sustainable Agriculture Network certification 164
Ralstonia solanacearum 149
RCA. *see* Rolling-circle amplification (RCA)
Red Dacca 250
Red rust thrips 185
Re-engineering approach 179
Regulation (EEC) No. 2092 319
RFLP markers 5
Ripening process control sheet (RPCS) 273, 276–277
Ripening systems 76
cycles 259–260
description 257
factors affecting 258–259
quality control 273–277
supply chain 261–266
discharge 264
distribution 265–266
growing 261–262
loading 264

packing 262–263
pre-cooling 263
ripening facility 265
transport 263
technologies and rooms 266–271
Rodale, J. L. 329
'The Rodale Book of Composting' 329
Rodale Press 329
Rolling-circle amplification (RCA) 46
RPCS. *see* Ripening process control sheet (RPCS)

SA8000 163, 165, 187, 290
Scale insects 176, 185, 339–340
Shoot-tip culture 129
Sigatoka disease. *see* Yellow Sigatoka
Silent Spring 160
Silver, Frank 25
SIMBA model 186, 229
Smallholder farmers 77
Smallholder intercropping systems 38
Smallholder monocropping systems 38
Smith, E. F. 45
Socio-economic LCA focus 286
Soil
microbiology and ecology 182
moisture 41, 71
nutrient deficiencies 71
preparation 150
Soil-borne pathogens 146
'Soil Degradation and the Rise of Organic Nutrition' 322
Soil-free potting mix 135
Solar radiation 108
Somaclonal variation 132, 133, 147
Somatic embryogenesis 126
Sooty mold 339
South Johnstone Research Facility 152
SSR markers 6, 9
Standard Fruit. *see* Dole
Steering Committee of the State-of-Knowledge Assessment of Standards and Certification 2012 167
Stop climate change seals 291
Student's *t*-tests 111
Subsistence and commercial 36
Sucker management
direction of selection 205, 207
stage of parent plant development 205
Sucriers. *see* Oritos
Sul-Po-Mag 326
Sustainable banana cultivation
banana producers in Philippines 309–310
description 303
and diversity in production 308–309
in food security 308
genetic diversity 309
in mixed cropping systems 308
limits of standards 307
range of standards 304–307
see also Banana cultivation; Organic banana cultivation
Swennen, Rony 29
Sword sucker 119, 127
Synthetic urea 326
System analysis 182

Tariff barriers 219
Tarp system. *see* Del Monte system
TDZ. *see* Thidiazuron (TDZ)
Temperature 75, 108, 135–136, 196
Temporary immersion systems (TISs) 132
Thiabendazole 334
Thidiazuron (TDZ) 123, 131
Thyrse 100–102
Time 26
Tin Pan Alley 25
TISs. *see* Temporary immersion systems (TISs)
Tissue culture 53
Training Manual for Organic Agriculture in the Tropics 321
Transitional and transformational approaches 188
Triploid AAA genome cultivars 6
True plantains 65
Turbana label 306
Twisting suckers 127

Uganda, banana production in 39–40
Ultraviolet (UV) radiation 210
UN. *see* United Nations (UN)
Underground corm 118
United Fruit. *see* Chiquita
United Nations Conference on Environment and Development (1992) 281
United Nations (UN) 160, 164
Universal Declaration of Human Rights 165
'Unti Belanda' 15
UPD 258, 273, 274
U.S. Department of Agriculture (USDA) 319, 331
USDA Organic Standards 319
UV. *see* Ultraviolet (UV) radiation

Vacuum bag packing 335
Value chain actors 68
Value chain analysis 282
Ventilation 276

Wageningen University and Research 343
Wages in banana plantations 68–69
Washington State Department of Agriculture (WSDA) 321
Waste 173
Water footprint 182, 188
Water sucker 119
WBF. *see* World Banana Forum (WBF)

West African rainforest plantains 13–14
White fluorescent lighting 126
Whitman, Ed 26
Wild bananas 93–94, 103
Williams 133, 136, 153, 196, 204, 217
World Animal Health 160
World Banana Forum (WBF) 293, 298
World Health Organization 159, 252, 321
World Trade Organization (WTO) 160, 164
WSDA. *see* Washington State Department of Agriculture (WSDA)
WTO. *see* World Trade Organization (WTO)

Xanthomonas campestris pv. *Musacearum* 47

Yellow Sigatoka 218

Zemurray, Samuel 26, 27

www.ingramcontent.com/pod-product-compliance
Ingram Content Group UK Ltd.
Pitfield, Milton Keynes, MK11 3LW, UK
UKHW021826100926
8556IPUK00003B/171

9 781786 761569

Decision-making process 290
'Decree 900' 26
Defoliation treatments 110–112
De Laat Koudetechniek BV 271
Del Monte system 263
Department of Agriculture 230
Dessert bananas 35, 50, 65
Di-ammonium phosphate 326
Diaspis boisduvalii 339
Diploid AA cultivars 5–6
Disc–on-stump traps 44
Diseases
 of banana 45–47
 management 47–49
  avoiding pathogens to new locations 47–48
  Black Sigatoka 331–334
  Crown rot and mould 334–337
  enforcement procedure 49
  good crop husbandry 49
  inoculum management 49
Dispersal of bananas
 into Africa 13–14
 into Oceania 12
 in South and Southeast Asia 14–15
Dole 23, 25–28, 162, 237
Domestication of bananas 10–11
Dow Chemical 162
DSSAT framework 229
'Ducasse' 153
Dulles, John Foster 26
Dusban 44
Dwarf Cavendish 101, 196, 199, 204, 216–217, 342
Dwarf off-types 133, 136

EAC. *see* East African Community (EAC)
EAHB. *see* East African highland bananas (EAHB)
Early banana cultivation
 archaeological evidence for 11–12
 domestication and 10–11
Earth Day 160
Earth Summit 160
East African Community (EAC) 80
East African highland bananas (AAA-EAHB) 35
East African highland bananas (EAHB) 13, 34–35, 43, 46, 52, 65, 107
Ecological intensification (EI) 161, 177, 179
'The Ecology of Commerce: A Declaration of Sustainability' 322
Eco-OK labelling 164
ECO-TOP ripening system 271
Ecuador, Banana value chain in 296–297
Edible seedless bananas 94
EI. *see* Ecological intensification (EI)
Eisenhower (Dwight D. - US President) 26
'El Pulpo' 23
Endophytic bacteria 130
Environmental Protection Agency (EPA) 160, 338
'Enyeru' 107
Epigenetic changes 133
Equalisation period 275
EU Commission Regulation No. 354 319
EU Commission Regulation No. 889 319
EU Organic 163
Eurasia Connection 263
EurepGAP 160
European Joint Programming Initiatives 282
European Union (EU) Organic Agriculture label 165, 166–167
EX-ACT BVC5. *see* EX-Ante Carbon-Balance Tool for Banana Value Chain (EX-ACT BVC5)
EX-Ante Carbon-Balance Tool for Banana Value Chain (EX-ACT BVC5) 293–294, 295

FairTrade label 29, 160, 163, 165, 166–167, 170–171, 182, 187, 188, 290, 304–307
FAO. *see* Food and Agriculture Organization (FAO)
Farmer networks 68
Farmers and marketing strategy 55
Farm Management Plan 324
FDA. *see* Food and Drug Administration (FDA)
'Female' flowers 102
Fencing 146
Fertigation techniques 230–231
Fertilizer band 230
FHIA. *see* Fundación Hondureña De Investigación Agrícola (FHIA)
FHIA 17 50
FHIA 23 50
Field genebanks 144, 152
Finger diameter 237
First ratoon (RI) 205, 207
Flowering 111–112
Flower types 102
FOC race 1 46
FOC-resistant cultivars 29
FOC SR4. *see* Fusarium wilt Subtropical Race 4 (FOC SR4)
FOC-TR4. *see* Fusarium wilt tropical race 4 (FOC-TR4)
Fogging chamber 248
Followers 119
FONTAGRO project 174, 179
Food and Agriculture Organization (FAO) 159, 160, 161, 252, 282, 293, 304, 305, 307, 321
Food and Drug Administration (FDA) 252, 320
Food intercropping systems 35
The Food Safety Modernization Act (2011) 252
Ford, Henry 25
French Agricultural Research Centre for International Development (CIRAD) 322, 343
Fruit fill, degree of 237

Fruit selection processes 247–249
Full-leaf defoliation 112
Fundación Hondureña De Investigación Agrícola (FHIA) 29, 144
Fusarium oxysporum 45, 73, 119, 149, 153, 342
Fusarium wilt 23, 25, 26, 30, 45–46, 47–48, 65, 73, 73–74, 119, 121, 123, 135, 144, 218
Fusarium wilt Subtropical Race 4 (FOC SR4) 217
Fusarium wilt tropical race 4 (FOC-TR4) 21–22, 28, 29, 46, 74, 133, 162, 303, 310, 343
Fyffes 306

GAP. *see* Good agricultural practices (GAP)
Gassing procedure 275–276
Gates Foundation 29
GCTCV. *see* Giant Cavendish Tissue Cultured Variants (GCTCV)
Gender and banana plantations 68
Gene sequencing and manipulation 182
Genetic admixture 10
Germplasm conservation 124
  case study 152–155
  choosing site 145–146
  crop management 150
  description 143–144
  dissemination of quality planting stock 152
  objectives of field genebank 144–145
  planting 149–150
  quality control 151–152
  sourcing material 147–149
GHG. *see* Greenhouse gas (GHG) emissions
Giant Cavendish Tissue Cultured Variants (GCTCV) 342
GlobalGap 160, 163, 164, 165, 167, 170, 172, 182, 184
Global Good Agricultural Practices (GLOBALG.A.P.) 304, 306
Global warming 199, 287
GLRA. *see* Great Lakes Region of Africa (GLRA)
GMO crops 184
'Gold standard' BSV diagnostic tool 46
Good agricultural practices (GAP)
  in banana production 161–163
  description 159–161
  improving 172–182
    bunch weight 174–175
    crop nutrient and residue management 175–176
    Fusarium Tropical Race 4 176–182
    insect pests of the bunch 176
    stand density 175
  related certification schemes 163–166
  scientific frontiers and implications 182–187
    banana plant and rhizosphere microbiome 184–185
    biodiversity for ecosystem services 185–186
    biotic and abiotic stresses 183–184
    integrated banana health and productivity 186–187
    managing soil microbial community 184
  and sustainable banana production 166–172
'Good for Development' certification framework 188
Gourmet Gardens Ltd 55
Government of Uganda 52
Grand Nain 133, 196, 199, 204
Great Lakes Region of Africa (GLRA) 34
Greenhouse gas (GHG) emissions 182, 281, 283–284, 287, 295
Greenhouse screening 185
Gros Michel 15, 23, 25, 27, 28, 29–30, 50, 65, 97, 102, 185, 217
'Growth Stages of Mono-and Dicotyledonous Plants' 322

HACCP. *see* Hazard analysis of critical control points (HACCP)
'*Hacer la césarea*' 197
Harvesting
  bunch movement 239–241
  description 235–236
  determining criteria 237
  techniques 237–239
Hazard analysis of critical control points (HACCP) 163
Healthy leaves removal 207
Helicotylenchus multicinctus 42, 72
Hermaphrodite flowers 94, 102
High Noon (SH-3640) 217
High-ridge Banana Growers and Marketing Association 79
Home garden systems 35–36
Honduran Foundation of Agricultural Investigation 343
Horn plantain 104, 110
'The Hotel Banana' 23
Humic-acid-based soil amendments 330

ICM. *see* Integrated crop management (ICM)
IFOAM. *see* International Federation of Organic Agriculture Movements (IFOAM)
'The IFOAM NORMS for Organic Production and Processing' 321
IIED. *see* International Institute for Environment and Development (IIED)
IISD. *see* International Institute for Sustainable Development (IISD)
IITA. *see* International Institute of Tropical Agriculture (IITA)
Imazalil 334
Immunocapture (IC)-PCR 46
INIBAP. *see* International Network for the Improvement of Banana and Plantain (INIBAP)